GENERAL RELATI

GENERAL RELATIVITY
An introduction to the theory of the gravitational field
2nd edition

HANS STEPHANI

SEKTION PHYSIK, FRIEDRICH SCHILLER UNIVERSITÄT,
JENA, G. D. R.

EDITED BY JOHN STEWART

TRANSLATED BY MARTIN POLLOCK

INSTITUTE OF ASTRONOMY, UNIVERSITY OF CAMBRIDGE

AND JOHN STEWART

DEPARTMENT OF APPLIED MATHEMATICS AND THEORETICAL
PHYSICS, UNIVERSITY OF CAMBRIDGE

Published by the Press Syndicate of the University of Cambridge
The Pitt Building, Trumpington Street, Cambridge CB2 1RP
40 West 20th Street, New York, NY 10011-4211, USA
10 Stamford Road, Oakleigh, Melbourne 3166, Australia

Originally published in German as *Allgemeine Relativitätstheorie*
by VEB Deutscher Verlag der Wissenschaften, Berlin, G.D.R.
© VEB Deutscher Verlag der Wissenschaften, first German edition 1977
second German edition 1980
third German edition 1989

First published in English as *General Relativity*
by Cambridge University Press 1982
second English edition 1990

English edition © Cambridge University Press 1982, 1990
Reprinted 1990, 1993, 1994

Printed in the United States of America

Library of Congress Cataloging-in-Publication Data
Staphani, Hans.
[Allgemeine Relativitätstheorie. English]
General relativity – an introduction to the theory of the
gravitational field / Hans Stephani : edited by John Stewart ;
translated by Martin Pollock and John Stewart. – 2nd ed.
p. cm.
Translation of: Allgemeine Relativitätstheorie.
Includes bibliographical references.
ISBN 0-521-37066-3. – ISBN 0-521-37941-5 (pbk.)
1. Gravitational fields.   2. General relativity (Physics)
I. Stewart, John M.   II. Title.
QC178.S8213   1990
530.1'1 – dc20   89-22351
                                                       CIP

British Library Cataloguing in Publication Data
Stephani, Hans
General relativity. – 2nd ed.
1. Gravitation. General theory of relativity
I. Title   II. Stewart, John, *1943–*   III. Allgemeine
Relativitätstheorie. *English*
531'.14

ISBN 0-521-37066-3 hardback
ISBN 0-521-37941-5 paperback

Contents

Foreword	*page*	xi
Editor's preface		xii
Notations, conventions and important formulae		xiii

Introduction   1

1   The force-free motion of particles in Newtonian mechanics   2
1.1   Coordinate systems   2
1.2   Equations of motion   4
1.3   The geodesic equation   6
1.4   Geodesic deviation   8

Foundations of Riemannian geometry   12

2   Why Riemannian geometry?   12
3   Riemannian space   14
3.1   The metric   14
3.2   Geodesics and Christoffel symbols   15
3.3   Coordinate transformations   17
3.4   Special coordinate systems   20
3.5   The physical meaning and interpretation of coordinate systems   24
4   Tensor algebra   25
4.1   Scalars and vectors   25
4.2   Tensors and other geometrical objects   27
4.3   Tensor algebra   30
4.4   Symmetries of tensors   33
4.5   Algebraic properties of second-rank tensors   34
4.6*   Tetrad and spinor components of tensors   37
5   The covariant derivative and parallel transport   42
5.1   Partial and covariant derivatives   42
5.2   The covariant differential and local parallelism   45
5.3   Parallel displacement along a curve and the parallel propagator   46

5.4	Fermi–Walker transport	47
5.5	The Lie derivative	49
6	The curvature tensor	51
6.1	Intrinsic geometry and curvature	51
6.2	The curvature tensor and global parallelism of vectors	52
6.3	The curvature tensor and second derivatives of the metric tensor	54
6.4	Properties of the curvature tensor	56
7	Differential operators, integrals and integral laws	60
7.1	The problem	60
7.2	Some important differential operators	61
7.3	Volume, surface and line integrals	62
7.4	Integral laws	64
7.5	Integral conservation laws	66
8	Fundamental laws of physics in Riemannian spaces	67
8.1	How does one find the fundamental physical laws?	67
8.2	Particle mechanics	70
8.3	Electrodynamics in vacuo	73
8.4	Geometrical optics	78
8.5	Thermodynamics	80
8.6	Perfect fluids and dust	83
8.7	Other fundamental physical laws	84

Foundations of Einstein's theory of gravitation — 86

9	The fundamental equations of Einstein's theory of gravitation	86
9.1	The Einstein field equations	86
9.2	The Newtonian limit	89
9.3	The equations of motion of test particles	91
9.4	A variational principle for Einstein's theory	95
10	The Schwarzschild solution	99
10.1	The field equations	99
10.2	The solution of the vacuum field equations	102
10.3	General discussion of the Schwarzschild solution	104
10.4	The motion of the planets and perihelion precession	105
10.5	The propagation of light in the Schwarzschild field	108
10.6	Further aspects of the Schwarzschild metric	113
10.7	Experiments to verify the Schwarzschild metric	114
11	The interior Schwarzschild solution	119
11.1	The field equations	119

11.2	The general solution of the field equations	120
11.3	Matching conditions and connection to the exterior Schwarzschild solution	122
11.4	A discussion of the interior Schwarzschild solution	125
12	The Reissner–Nordström (Reissner–Weyl) solution	126

Linearized theory of gravitation, far fields and gravitational waves — **128**

13	The linearized Einstein theory of gravitation	128
13.1	Justification for a linearized theory and its realm of validity	128
13.2	The fundamental equations of the linearized theory	129
13.3	A discussion of the fundamental equations and a comparison with special-relativistic electrodynamics	130
13.4	The far field due to a time-dependent source	132
13.5	A discussion of the properties of the far field (linearized theory)	136
14	Far fields due to arbitrary matter distributions and balance equations for momentum and angular momentum	138
14.1	What are far fields?	138
14.2	The energy–momentum pseudo-tensor for the gravitational field	141
14.3	The balance equations for momentum and angular momentum	144
14.4	Is there an energy law for the gravitational field?	147
15	Gravitational waves	149
15.1	Do gravitational waves really exist?	149
15.2	Plane gravitational waves in the linearized theory	151
15.3	Plane waves as exact solutions of Einstein's equations	154
15.4	The experimental evidence for gravitational waves	158
16*	The Cauchy problem for the Einstein field equations	160
16.1	The problem	160
16.2	Three-dimensional hypersurfaces and reduction formulae for the curvature tensor	160
16.3	The Cauchy problem for the Einstein vacuum field equations	164
16.4	The characteristic initial value problem	166
16.5	Matching conditions at the boundary surface of two metrics	168

Invariant characterization of exact solutions — 171

- 17 Preferred vector fields and their properties — 171
- 17.1 Special simple vector fields — 171
- 17.2 Timelike vector fields — 175
- 17.3* Null vector fields — 178
- 18* The Petrov classification — 183
- 18.1 What is the Petrov classification? — 183
- 18.2 The algebraic classification of electromagnetic fields — 184
- 18.3 The physical interpretation of null electromagnetic fields — 187
- 18.4 The algebraic classification of gravitational fields — 189
- 18.5 The physical interpretation of degenerate vacuum gravitational fields — 192
- 19 Killing vectors and groups of motion — 194
- 19.1 The problem — 194
- 19.2 Killing vectors — 195
- 19.3 Killing vectors of some simple spaces — 196
- 19.4 Relations between the curvature tensor and Killing vectors — 198
- 19.5 Groups of motion — 200
- 19.6 Killing vectors and conservation laws — 204
- 20* The embedding of Riemannian spaces in flat spaces of higher dimension — 210
- 21 A survey of some selected classes of solutions — 211
- 21.1 Vacuum solutions — 212
- 21.2 Solutions with special symmetry properties — 213

Gravitational collapse and black holes — 221

- 22 The Schwarzschild singularity — 221
- 22.1 How does one examine the singular points of a metric? — 221
- 22.2 Radial geodesics in the neighbourhood of the Schwarzschild singularity — 222
- 22.3 The Schwarzschild solution in other coordinate systems — 224
- 22.4 The Schwarzschild solution as a black hole — 227
- 23 Gravitational collapse – the possible life history of a spherically symmetric star — 230
- 23.1 The evolutionary phases of a spherically symmetric star — 230
- 23.2 The critical mass of a star — 231
- 23.3 Gravitational collapse — 235
- 24 Rotating black holes — 242
- 24.1 The Kerr solution — 242

24.2	Gravitational collapse – the possible life history of a rotating star	246
24.3	Some properties of black holes	247
24.4	Can and do black holes exist?	248

Cosmology 251

25	Robertson–Walker metrics and their properties	251
25.1	The cosmological principle and Robertson–Walker metrics	251
25.2	The motion of particles and photons in Robertson–Walker metrics	253
25.3	Distance measurement and horizons in Robertson–Walker metrics	256
25.4	Physics in closed universes	260
26	The dynamics of Robertson–Walker metrics and the Friedmann universe	265
26.1	The Einstein field equations for Robertson–Walker metrics	265
26.2	The most important Friedmann universes	267
26.3	Consequences of the field equations for models with arbitrary equation of state having positive pressure and positive rest mass density	271
27	Our universe as a Friedmann model	273
27.1	Red shift and mass density	273
27.2	The earliest epochs of our universe and the cosmic background radiation	275
27.3	A Schwarzschild cavity in the Friedmann universe	279
28	General cosmological models	283
28.1	What is a cosmological model?	283
28.2	Solutions of Bianchi type I with dust	284
28.3	The Gödel universe	288
28.4	Singularity theorems	289

Non-Einsteinian theories of gravitation 292

29	Classical field theories	292
29.1	Why and how can one generalize the Einstein theory?	292
29.2	Possible tests of gravitational theories and the PPN formalism	295
30	Relativity theory and quantum theory	297
30.1	The problem	297

30.2	Unified quantum field theory and the quantization of the gravitational field	298
30.3	Semiclassical gravity	299
30.4*	Quantization in a given classical gravitational field. The thermodynamics of black holes	300
Bibliography		309
Index		319

Foreword

This book is designed to provide an introduction to the foundations of Einstein's theory and also to survey the questions it raises, its concepts and its methods. Because of the rapid development of relativity physics in recent years it has been impossible to avoid some restriction and selection of the subject matter; it is hoped, however, that the gap can be filled as far as possible by means of the bibliography at the end of every major section. Several rather more exacting sections, which the reader may omit at a first reading, are denoted by an asterisk. The reader is assumed to be familiar with theoretical mechanics, electrodynamics and special relativity. The basic ideas of Riemannian geometry which are necessary for the theory of general relativity are described in the first chapters.

My thanks go to all colleagues of the Jena research group, led by Professor Schmutzer, with whom and from whom I have learnt the theory of relativity, and to all authors of books and articles, whether mentioned by name or not, whose ideas this book contains. I am especially indebted to my colleagues Dr. G. Kluge and Dr. D. Kramer for numerous critical remarks on the form of the manuscript, and to Professor Dr. E. Schmutzer and Dr. F. Gehlhar who suggested changes. I also have to thank Frau U. Kaschlik for a careful typing of the manuscript, Herr Th. Elster for help with the corrections and not least the publisher for a pleasant collaboration. In the preparation of the third (German) edition the additional references collected by Dr. J. Stewart for the English edition have been incorporated.

Jena   HANS STEPHANI

Editor's preface

Several years ago I was asked by an astrophysicist to recommend a textbook on relativity which, although physically oriented, contained a clear, unbiased description of mainstream relativity. On the same day Cambridge University Press sent me Dr. Hans Stephani's *Allgemeine Relativitätstheorie* to review, thus answering the astrophysicist's question. This book is a translation of the first (1977) edition with all of the amendments and corrections of the second (1980) and third (1988) editions included. In a few places I have added comments in footnotes. Further, the bibliography has been updated to 1987/88, and where possible English translations have been cited.

JOHN STEWART

Notations, conventions and important formulae

Minkowski space: $ds^2 = \eta_{ab} dx^a dx^b = dx^2 + dy^2 + dz^2 - c^2 dt^2$.

Riemannian space: $ds^2 = g_{ab} dx^a dx^b = -c^2 d\tau^2$.

$$g = |g_{ab}|, \quad g^{ab} g_{bm} = \delta^a_m = g^a_m.$$

ϵ-pseudo-tensor: ϵ^{abmn}; $\epsilon^{1234} = 1/\sqrt{-g}$,

$$\epsilon_{abcd}\epsilon^{abnm} = -2(g^n_c g^m_d - g^m_c g^n_d).$$

Dualization of an antisymmetric tensor: $\tilde{F}^{ab} = \frac{1}{2}\epsilon^{abmn} F_{mn}$.

Christoffel symbols: $\Gamma^a_{mn} = \frac{1}{2} g^{ab}(g_{bm,n} + g_{bn,m} - g_{mn,b})$.

Covariant derivative: $DT^a/Dx^m = T^a_{;m} = T^a_{,m} + \Gamma^a_{mn} T^n$,

$$DT_a/Dx^m = T_{a;m} = T_{a,m} - \Gamma^n_{am} T_n.$$

Geodesic equation: $\dfrac{D^2 x^i}{D\lambda^2} = \dfrac{d^2 x^i}{d\lambda^2} + \Gamma^i_{nm} \dfrac{dx^n}{d\lambda} \dfrac{dx^m}{d\lambda} = 0$.

Parallel transport along the curve $x^i(\lambda)$: $DT^a/D\lambda = T^a_{;b} dx^b/d\lambda = 0$.

Fermi-Walker transport: $\dfrac{DT^n}{D\tau} - \dfrac{1}{c^2} T_a \left(\dfrac{dx^n}{d\tau} \dfrac{D^2 x^a}{D\tau^2} - \dfrac{dx^a}{d\tau} \dfrac{D^2 x^n}{D\tau^2} \right) = 0$.

Lie derivative in the direction of the vector field $a^k(x^i)$:

$$\pounds_a T^n = T^n_{,k} a^k - T^k a^n_{,k} = T^n_{;k} a^k - T^k a^n_{;k}.$$

$$\pounds_a T_n = T_{n,k} a^k + T_k a^k_{,n} = T_{n;k} a^k + T_k a^k_{;n}.$$

Killing equation: $\xi_{i;n} + \xi_{n;i} = \pounds_\xi g_{in} = 0$.

Divergence of a vector field: $a^i_{;i} = \dfrac{1}{\sqrt{-g}} (\sqrt{-g}\, a^i)_{,i}$.

Maxwell equations: $F^{mn}_{;n} = \dfrac{1}{\sqrt{-g}} (\sqrt{-g} F^{mn})_{,n} = \dfrac{1}{c} j^m$,

$$\tilde{F}^{mn}_{;n} = 0.$$

Curvature tensor:
$$a_{m;s;q} - a_{m;q;s} = a_b R^b{}_{msq},$$
$$R^b{}_{msq} = \Gamma^b{}_{mq,s} - \Gamma^b{}_{ms,q} + \Gamma^b{}_{ns}\Gamma^n{}_{mq} - \Gamma^b{}_{nq}\Gamma^n{}_{ms},$$
$$R_{amsq} = \tfrac{1}{2}(g_{aq,ms} + g_{ms,aq} - g_{as,mq} - g_{mq,as}) + \text{non-linear terms}.$$

Ricci tensor: $R_{mq} = R^s{}_{msq} = -R^s{}_{mqs}; \quad R^m{}_m = R.$

Field equations: $G_{ab} = R_{ab} - \dfrac{R}{2} g_{ab} = \kappa T_{ab}.$

Perfect fluid: $T_{ab} = (\mu + p/c^2)u_a u_b + p g_{ab}.$

Schwarzschild metric: $\mathrm{d}s^2 = \dfrac{\mathrm{d}r^2}{1-2M/r} + r^2(\mathrm{d}\vartheta^2 + \sin^2\vartheta\,\mathrm{d}\varphi^2)$
$$- (1-2M/r)c^2\mathrm{d}t^2.$$

Robertson–Walker metric:
$$\mathrm{d}s^2 = K^2(ct)\left[\dfrac{\mathrm{d}r^2}{1-\epsilon r^2} + r^2(\mathrm{d}\vartheta^2 + \sin^2\vartheta\,\mathrm{d}\varphi^2)\right] - c^2\mathrm{d}t^2.$$

Hubble constant: $H(ct) = \check{K}/K.$

Acceleration parameter: $q(ct) = -K\ddot{K}/K^2.$

$\kappa = 2.07 \times 10^{-48} \mathrm{g}^{-1}\mathrm{cm}^{-1}\mathrm{s}^2, \quad cH = 55 \text{ km/s Mpc}.$

$2M_{\text{Earth}} = 0.8876 \text{ cm}, \quad 2M_{\text{Sun}} = 2.9533 \times 10^5 \text{ cm}.$

Introduction

The general theory of relativity came into being historically as an extension of the special theory of relativity. However the questions to which it is addressed are somewhat incompletely described by its name. As in every living, evolving science there is no agreement amongst scientists as to which of the viewpoints and ideas discussed were particularly important when the theory was in its infancy, which are absolutely necessary for a logical structure, or which will prove fruitful for its future development. Three groups of questions, however, played a special rôle and led ultimately to the general theory of relativity.

(1) For a description of nature and its laws one should be able to use *arbitrary coordinate systems,* and in accordance with the *principle of covariance* the form of the laws of nature should not depend essentially upon the choice of the coordinate system. This requirement, in the first place purely mathematical, acquires a physical meaning through the substitution of 'arbitrary coordinate system' by 'arbitrarily moving observer.' The laws of nature should be independent of the state of motion of the observer, as analogously, they are the same in the special theory of relativity for all inertial systems, that is, for all observers moving with constant speed relative to one another. To this group belongs also the question, raised in particular by Ernst Mach, of whether an absolute acceleration (including an absolute rotation) can really be defined meaningfully, or whether every measurable rotation implies a rotation relative to the fixed stars (*Mach's Principle*).

(2) The Newtonian theory of gravitation is inconsistent with the special principle of relativity. In it gravitational effects propagate with an infinitely large velocity. A new, better formulation of the *field equations of gravitation* should therefore be found, which includes also the influence of gravity upon other physical processes and which agrees with experiment.

(3) In astrophysics and *cosmology* large masses are involved; gravitational forces dominate short-range nuclear forces. Thus a theory of gravitation has to be found which correctly reflects the dynamical behaviour of the whole Universe and which at the same time is valid for stellar evolution.

Introduction

The general theory of relativity began with the formulation of the fundamental equations by Albert Einstein in 1915, followed by a series of articles on the foundations of the theory and on its possible experimental confirmation. In spite of the success of the theory (precession of the perihelion on Mercury, deflection of light by the Sun, explanation of the cosmological redshift), it has retained for a long time the reputation of an esoteric science for specialists and outsiders, perhaps because of the mathematical difficulties, the new concepts and the paucity of applications (for example, in comparison with quantum theory, which came into existence at almost the same time). Through the development of new methods of obtaining solutions and the physical interpretation of the theory, and even more through the surprising astrophysical discoveries (pulsars, cosmic background radiation), and the improved possibilities of demonstrating general relativistic effects, in the course of the last thirty years the general theory of relativity has become a true physical science, with many associated experimental questions and observable consequences.

The general theory of relativity is the theory of the gravitational field; the description of its language and concepts, and its methods and conclusions, form the main content of this book.

Modern theoretical physics uses and needs ever more complicated mathematical tools – this statement, with its often unwelcome consequences for the physicist, is true also for the theory of gravitation. The language of the general theory of relativity is differential geometry, and we must learn it, if we wish to ask and answer precisely physical questions. This book therefore begins with seven chapters in which the essential concepts and formulae of Riemannian geometry are described. Here suffix notation will be used in preference to the modern coordinate-free notation in order to make the book easier to read for non-mathematicians.

1 The force-free motion of particles in Newtonian mechanics

1.1 *Coordinate systems*

In theoretical mechanics one usually meets only a few simple coordinate systems for describing the motion of a particle. For the purposes of mechanics one can characterize the coordinate system best via the specification of the connection between the infinitesimal separation ds of two points and the difference of their coordinates. In describing the motion in three-dimensional space one chooses Cartesian coordinates x, y, z with

1 Force-free motion in Newtonian mechanics

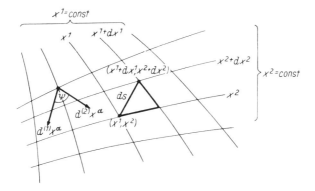

Fig. 1.1. Measurement of lengths and angles by the use of the metric tensor.

$$ds^2 = dx^2 + dy^2 + dz^2, \tag{1.1}$$

cylindrical coordinates ρ, φ, z with

$$ds^2 = d\rho^2 + \rho^2 d\varphi^2 + dz^2 \tag{1.2}$$

or spherical coordinates r, ϑ, φ with

$$ds^2 = dr^2 + r^2 d\vartheta^2 + r^2 \sin^2 \vartheta \, d\varphi^2. \tag{1.3}$$

If the motion is restricted to a surface which does not change with time, for example, a sphere, then one would use the corresponding two-dimensional section ($dr = 0$) of spherical coordinates

$$ds^2 = r^2 d\vartheta^2 + r^2 \sin^2 \vartheta \, d\varphi^2. \tag{1.4}$$

For other arbitrary coordinate systems ds^2 is also a quadratic function of the coordinate differentials:

$$ds^2 = g_{\alpha\beta}(x^\nu) dx^\alpha dx^\beta; \quad \alpha, \beta, \nu = 1, 2, 3. \tag{1.5}$$

Here and in all following formulae indices occurring twice are to be summed, from one to three for a particle in three-dimensional space and from one to two for a particle on a plane.

The form (1.5) is called the *fundamental metric form;* the position-dependent coefficients $g_{\alpha\beta}$ form the components of the *metric tensor*. It is symmetric: $g_{\alpha\beta} = g_{\beta\alpha}$. The name 'metric tensor' refers to the fact that by its use the quantities length and angle which are fundamental to geometrical measurement can be defined and calculated. The displacement ds of two points with coordinates (x^1, x^2) and $(x^1 + dx^1, x^2 + dx^2)$ is given by (1.5), and the angle ψ between two infinitesimal vectors $\mathrm{d}^{(1)}x^\alpha$ and $\mathrm{d}^{(2)}x^\alpha$ diverging from a point can be calculated as

$$\cos \psi = \frac{g_{\alpha\beta}\mathrm{d}^{(1)}x^\alpha \mathrm{d}^{(2)}x^\beta}{\sqrt{g_{\rho\sigma}\mathrm{d}^{(1)}x^\rho \mathrm{d}^{(1)}x^\sigma}\sqrt{g_{\mu\nu}\mathrm{d}^{(2)}x^\mu \mathrm{d}^{(2)}x^\nu}}. \qquad (1.6)$$

Formula (1.6) is nothing other than the familiar vector relation $\mathbf{ab} = |\mathbf{a}||\mathbf{b}|\cos(\mathbf{a},\mathbf{b})$ applied to infinitesimal vectors.

If the matrix of the metric tensor is diagonal, that is to say, $g_{\alpha\beta}$ differs from zero only when $\alpha=\beta$, then one calls the coordinate system orthogonal. As (1.6) shows, the coordinate lines $x^\alpha =$ constant are then mutually perpendicular.

If the determinant of $g_{\alpha\beta}$ is non-zero, the matrix possesses an inverse matrix $g^{\beta\mu}$ given by

$$g_{\alpha\beta}g^{\beta\mu} = \delta_\alpha^\mu = g_\alpha^\mu. \qquad (1.7)$$

The immediate significance of the fundamental metric form (1.5) for mechanics rests on its simple connection with the square of the speed of the particle:

$$v^2 = \left(\frac{\mathrm{d}s}{\mathrm{d}t}\right)^2 = g_{\alpha\beta}\frac{\mathrm{d}x^\alpha}{\mathrm{d}t}\frac{\mathrm{d}x^\beta}{\mathrm{d}t}, \qquad (1.8)$$

which we need for the construction of the kinetic energy as one part of the Lagrangian.

1.2 Equations of motion

We can obtain the equations of motion most quickly from the Lagrangian L, which for force-free motion is identical with the kinetic energy of the particle

$$L = \frac{m}{2}v^2 = \frac{m}{2}g_{\alpha\beta}\frac{\mathrm{d}x^\alpha}{\mathrm{d}t}\frac{\mathrm{d}x^\beta}{\mathrm{d}t} = \frac{m}{2}g_{\alpha\beta}\dot{x}^\alpha \dot{x}^\beta. \qquad (1.9)$$

The corresponding Lagrange equations (of the second kind)

$$\frac{\mathrm{d}}{\mathrm{d}t}\frac{\partial L}{\partial \dot{x}^\nu} - \frac{\partial L}{\partial x^\nu} = 0 \qquad (1.10)$$

are easily set up. If we write partial derivatives with respect to the coordinates in the symbolic manner

$$\frac{\partial}{\partial x^\nu}(\ldots) = (\ldots)_{,\nu}, \qquad (1.11)$$

1 Force-free motion in Newtonian mechanics

then

$$\frac{\partial L}{\partial \dot{x}^\nu} = mg_{\alpha\nu}\dot{x}^\alpha, \qquad \frac{\partial L}{\partial x^\nu} = L_{,\nu} = \frac{m}{2}g_{\alpha\beta,\nu}\dot{x}^\alpha\dot{x}^\beta, \qquad (1.12)$$

and from (1.10) it follows immediately that

$$g_{\alpha\nu}\ddot{x}^\alpha + g_{\alpha\nu,\beta}\dot{x}^\alpha\dot{x}^\beta - \tfrac{1}{2}g_{\alpha\beta,\nu}\dot{x}^\alpha\dot{x}^\beta = 0. \qquad (1.13)$$

If we first write the second term in this equation in the form

$$g_{\alpha\nu,\beta}\dot{x}^\alpha\dot{x}^\beta = \tfrac{1}{2}(g_{\alpha\nu,\beta} + g_{\beta\nu,\alpha})\dot{x}^\alpha\dot{x}^\beta, \qquad (1.14)$$

then multiply (1.13) by $g^{\mu\nu}$ and sum over ν, then because of (1.7) we obtain

$$\ddot{x}^\mu + \Gamma^\mu_{\alpha\beta}\dot{x}^\alpha\dot{x}^\beta = 0, \qquad (1.15)$$

where the abbreviation

$$\Gamma^\mu_{\alpha\beta} = \tfrac{1}{2}g^{\mu\nu}(g_{\alpha\nu,\beta} + g_{\beta\nu,\alpha} - g_{\alpha\beta,\nu}) \qquad (1.16)$$

has been used.

Equations (1.15) are the required equations of motion of a particle. In the course of their derivation we have also come across the *Christoffel symbols* $\Gamma^\mu_{\alpha\beta}$, defined by (1.16), which play a great rôle in differential geometry. As is evident from (1.16), they possess the symmetry

$$\Gamma^\mu_{\alpha\beta} = \Gamma^\mu_{\beta\alpha}, \qquad (1.17)$$

and hence there are eighteen distinct Christoffel symbols in three-dimensional space, and six for two-dimensional surfaces.

On contemplating (1.15) and (1.16), one might suppose that the Christoffel symbols lead to a particularly simple way of constructing the equations of motion. This supposition is, however, false; on the contrary, one uses the very equations of motion in order to construct the Christoffel symbols. We shall illustrate this method by means of an example. In spherical coordinates (1.3), $x^1 = r$, $x^2 = \vartheta$, $x^3 = \varphi$, the Lagrangian

$$L = \frac{m}{2}(\dot{r}^2 + r^2\dot{\vartheta}^2 + r^2\sin^2\vartheta\,\dot{\varphi}^2) \qquad (1.18)$$

implies the following Lagrange equations of the second kind:

$$\left.\begin{array}{l} \ddot{r} - r\dot{\vartheta}^2 - r\sin^2\vartheta\,\dot{\varphi}^2 = 0, \\[4pt] \ddot{\vartheta} + \dfrac{2}{r}\dot{r}\dot{\vartheta} - \sin\vartheta\cos\vartheta\,\dot{\varphi}^2 = 0, \\[4pt] \ddot{\varphi} + \dfrac{2}{r}\dot{r}\dot{\varphi} + 2\cot\vartheta\,\dot{\varphi}\dot{\vartheta} = 0. \end{array}\right\} \qquad (1.19)$$

and

Comparison with (1.15) shows that (noticing that, because of the symmetry relation (1.17), mixed terms in the speeds $\dot r, \dot\vartheta, \dot\varphi$ always occur twice) only the Christoffel symbols

$$
\begin{aligned}
&\Gamma^1_{22}=-r, &&\Gamma^1_{33}=-r\sin^2\vartheta, \\
&\Gamma^2_{12}=\Gamma^2_{21}=\frac{1}{r}, &&\Gamma^2_{33}=-\sin\vartheta\cos\vartheta, \\
&\Gamma^3_{13}=\Gamma^3_{31}=\frac{1}{r}, &&\Gamma^3_{23}=\Gamma^3_{32}=\cot\vartheta
\end{aligned}
\qquad (1.20)
$$

differ from zero.

In the case of free motion of a particle in three-dimensional space the physical content of the equations of motion is naturally rather scanty; it is merely a complicated way of writing the law of inertia – we know beforehand that the particle moves in a straight line in the absence of forces. In the two-dimensional case, for motion over an arbitrary surface, the path of the particle can be rather complicated. As we shall show in the following section, however, a simple geometrical interpretation of the equations of motion (1.15) is then possible.

1.3 The geodesic equation

In three-dimensional space the path of a force-free particle, the straight line, has the property of being the shortest curve between any two points lying on it. We are going to generalize this relation, and therefore ask for the shortest curve connecting two points in a three-dimensional or two-dimensional space; that is, for that curve whose arc-length s is a minimum for given initial-point and end-point:

$$ s = \int_{P_I}^{P_E} ds = \text{extremum}. \qquad (1.21) $$

In order to describe this curve we need an initially arbitrary parameter λ, which for all curves under comparison has the same value at the endpoints P_E and P_I, and if for the differential arc-length ds we substitute the expression (1.5), then (1.21) implies

$$ s = \int_{\lambda_I}^{\lambda_E} \frac{ds}{d\lambda} d\lambda = \int_{\lambda_I}^{\lambda_E} \sqrt{g_{\alpha\beta}\frac{dx^\alpha}{d\lambda}\frac{dx^\beta}{d\lambda}}\, d\lambda = \text{extremum}, \qquad (1.22) $$

from which we shall determine the required shortest connecting curve, the geodesic, in the form $x^\alpha(\lambda)$.

1 Force-free motion in Newtonian mechanics

The variational problem (1.22) has precisely the mathematical form of Hamilton's Principle with the Lagrangian,

$$L = \sqrt{g_{\alpha\beta} x'^{\alpha} x'^{\beta}} = \sqrt{F}, \quad x'^{\alpha} \equiv \frac{dx^{\alpha}}{d\lambda}, \qquad (1.23)$$

and the parameter λ instead of the time t. Thus the geodesic must obey the associated Lagrange equations of the second kind

$$\frac{d}{d\lambda} \frac{\partial L}{\partial x'^{\nu}} - \frac{\partial L}{\partial x^{\nu}} = \frac{d}{d\lambda}\left(\frac{g_{\alpha\nu} x'^{\alpha}}{\sqrt{F}}\right) - \frac{1}{2\sqrt{F}} g_{\alpha\beta,\nu} x'^{\alpha} x'^{\beta}$$

$$= \frac{1}{2F\sqrt{F}}\left[-\frac{dF}{d\lambda} g_{\alpha\nu} x'^{\alpha} + 2F \frac{d}{d\lambda}(g_{\alpha\nu} x'^{\alpha}) - F g_{\alpha\beta,\nu} x'^{\alpha} x'^{\beta}\right]$$

$$= 0. \qquad (1.24)$$

We can simplify this differential equation for the geodesic by choosing the parameter λ appropriately (only for this extremal curve, not for the comparison curves); we demand that λ be proportional to the arc-length s. From (1.22) and (1.23) it follows that F = constant, and from (1.24) we get the differential equation of the geodesic

$$\frac{d^2 x^{\mu}}{d\lambda^2} + \Gamma^{\mu}_{\alpha\beta} \frac{dx^{\alpha}}{d\lambda} \frac{dx^{\beta}}{d\lambda} = 0. \qquad (1.25)$$

This differential equation not only has the same form as the equation of motion (1.15), it is also completely equivalent to it, since, of course, for a force-free motion the magnitude $v = ds/dt$ of the speed is constant because of the law of conservation of energy, and consequently the time t is one of the allowable possibilities in (1.23) for the parameter λ which is proportional to the arc-length s.

If we choose as parameter λ the arc-length s itself, then we can recapitulate our result in the following law:

A force-free particle moves on a geodesic

$$\frac{d^2 x^{\mu}}{ds^2} + \Gamma^{\mu}_{\alpha\beta} \frac{dx^{\alpha}}{ds} \frac{dx^{\beta}}{ds} = 0, \qquad (1.26)$$

of the three-dimensional space or of the surface to which it is constrained. Its path is therefore always the shortest curve between any two points lying on it; for example, on the spherical surface the paths are great circles.

In the general theory of relativity we shall meet the problem of how to set up the equation of motion of a point mass in an arbitrary gravitational field. It will turn out that the formulation of the equation of motion for

1.4 Geodesic deviation

In this section we shall turn to a question whose answer requires the help of Riemannian geometry which we shall indeed use. The reader is therefore asked for indulgence if some of the formalism appears rather vague and the calculations inadequately motivated. The reader is recommended to read this section again after mastering chapter 5.

If the surface to which the particle is constrained is a plane, or a surface which is due to the distortion of a plane or part of a plane (e.g. cylinder, cone) then the geodesics are straight lines of this plane, and the equations of motion of the point mass are very simple to integrate. With the use of unsuitable coordinates, however, the geodesic equation (1.26) can be rather complicated. In such a case how can one tell from the equation of motion, that is, from the Christoffel symbols $\Gamma^\mu_{\alpha\beta}$, that motion on a plane is being described?

To answer this question we examine a family $x^\alpha(s, p)$ of geodesics on a surface. Here the parameter p labels the different geodesics and the arc-length s is the parameter along the curves fixing the different points of the same geodesic.

A family of straight lines in the plane is now distinguished by the displacement of two neighbouring geodesics, as measured between points with the same value of the parameter s, being a *linear* function of arc-length s. This is a hint that also in the general case of geodesics on an arbitrary surface we should examine the behaviour of the separation of neighbouring geodesics and from this draw conclusions about the properties of the surface.

We first form the partial derivatives

$$\frac{\partial x^\alpha}{\partial s} = t^\alpha, \qquad \frac{\partial x^\alpha}{\partial p} = V^\alpha, \qquad \frac{\partial t^\alpha}{\partial p} \equiv \frac{\partial V^\alpha}{\partial s}. \qquad (1.27)$$

The unit tangent vector t^α points in the direction of the velocity, and $V^\alpha dp$ is just the displacement vector of two neighbouring geodesics. In order to see whether we are dealing with a plane or not, it is, however, insufficient to simply form $\partial^2 V^\alpha/\partial s^2$. Indeed, even for a straight line in the plane, the fact that the tangent vector $t^\mu = dx^\mu/ds$ is constant, that is, independent of s, is not expressed in an arbitrary coordinate system by $dt^\mu/ds = 0$, but, as a glance at the geodesic equation (1.26) shows, by

1 Force-free motion in Newtonian mechanics

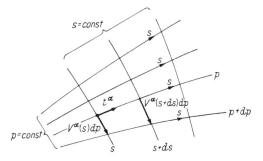

Fig. 1.2. The family of geodesics $x^\alpha(s, p)$.

$$\frac{D}{Ds}t^\mu \equiv \frac{dt^\mu}{ds} + \Gamma^\mu_{\alpha\beta} t^\alpha \frac{dx^\beta}{ds} = 0. \tag{1.28}$$

We interpret (1.28) as the defining equation for the operator D/Ds, valid for every parameter s and applicable to every vector t^μ. If according to this prescription we form the expressions

$$\begin{aligned}\frac{D}{Dp}t^\alpha &= \frac{\partial^2 x^\alpha}{\partial s\, \partial p} + \Gamma^\alpha_{\mu\nu} t^\mu V^\nu, \\ \frac{D}{Ds}V^\alpha &= \frac{\partial^2 x^\alpha}{\partial s\, \partial p} + \Gamma^\alpha_{\mu\nu} V^\mu t^\nu,\end{aligned} \tag{1.29}$$

then we can at once read off the relation analogous to (1.27),

$$\frac{D}{Dp}t^\alpha = \frac{D}{Ds}V^\alpha. \tag{1.30}$$

We shall now calculate the quantity

$$\frac{D^2 V^\alpha}{Ds^2} = \frac{D}{Ds}\left(\frac{D}{Dp}t^\alpha\right) \tag{1.31}$$

in order to discuss with its help the behaviour of the separation of two neighbouring geodesics. Our first goal is to express the right-hand side in terms of the Christoffel symbols. Substitution of the defining equation (1.28) gives us immediately

$$\begin{aligned}\frac{D^2 V^\alpha}{Ds^2} &= \frac{D}{Ds}\left(\frac{\partial t^\alpha}{\partial p} + \Gamma^\alpha_{\mu\nu} t^\mu V^\nu\right) \\ &= \frac{\partial^2 t^\alpha}{\partial s\, \partial p} + \Gamma^\alpha_{\mu\nu,\beta} t^\beta t^\mu V^\nu + \Gamma^\alpha_{\mu\nu}\left(\frac{\partial t^\mu}{\partial s} V^\nu + t^\mu \frac{\partial V^\nu}{\partial s}\right) \\ &\quad + \Gamma^\alpha_{\rho\tau}\left(\frac{\partial t^\rho}{\partial p} + \Gamma^\rho_{\mu\nu} t^\mu V^\nu\right) t^\tau.\end{aligned} \tag{1.32}$$

We can simplify this equation by invoking the relation

$$0 = \frac{D}{Dp}\frac{Dt^\alpha}{Ds} = \frac{D}{Dp}\left(\frac{\partial t^\alpha}{\partial s} + \Gamma^\alpha_{\mu\nu}t^\mu t^\nu\right),$$

which follows from the geodesic equation (1.28). This leads to

$$\frac{\partial^2 t^\alpha}{\partial s\, \partial p} = -\Gamma^\alpha_{\mu\nu,\beta}V^\beta t^\mu t^\nu - \Gamma^\alpha_{\mu\nu}\left(\frac{\partial t^\mu}{\partial p}t^\nu + \frac{\partial t^\nu}{\partial p}t^\mu\right)$$
$$-\Gamma^\alpha_{\rho\tau}\left(\frac{\partial t^\rho}{\partial s} + \Gamma^\rho_{\mu\nu}t^\mu t^\nu\right)V^\tau,$$

which we can substitute into (1.32). Bearing in mind (1.27) and (1.17), we then find that

$$\frac{D^2 V^\alpha}{Ds^2} = t^\beta t^\mu V^\nu (\Gamma^\alpha_{\mu\nu,\beta} - \Gamma^\alpha_{\mu\beta,\nu} + \Gamma^\alpha_{\rho\beta}\Gamma^\rho_{\mu\nu} - \Gamma^\alpha_{\rho\nu}\Gamma^\rho_{\mu\beta}). \tag{1.33}$$

The right-hand side of this equation gives us a measure of the change in separation of neighbouring geodesics, or, in the language of mechanics, of the relative acceleration of two particles moving towards one another on neighbouring paths ($V^\alpha dp$ is their separation, ds is proportional to dt for force-free motion). It is also called the *geodesic deviation*.

When the geodesics are straight lines in a plane, the right-hand side should vanish; it is therefore – geometrically speaking – also a measure of the curvature of the surface, of the deviation of the surface from a plane. This intuitive basis also makes understandable the name 'curvature tensor' for the expression $R^\alpha_{\mu\beta\nu}$, defined by

$$R^\alpha{}_{\mu\beta\nu} = \Gamma^\alpha_{\mu\nu,\beta} - \Gamma^\alpha_{\mu\beta,\nu} + \Gamma^\alpha_{\rho\beta}\Gamma^\rho_{\mu\nu} - \Gamma^\alpha_{\rho\nu}\Gamma^\rho_{\mu\beta}. \tag{1.34}$$

It can be determined from the Christoffel symbols by calculation or by measurement of the change in separation of neighbouring paths. If it vanishes, the surface is a plane and the paths are straight lines.

As an illustration we calculate the curvature tensor of the spherical surface

$$ds^2 = K^2(d\vartheta^2 + \sin^2\vartheta\, d\varphi^2)$$
$$= K^2[(dx^1)^2 + \sin^2 x^1 (dx^2)^2] = g_{\alpha\beta}dx^\alpha dx^\beta. \tag{1.35}$$

The only non-vanishing Christoffel symbols are

$$\Gamma^1_{22} = -\sin\vartheta\cos\vartheta, \qquad \Gamma^2_{12} = \Gamma^2_{21} = \cot\vartheta, \tag{1.36}$$

and from them we find after a simple calculation the components of the curvature tensor

1 Force-free motion in Newtonian mechanics

$$R^1{}_{221} = -R^1{}_{212} = -\sin^2\vartheta = -\frac{1}{K^2}g_{22},$$

$$R^2{}_{121} = -R^2{}_{112} = 1 = \frac{1}{K^2}g_{11},\qquad(1.37)$$

and $\qquad R^\alpha{}_{\mu\beta\nu} = 0 \quad$ otherwise.

This result can be summarized by the formula

$$R^\alpha{}_{\mu\beta\nu} = \frac{1}{K^2}(\delta^\alpha_\beta g_{\mu\nu} - \delta^\alpha_\nu g_{\mu\beta}).\qquad(1.38)$$

It expresses the fact that apart from the coordinate-dependent metric, the curvature tensor depends only upon the radius of the sphere K. For $K\to\infty$ the curvature tensor vanishes.

One can show that the curvature tensor of an arbitrary two-dimensional surface always has the form (1.38), with of course a position-dependent K.

Foundations of Riemannian geometry

2 Why Riemannian geometry?

It is one of the most important results of special relativity that basic physical laws are most simply expressed when they are formulated not in three-dimensional space but in four-dimensional space-time

$$ds^2 = \eta_{ab} dx^a dx^b = dx^2 + dy^2 + dz^2 - c^2 dt^2,$$

$$\eta_{ab} = \eta^{ab} = \begin{pmatrix} 1 & & & \\ & 1 & & \\ & & 1 & \\ & & & -1 \end{pmatrix}, \quad a, b = 1, \ldots, 4. \tag{2.1}$$

We shall now show that it is worthwhile replacing this Minkowski space by a yet more complicated mathematical space-time structure.

If we were to examine a circular disk which is at rest in an inertial system, from the standpoint of a coordinate system rotating around the axis of the disk, and try to measure the geometrical properties of the disk with the help of rulers, then the following result would be plausible: rulers laid out in the radial direction are not influenced by the rotation of the disk, and the radius of the circle is unchanged; rulers laid out along the periphery of the disk are shortened by the Lorentz contraction, the circumference of the circle being thereby decreased. The rotating observer thus establishes that the ratio of the circumference of the circle to its diameter is less than π; he finds geometrical relations similar to those on the curved surface of a sphere. Naturally the application of the Lorentz transformation to rotating systems, and the definition of simultaneity hidden in this measuring procedure, are questionable. But that would only support the result of our 'gedanken' investigation, that in going over to observers (coordinate systems) in arbitrary motion real changes in the space-time structure (the behaviour of rulers and clocks) can arise.

Even more significant physically is the indication to be deduced from the investigation of the equation of motion of a particle moving in a

2 Why Riemannian geometry?

gravitational field $g(r, t)$. If we write this equation in a Cartesian coordinate system, whose origin moves with acceleration a with respect to an inertial system, and which rotates with angular velocity ω, then the familiar equation

$$m\ddot{r} = mg - ma - 2m\omega \times \dot{r} - m\omega \times (\omega \times r) - m\dot{\omega} \times r \qquad (2.2)$$

holds.

All the terms of this equation of motion have the mass m as a factor. From the standpoint of Newtonian mechanics this factor possesses two distinct physical meanings: the force mg, which acts upon a body in the gravitational field, is proportional to the gravitational mass m_G, whilst all other terms in (2.2) are an expression of the inertial behaviour of the body (which is the same for all kinds of forces) and consequently contain the inertial mass m_I. It was one of the most important discoveries of mechanics that for all bodies these two parameters are the same: all bodies fall equally fast, and hence

$$m_G = m_I = m. \qquad (2.3)$$

The numerical identity of inertial and gravitational mass (the current value is $|m_G - m_I|/(m_G + m_I) \leq 10^{-12}$, Braginsky (1972)) also points to a more essential identity. In the language of (2.2) gravitation is perhaps just as much an apparent force as the Coriolis force or the centrifugal force. One could therefore suppose that the particle moves weightlessly in reality, and that also the gravitational force can be eliminated by a suitable choice of coordinate system.

As an exact consideration shows, the gravitational force can really only be transformed away locally, that is, over a spatial region within which the gravitational field can be regarded as homogeneous: inside an Earth satellite or a falling box bodies move force-free for the co-moving observer. Globally, however, this is not attainable through a simple coordinate transformation (by changing to a moving observer): there is no Cartesian coordinate system in which two distantly separated satellites simultaneously move force-free.

If we therefore wish to adhere to the view that in spite of the existing gravitational field the particle moves force-free, and in the sense of chapter 2 translate 'force-free' by 'along a geodesic,' then we must alter the geometry of the space. Just as the geodesics on a surface fail to be straight lines only if the surface is curved and the curvature tensor defined by (1.33) and (1.34) does not vanish, so the planetary orbits are only geodesics of the space if this space is curved.

Foundations of Riemannian geometry

In fact this idea of Einstein's, to regard the gravitational force as a property of the space and thereby to geometrize it, turns out to be extraordinarily fruitful. In the following chapters we shall therefore describe in detail the properties of such curved spaces.

3 Riemannian space

3.1 *The metric*

The geometrical background to the special theory of relativity is the pseudo-Euclidean space (2.1) with one timelike and three spacelike coordinates. In the generalization which we shall develop we also start with a four-dimensional manifold; that is, we shall assume that every point (within a small finite neighbourhood) can be fixed uniquely by the specification of four coordinates x^n. It can of course occur that it is not possible to cover the whole space-time with a single coordinate system. In order to be able to study physics in this manifold, we must be able to measure the spatial and temporal separations of neighbouring points. As the generalization of (2.1) and (1.5) we therefore introduce the metric

$$ds^2 = g_{nm}(x^i)dx^n dx^m. \tag{3.1}$$

(Summation occurs from 1 to 4 over Latin indices appearing twice.) This *fundamental metric form* indicates how one measures on the small scale (in the infinitesimal neighbourhood of a point) the interval ds between the points (x^n) and $(x^n + dx^n)$ and the angle between two directions dx^n and $d\bar{x}^n$

$$\cos(dx^n, d\bar{x}^m) = \frac{g_{nm}dx^n d\bar{x}^m}{\sqrt{ds^2 d\bar{s}^2}} \tag{3.2}$$

(compare also section 1.1). The *metric tensor* (the metric) g_{mn} characterizes the space completely (locally). It is symmetric, its determinant g is in general different from zero, and it possesses therefore an inverse g^{in}:

$$g_{mn} = g_{nm}, \qquad |g_{nm}| = g \neq 0, \qquad g^{in}g_{nm} = \delta^i_m = g^i_m. \tag{3.3}$$

A space with the properties (3.1) and (3.3) is accordingly a generalization as much of the two-dimensional surfaces as also of the four-dimensional uncurved (flat) Minkowski space. If ds^2 is positive definite, that is, zero only for $dx^i = 0$ and positive otherwise (and if the parallel transport of a vector is defined as in chapter 5), then we are dealing with a *Riemannian space* in the narrower sense. But, as we know from the special theory of relativity, the physical space-time must have an extra structure: we can distinguish between timelike and spacelike intervals, between clocks and rulers, and there is a light cone with $ds = 0$. Our space is therefore a *pseudo-Riemannian space,* ds^2 can be positive (spacelike), negative (timelike) or null (lightlike); it is a Lorentzian metric. Nevertheless we shall usually use the term Riemannian space (in the broader sense) for it.

In section 3.4 we shall describe how one takes into account the requirement that there exist one timelike and three spacelike directions.

3.2 Geodesics and Christoffel symbols

On a two-dimensional surface we can define geodesics by making them the shortest curve between two points:

$$\int ds = \text{extremum}. \tag{3.4}$$

In a pseudo-Riemannian space, in which ds^2 can also be zero or negative, we encounter difficulties in the application of (3.4), especially for curves with $ds = 0$ (null lines). We therefore start here from the variational principle

$$\int L \, d\lambda = \int \left(\frac{ds}{d\lambda}\right)^2 d\lambda = \int g_{mn} \frac{dx^n}{d\lambda} \frac{dx^m}{d\lambda} d\lambda = \text{extremum}, \tag{3.5}$$

which, as we have shown in section 1.3, is equivalent to (3.4) for $ds \neq 0$. The Lagrange equations of the second kind for the Lagrangian $L = (ds/d\lambda)^2$ give (cf. (1.3))

$$\frac{d^2 x^m}{d\lambda^2} + \Gamma^m_{ab} \frac{dx^a}{d\lambda} \frac{dx^b}{d\lambda} = 0 \tag{3.6}$$

as differential equations of the geodesics. There are four second-order differential equations for the four functions $x^m(\lambda)$, and accordingly geodesics are locally uniquely determined if the initial-point and the initial direction or the initial-point and end-point are given. When later we speak briefly of the separation of two points, we always mean the arc-length of the connecting geodesics. The Christoffel symbols occurring in (3.6) are of course defined as in (1.16) by

$$\Gamma^m_{ab} = \tfrac{1}{2} g^{mn}(g_{an,b} + g_{bn,a} - g_{ab,n}). \tag{3.7}$$

The Lagrangian $L = g_{mn} dx^m dx^n/d\lambda^2$ is a homogeneous function of second degree in the 'velocities' $\dot{x}^n = dx^n/d\lambda$

$$\frac{\partial L}{\partial \dot{x}^n}\dot{x}^n = 2L. \tag{3.8}$$

Because of (3.8) and the Lagrange equation of the second kind we have

$$\frac{dL}{d\lambda} = \frac{\partial L}{\partial x^n}\dot{x}^n + \frac{\partial L}{\partial \dot{x}^n}\ddot{x}^n = \left(\frac{d}{d\lambda}\frac{\partial L}{\partial \dot{x}^n}\right)\dot{x}^n + \frac{\partial L}{\partial \dot{x}^n}\ddot{x}^n$$

$$= \frac{d}{d\lambda}\left(\frac{\partial L}{\partial \dot{x}^n}\dot{x}^n\right) = 2\frac{dL}{d\lambda},$$

and therefore

$$\frac{dL}{d\lambda} = 0, \quad L = g_{nm}\frac{dx^m}{d\lambda}\frac{dx^n}{d\lambda} = \left(\frac{ds}{d\lambda}\right)^2 = C = \text{constant}. \tag{3.9}$$

The constant C can be positive, negative or zero, and we distinguish correspondingly spacelike, timelike and null geodesics. We shall meet timelike geodesics again as paths for particles, and null geodesics as light rays. Because of (3.9) the (affine) parameter λ along a geodesic is clearly determined uniquely up to a linear transformation $\lambda' = a\lambda + b$; for timelike curves we shall identify λ with the proper time τ.

Christoffel symbols are important quantities in Riemannian geometry. We therefore want to investigate more closely their relation to the partial derivatives of the metric tensor given by (3.7). Because of the symmetry $g_{mn} = g_{nm}$ of the metric tensor, the Christoffel symbols too are symmetric in the lower indices:

$$\Gamma^m_{ab} = \Gamma^m_{ba}. \tag{3.10}$$

In four dimensions there are $\binom{5}{2} = 10$ different components of the metric tensor, and therefore because of the additional freedom provided by the upper indices, $4 \times 10 = 40$ distinct Christoffel symbols. But this number is the same as the number of partial derivatives $g_{mn,a}$ of the metric tensor, and it should therefore be possible to express the partial derivatives through the Christoffel symbols, thus solving (3.7). In fact because of (3.3) we have

$$g_{mi}\Gamma^m_{ab} = \tfrac{1}{2}(g_{ai,b} + g_{bi,a} - g_{ab,i}),$$

$$g_{ma}\Gamma^m_{ib} = \tfrac{1}{2}(g_{ia,b} + g_{ab,i} - g_{ib,a}),$$

and adding the two equations we get

$$g_{ia,b} = g_{mi}\Gamma^m_{ab} + g_{ma}\Gamma^m_{ib}. \tag{3.11}$$

3 Riemannian space

The partial derivatives of the determinant g of the metric tensor can also be calculated in a simple manner from the Christoffel symbols. The chain rule implies that

$$\frac{\partial g}{\partial x^b} = \frac{\partial g}{\partial g_{ia}} g_{ia,b}. \tag{3.12}$$

If one now introduces the expansion of the determinant $g = |g_{mn}|$ along the ith row by

$$g = \sum_a g_{ia} G_{ia} \quad \text{(no summation over } i!\text{)}$$

and uses the fact that the elements g^{ia} of the inverse matrix can be expressed through the co-factors G_{ia} according to

$$gg^{ia} = G_{ia},$$

then one finds that

$$\frac{\partial g}{\partial g_{ia}} = gg^{ia}. \tag{3.13}$$

If one also takes into account (3.11), then it follows that

$$\frac{\partial g}{\partial x^b} = gg^{ia} g_{ia,b} = g(\Gamma^a_{ab} + \Gamma^i_{ib}) = 2g\Gamma^a_{ab},$$

and from this finally that

$$\frac{\partial \ln \sqrt{-g}}{\partial x^b} = \frac{1}{2g} \frac{\partial g}{\partial x^b} = \Gamma^a_{ab}. \tag{3.14}$$

In writing the formula thus we have already assumed that g is negative (cf. section 4.2).

3.3 Coordinate transformations

Naturally the physical structure of our space-time manifold is not allowed to depend upon the choice of the coordinates with which we describe it. We now investigate which properties of the metric tensor and of the Christoffel symbols are derivable from this requirement, that is to say, how these quantities behave under a coordinate transformation.

All coordinate transformations of the old coordinates x^n into new coordinates $x^{n'}$ are permitted which guarantee a one-to-one relationship of the form

18　　　　　　　　*Foundations of Riemannian geometry*

$$x^{n'} = x^{n'}(x^n), \qquad \left|\frac{\partial x^{n'}}{\partial x^n}\right| \neq 0. \tag{3.15}$$

We have in (3.15) made use of the convention of distinguishing the new coordinates from the old not by a new symbol (perhaps y^n) but by a dash on the index. This convention is advantageous for many calculations of a general kind, although we shall occasionally deviate from it. With the abbreviation

$$A^{n'}_n = \frac{\partial x^{n'}}{\partial x^n}, \tag{3.16}$$

we obtain from (3.15) the transformation law for the coordinate differentials

$$dx^{n'} = A^{n'}_n dx^n. \tag{3.17}$$

The inverse transformation to (3.15),

$$x^n = x^n(x^{n'}),$$

implies analogously

$$dx^n = A^n_{n'} dx^{n'}, \qquad A^n_{n'} = \frac{\partial x^n}{\partial x^{n'}}, \tag{3.18}$$

and, from (3.17) and (3.18),

$$A^{n'}_n A^n_{m'} = \delta^{n'}_{m'}, \qquad A^n_{n'} A^{n'}_m = \delta^n_m. \tag{3.19}$$

If one performs two successive transformations

$$x^{n''} = x^{n''}(x^{n'}), \qquad x^{n'} = x^{n'}(x^n),$$

then the differentials are

$$dx^{n'} = A^{n'}_n dx^n,$$

$$dx^{n''} = A^{n''}_{n'} dx^{n'},$$

and for the overall transformation,

$$dx^{n''} = A^{n''}_n dx^n, \qquad A^{n''}_n = A^{n''}_{n'} A^{n'}_n. \tag{3.20}$$

Equations (3.17)–(3.20) show that the coordinate transformations are described locally by the 4×4 matrices $A^{n'}_n$, which form a group (matrix multiplication as the group operation).

Occasionally one transforms the coordinate differentials according to the prescription (3.17) *without* the $A^{n'}_n$ being the partial derivatives (3.16) of four functions $x^{n'}$; that is, there exists no associated coordinate transformation $x^{n'}(x^n)$. One then speaks of the introduction of an anholonomic basis (of differentials).

3 Riemannian space

We obtain the prescription for the transformation of the components of the metric tensor from the requirement that lengths and angles should not change under a coordinate transformation; that is, ds^2 is an invariant

$$ds'^2 \equiv g_{n'm'}dx^{n'}dx^{m'} = ds^2 = g_{nm}dx^n dx^m = g_{nm}A_{n'}^n A_{m'}^m dx^{n'}dx^{m'}.$$

Since this equation must hold for arbitrary choice of the $dx^{n'}$, it follows that

$$g_{n'm'} = g_{nm}A_{n'}^n A_{m'}^m. \tag{3.21}$$

The behaviour of the Christoffel symbols under transformations is most easily calculated using the geodesic equation (3.6). Since the variational principle (3.5) was formulated with the help of the invariant quantities ds and $d\lambda$, the geodesic equation must have the form (3.6) in the new coordinates $x^{n'}$ as well

$$\frac{d^2 x^{m'}}{d\lambda^2} + \Gamma^{m'}_{a'b'} \frac{dx^{a'}}{d\lambda} \frac{dx^{b'}}{d\lambda} = 0. \tag{3.22}$$

(The property of a curve, that it be the shortest curve between two points, is independent of the choice of coordinates.) If we substitute into equation (3.22) the equation

$$\frac{d^2 x^{m'}}{d\lambda^2} = A_m^{m'} \frac{d^2 x^m}{d\lambda^2} + A_{m,b}^{m'} \frac{dx^b}{d\lambda} \frac{dx^m}{d\lambda}$$

$$= -A_m^{m'} \Gamma^m_{ab} \frac{dx^a}{d\lambda} \frac{dx^b}{d\lambda} + A_{a,b}^{m'} \frac{dx^b}{d\lambda} \frac{dx^a}{d\lambda},$$

which follows from (3.17) and

$$\frac{dx^{m'}}{d\lambda} = A_m^{m'} \frac{dx^m}{d\lambda},$$

and transform everything to dashed coordinates, we obtain finally

$$\Gamma^{m'}_{a'b'} = A_m^{m'} A_{a'}^a A_{b'}^b \Gamma^m_{ab} - A_{a,b}^{m'} A_{a'}^a A_{b'}^b. \tag{3.23}$$

In this transformation formula it should be noted that the new Christoffel symbols are not homogeneous linear functions of the old. It is therefore quite possible that in a Riemannian space the Christoffel symbols are non-zero in one coordinate system, whilst in another coordinate system they vanish identically. Thus in the usual three-dimensional space the Christoffel symbols are identically zero in Cartesian coordinates, whereas in spherical coordinates they have the values (1.20). We shall answer in section 6.2 the question of whether the Christoffel symbols can always be made to vanish.

3.4 Special coordinate systems

For many calculations and considerations it is convenient to use a special coordinate system. But one must examine in each individual case whether a coordinate system with the desired properties really does exist; that is, whether it is possible for a given metric $g_{a'b'}$ to define the four functions $x^n(x^{n'})$ so that the transformed metric g_{ab} fulfils the chosen requirements.

Orthogonal coordinates If the matrix g_{ab} has only diagonal elements,

$$ds^2 = g_{11}(dx^1)^2 + g_{22}(dx^2)^2 + g_{33}(dx^3)^2 + g_{44}(dx^4)^2, \qquad (3.24)$$

then we are dealing with orthogonal coordinates, and the coordinate lines (lines along which only one coordinate varies at any given time) form right-angles with one another. In the three-dimensional Euclidean space one uses and prefers such coordinates; for example, spherical coordinates or cylindrical coordinates. As a more exact analysis shows, such orthogonal coordinate systems do not in general exist in a four-dimensional Riemannian space, since the system of differential equations,

$$g_{a'b'}\frac{\partial x^{a'}}{\partial x^a}\frac{\partial x^{b'}}{\partial x^b} = 0 \quad \text{for} \quad a \neq b, \qquad (3.25)$$

has no solutions $x^{a'}(x^a)$ which satisfy the conditions (3.15) for arbitrarily given functions $g_{a'b'}$. This result is plausible, since (3.25) is a system of six differential equations for four functions.

Time-orthogonal coordinates We shall customarily choose time as the fourth coordinate: $x^4 = ct$; time-orthogonal coordinates exist when $g_{4\alpha} = 0$. If, moreover, g_{44} has the value (± 1), then we are dealing with *Gaussian coordinates* often called *synchronous* coordinates. Since it follows from $g_{4\alpha} = 0$ that also $g^{4\alpha} = 0$ (and vice versa), in going over to time-orthogonal coordinates we have to satisfy the system

$$g^{4\alpha} = A^\alpha_{a'} A^4_{b'} g^{a'b'} = \frac{\partial x^\alpha}{\partial x^{a'}}\frac{\partial x^4}{\partial x^{b'}} g^{a'b'} = 0, \quad \alpha = 1, 2, 3. \qquad (3.26)$$

One can see that it is still possible to specify arbitrarily the function $x^4(x^{b'})$ and then for every one of the functions $x^\alpha(x^{a'})$ a partial differential equation has to be solved, the existence of the solution being guaranteed by general laws. Time-orthogonal coordinates,

$$ds^2 = g_{\alpha\beta} dx^\alpha dx^\beta + g_{44}(dx^4)^2, \qquad (3.27)$$

3 Riemannian space

can therefore always be introduced (the fact that x^4 has the name 'time' here plays no rôle at all), and also it is still possible to satisfy the additional condition $|g_{44}| = 1$ by choice of the function $x^4(x^{b'})$.

Co-moving coordinates Later applications in Riemannian spaces often deal with a velocity field $u^{\bar{z}} = dx^{\bar{z}}/d\lambda$ (a flux of bodies, or of observers). Since λ is a coordinate-independent parameter, the components of this velocity transform like the coordinate differential

$$u^n = A^n_{n'} u^{n'} = \frac{\partial x^n}{\partial x^{n'}} u^{n'}. \tag{3.28}$$

By means of a coordinate transformation it is always possible to make the three spatial components u^α of the velocity zero, since the differential equations

$$u^1 = \frac{\partial x^1}{\partial x^{1'}} u^{1'} + \frac{\partial x^1}{\partial x^{2'}} u^{2'} + \frac{\partial x^1}{\partial x^{3'}} u^{3'} + \frac{\partial x^1}{\partial x^{4'}} u^{4'} = 0, \ldots \tag{3.29}$$

always have a solution $x^\alpha(x^{n'})$. In the resulting coordinate system, in which the velocity has the form $u^n = (0, 0, 0, u^4)$, the particles do not change their position; the coordinates move with the particles (one can visualize the coordinate values attached to the particles as names). Although the coordinate difference of two particles never alters, their separation can vary because of the time-dependence of the metric.

Local Minkowski system At an arbitrarily given point, which in the following we shall identify with the origin of the coordinates, let the coordinate lines form right-angles with one another. That this is possible is intuitively obvious, and mathematically provable since, with the help of suitable transformation matrices $A^{a'}_a$, one can transform the constant matrix $g_{a'b'}(O)$ to principal axes. Then the metric at the point O,

$$ds'^2 = g_{11}(dx^{1'})^2 + g_{22}(dx^{2'})^2 + g_{33}(dx^{3'})^2 + g_{44}(dx^{4'})^2,$$

can be further simplified to

$$ds^2 = \pm (dx^1)^2 \pm (dx^2)^2 \pm (dx^3)^2 \pm (dx^4)^2 \tag{3.30}$$

by a stretching of the coordinates

$$x^1 = \sqrt{|g_{11}|}\, x^{1'}, \ldots .$$

In the general case of an arbitrary metric no statement can be made about the signs occurring in (3.30). In order to make sure of the connection to

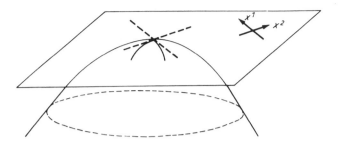

Fig. 3.1. Tangent plane and locally flat coordinate system.

the structure of Minkowski space we *demand* that the spaces used in the general theory of relativity have signature (+2); that is, at every point under transformation of the metric to the form (3.30) three positive signs occur and one negative. We call such spaces *normal hyperbolic pseudo-Riemannian spaces*. One can show that the signature is an invariant, that is to say, it is independent of the choice of the initial coordinate system and of the (not uniquely determined) coordinate transformations, which lead to (3.30) (law of inertia of the quadratic forms).

In physically important spaces (e.g. gravitational fields), there can be singular points, however, at which the metric cannot be brought to the normal form (3.30). Obviously at these points the structure of the space really does depart from that with which we are familiar.

Locally flat (geodesic) system After the introduction of a local Minkowski system the situation at a point is as in a flat four-dimensional space. One can also extend such a system into a (differential) neighbourhood of the point.

This can be illustrated by the example of an arbitrarily curved two-dimensional surface (cf. section 3.1). Suppose one sets up the tangent plane to the surface at the point O under consideration and projects the Cartesian coordinates of the plane onto the surface. Since surface and tangent plane touch, the resulting coordinate lines on the surface differ from the straight lines of the plane only in second order. Applying the same consideration to a four-dimensional space-time, we would project the quasi-Cartesian coordinates of the tangential Minkowski space onto the Riemannian space and expect a metric of the form

$$g_{nm}(x^a) = \eta_{nm} + \tfrac{1}{2} d_{nmab}(O) x^a x^b + \ldots . \tag{3.31}$$

We therefore call such a coordinate system locally flat.

3 Riemannian space

In fact one can always locally transform an arbitrary metric

$$\bar{g}_{nm}(\bar{x}^a) = \bar{g}_{nm}(O) + \bar{g}_{nm,i}(O)\bar{x}^i + \dots \quad (3.32)$$

into the form (3.31). For if one introduces new coordinates x^a by

$$\left.\begin{array}{l} x^a = \bar{x}^a + \tfrac{1}{2}\bar{\Gamma}^a_{mn}(O)x^{\bar{m}}x^{\bar{n}} + \dots, \\ \bar{x}^a = x^a - \tfrac{1}{2}\bar{\Gamma}^a_{mn}(O)x^m x^n + \dots, \end{array}\right\} \quad (3.33)$$

then, because of

$$\frac{\partial \bar{x}^a}{\partial x^n} = \delta^a_n - \bar{\Gamma}^a_{mn}(O)x^m + \dots,$$

the new metric tensor has the form (ignoring terms higher than linear in x^a)

$$g_{mn} = \bar{g}_{ab}(\delta^a_n - \bar{\Gamma}^a_{in}(O)x^i)(\delta^b_m - \bar{\Gamma}^b_{km}(O)x^k).$$

Its partial derivatives

$$\begin{aligned}\left.\frac{\partial g_{mn}}{\partial x^s}\right|_{x^i=0} &= \bar{g}_{ab}(O)(-\bar{\Gamma}^a_{sn}(O)\delta^b_m - \delta^a_n \bar{\Gamma}^b_{sm}(O)) + \bar{g}_{nm,s}(O) \\ &= -\bar{g}_{am}(O)\bar{\Gamma}^a_{sn}(O) - \bar{g}_{nb}(O)\bar{\Gamma}^b_{sm}(O) + \bar{g}_{nm,s}(O)\end{aligned}$$

all vanish, however, since the last row is zero because of (3.11). We have therefore arrived at a metric

$$g_{mn}(x^i) = g_{mn}(O) + \tfrac{1}{2}d_{nmab}(O)x^a x^b,$$

which can be changed into (3.31) by transformation to principal axes and stretching of the axes.

Since in a locally flat coordinate system the partial derivatives of the metric vanish at the point $x^a = 0$, and with them the Christoffel symbols, the geodesic equation (3.6) simplifies locally to

$$\frac{d^2 x^n}{d\lambda^2} = 0, \quad (3.34)$$

that is, the coordinate lines (e.g. x^1 variable; x^2, x^3 and x^4 constant) are geodesics. One therefore also calls such a coordinate system locally geodesic (at $x^a = 0$).

A locally flat coordinate system offers the best approximation to a Minkowski space that is possible in Riemannian geometry. How good this substitution of a curved space by the tangent space is depends upon the magnitudes of the coefficients d_{nmab} in (3.31), from which we can also expect to obtain a measure for the curvature of the space.

24   *Foundations of Riemannian geometry*

3.5 *The physical meaning and interpretation of coordinate systems*

Coordinates are names which we give to events in the Universe; they have in the first instance nothing to do with physical properties. For this reason all coordinate systems are also in principle equivalent, and the choice of a special system is purely a question of expediency. Just as in three-dimensional space for a problem with spherical symmetry one would use spherical coordinates, so, for example, for a static metric one will favour time-orthogonal coordinates (3.27). Because of the great mathematical difficulties in solving problems in the general theory of relativity, the finding of a coordinate system adapted to the problem is often the key to success.

In many applications one is interested in the outcome of measurements performed by a special observer (or a family of observers); then one will link the coordinate system with the observer and the objects which he studies (observer on the rotating Earth, in a satellite...). After having been thus fixed, the coordinate system naturally has a physical meaning, because it is tied to real objects.

In addition to the co-moving coordinates the locally flat coordinate system possesses a particular significance. For an observer at the preferred origin of the coordinate system, particles whose paths are geodesics move force-free, because of (3.34). But geodesics are paths of particles in the gravitational field (as we have made plausible and shall later prove); that is, for the observers just mentioned there exists (locally) no gravitational field: the locally flat coordinate system is the system of a freely falling observer at the point in question of the space-time. This is the best approximation to the Minkowski world, that is, to an inertial system, that Riemannian geometry offers. It is determined in this manner only up to four-dimensional rotations (Lorentz transformations).

Even when one has decided on a particular coordinate system, one should always try to state results in an invariant form; that is, a form independent of the coordinate system. To this end it is clearly necessary to characterize the coordinate system itself invariantly. We shall later familiarize ourselves with the necessary means to do this.

Finally a few remarks on the question of how one can determine the metric tensor g_{ab} when the coordinate system and auxiliary physical quantities have been specified. Specification of the coordinate system means physically that observers possessing rulers and clocks are distributed in the space. Locally, in the infinitesimal neighbourhood of a point, the question is very easy to answer. One takes a freely falling observer, who measures lengths and times in the manner familiar from special relativity, and one then knows the interval ds^2 of two points. One then transforms to the originally given coordinate system; that is, one expresses the result through the coordinates of the observers distributed in the space. Since ds^2 does not change, from $ds^2 = g_{ab}dx^a dx^b$ one can read off the g_{ab} for known ds^2 and dx^a.

In time measurement, which is especially important, one distinguishes between clocks which run (forwards) *arbitrarily* and thereby show coordinate times t (which therefore have no immediate physical significance), and standard clocks which show proper time τ, defined by $ds^2 = -c^2 d\tau^2$. For a clock at rest ($dx^\alpha = 0$) the two times are related by

$$d\tau^2 = -g_{44} dt^2. \tag{3.35}$$

Bibliography to section 3

Textbooks Eisenhart (1949), Fock (1964), Hicks (1971), Schmutzer (1968), Schouten (1954), Schutz (1980), Spivak (1979), Synge (1960)

Monographs and collected works Farnsworth *et al.* (1972)

4 Tensor algebra

In general relativity physical quantities and laws are required to have a simple and well-defined behaviour under coordinate transformations

$$dx^{a'} = \frac{\partial x^{a'}}{\partial x^a} dx^a = A_a^{a'} dx^a, \tag{4.1}$$

just as they do in special relativity. In contrast to Lorentz transformations

$$x^{a'} = L_a^{a'} x^a, \tag{4.2}$$

which are linear transformations of the coordinates with position-independent coefficients $L_a^{a'}$, we shall now be dealing with linear transformations of coordinate differentials with position-dependent coefficients $A_a^{a'}$. But if we restrict ourselves to the investigation of physical quantities at one point, without forming derivatives, then the differences from the rules used in calculating with Lorentz transformations will be trivial: they correspond to the difference between orthogonal and nonorthogonal Cartesian coordinates.

4.1 *Scalars and vectors*

Scalars (invariants) A scalar does not change under coordinate transformations

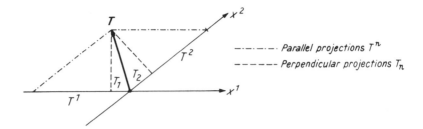

Fig. 4.1. Covariant and contravariant components of a vector **T**.

$$\varphi' = \varphi. \tag{4.3}$$

Its numerical value at a point remains constant even if the coordinates of this point change. For example, the interval ds of two points is a scalar.

Vectors The four quantities T^n are called the *contravariant components* of a vector if they transform like the coordinate differentials

$$T^{n'} = T^n A_n^{n'}. \tag{4.4}$$

This definition implies that the coordinates x^a themselves are not the components of a vector – in a Riemannian space there is no position vector.

Using the prescription

$$T_a = g_{an} T^n, \qquad T^n = g^{na} T_a, \tag{4.5}$$

with the contravariant components T^n (index upstairs) one can associate the *covariant components* T_a (index downstairs). Because of the transformation laws (4.4) and (3.21) and the relation (3.19), the relations

$$T_{a'} = g_{a'n'} T^{n'} = g_{am} A_{a'}^a A_n^m T^n A_n^{n'} = g_{an} T^n A_{a'}^a$$

hold, and therefore

$$T_{a'} = A_{a'}^a T_a. \tag{4.6}$$

Covariant and contravariant components describe the same vector, the difference between them being typical of non-orthogonal coordinates.

Figure 4.1 shows how in the x^1, x^2-plane one obtains the contravariant components by parallel projection onto, and the covariant components by dropping perpendiculars onto, the coordinate axes of a non-orthogonal Cartesian system (with $g_{11} = g_{22} = 1$). A recommended exercise is to examine the behaviour of the components under the transformation

$$x^{1'} = x^1 + bx^2, \qquad x^{2'} = x^2.$$

4.2 Tensors and other geometrical objects

'Geometrical object' is the collective name for all objects whose components Ω^k transform under a given coordinate transformation in such a way that the new components $\bar{\Omega}^k$ are unique functions of the old ones, of the transformation matrix $A^{a'}_a$, and of its derivatives:

$$\bar{\Omega}^k = \bar{\Omega}^k(\Omega^k; A^{a'}_a; A^{a'}_{a,m}; A^{a'}_{a,m,n}; \ldots). \tag{4.7}$$

In this section we shall encounter several geometrical objects which are especially important for physics.

Tensors The quantities $T^a{}_{bc}{}^d{}_{\ldots}$ are the components of a tensor if, with respect to every upper (contravariant) index, they transform like the contravariant components of a vector, and, to every lower (covariant) index, like the covariant components of a vector:

$$T^{a'}{}_{b'c'}{}^{d'}{}_{\ldots} = T^a{}_{bc}{}^d{}_{\ldots} A^{a'}_a A^b_{b'} A^c_{c'} A^{d'}_d \cdots. \tag{4.8}$$

The *rank* of a tensor is equal to the number of its indices.

Corresponding to the rule (4.5) for vectors, we can also transform between covariant and contravariant indices. For example, from (4.8) we can form the covariant tensor (tensor written out in purely covariant components)

$$T_{abcd} = g_{an} g_{dm} T^n{}_{bc}{}^m \tag{4.9}$$

(cf. (4.24)).

Evidently the g_{ab} are the covariant components of a second-rank tensor (cf. (3.21)), whose 'mixed' components coincide with the Kronecker symbol (because of (3.3)),

$$g^i_n = \delta^i_n. \tag{4.10}$$

We shall meet more examples of tensors later.

Tensor densities If we transform the determinant $g = |g_{ab}|$ of the metric tensor to another coordinate system, then we obtain

$$g' = |g_{a'b'}| = |g_{ab} A^a_{a'} A^b_{b'}| = |g_{ab}| |A^a_{a'}| |A^b_{b'}|;$$

that is,

$$g' = |A^a_{a'}|^2 g = \left| \frac{\partial x^a}{\partial x^{a'}} \right|^2 g. \tag{4.11}$$

The square of the functional determinant (Jacobian) $|\partial x^a/\partial x^{a'}|$ occurs in the transformation formula for g; we are dealing with a scalar density of weight 2.

In general we speak of a tensor density of weight W, whenever

$$T^{a'\cdots}{}_{b'}{}^{c'\cdots}{}_{\cdots} = |A_n^{n'}|^W T^{a\cdots}{}_b{}^{c\cdots}{}_{\cdots} A_a^{a'} A_{b'}^b A_c^{c'} \cdots . \qquad (4.12)$$

We can draw an important conclusion from equation (4.11); since we admit only normal hyperbolic Riemannian spaces, which at every point allow the introduction of a local Minkowski system with $g' = |\eta_{mn}| = -1$, then from (4.11) the sign of g does not change under an arbitrary coordinate transformation, and so g is always negative.

Pseudo-tensors In the transformation law of a pseudo-tensor (compared with that of a tensor) there occurs also the sign of one of the elements of the transformation matrix $A_{a'}^a$ or of a combination of its elements. A simple example is a pseudo-vector,

$$T^{n'} = \operatorname{sgn}|A_a^{a'}| T^n A_n^{n'}, \qquad (4.13)$$

in whose transformation law the sign of the functional determinant of the coordinate transformation occurs. Under coordinate transformations possessing locally the character of a rotation it behaves like a vector; under reflections it also changes its sign.

The ϵ-pseudo-tensor In special relativity the Levi-Civita symbol Δ^{abcd} is used repeatedly. It is so defined that under interchange of two arbitrary indices its sign changes, and

$$\Delta^{1234} = 1 \qquad (4.14)$$

always holds. Obviously because of this definition all the components with two identical indices vanish. One can show that this symbol is a pseudo-tensor with respect to Lorentz transformations; it will therefore also be referred to as the ϵ-pseudo-tensor.

Under arbitrary coordinate transformations the Levi-Civita symbol does not, however, behave like a tensor; (4.14) is, for example (applying the transformation formula (4.8)), not invariant with respect to the change of scale $x^{a'} = \text{constant } x^a$. One can, however, associate the Levi-Civita symbol with a tensor in a Riemannian space, and thereby generalize the ϵ-tensor.

We define this generalised ϵ-tensor so that upon introduction of a local Minkowski system it coincides at every point with the Levi-Civita symbol (M = Minkowski system):

4 Tensor algebra

$$\overset{M}{\epsilon}{}^{a'b'c'd'} = \Delta^{a'b'c'd'}. \tag{4.15}$$

If we transform this equation from the local Minkowski system $x^{n'}$ to an arbitrary coordinate system x^n, then we have

$$\epsilon^{abcd} = \pm \Delta^{a'b'c'd'} A^a_{a'} A^b_{b'} A^c_{c'} A^d_{d'}. \tag{4.16}$$

The ambiguity of sign appearing in (4.16) has its origin in the pseudo-tensorial property of the Levi-Civita symbol. For example, (4.15) must again hold for a pure reflection $x^{1'} = -x^1$, that is, for $A^1_{1'} = -1$, $A^2_{2'} = A^3_{3'} = A^4_{4'} = 1$, $A^a_{a'} = 0$ otherwise, which once again leads to a Minkowski system; the components of the ϵ-tensor must not change.

One can, however, bring (4.16) into a more easily manageable form. If one imagines that the indices $abcd$ in (4.16) are held fixed, for example, in the combination 1234, then the right-hand side of the formula

$$\epsilon^{1234} = \pm \Delta^{a'b'c'd'} A^1_{a'} A^2_{b'} A^3_{c'} A^4_{d'}$$

is (up to a sign) precisely the determinant $A^n_{n'}$. Since in every Minkowski system $g' = -1$, then because of (4.11) this determinant has the value

$$|A^n_{n'}| = \pm \frac{1}{\sqrt{-g}},$$

so that from (4.16), with

$$\epsilon^{abcd} = \pm \Delta^{abcd} |A^n_{n'}|,$$

it follows that

$$\epsilon^{abcd} = \frac{\Delta^{abcd}}{\sqrt{-g}}, \tag{4.17}$$

which we can also take as a definition of the ϵ-tensor. The tensorial (pseudo-tensorial) property of the ϵ-tensor can be easily verified in this equation by the use of (4.11) and the rules governing determinants.

In analogy to (4.5) we obtain the covariant components of the ϵ-tensor from

$$\epsilon_{abcd} = g_{an} g_{bm} g_{cp} g_{dq} \epsilon^{nmpq} = g_{an} g_{bm} g_{cp} g_{dq} \frac{\Delta^{nmpq}}{\sqrt{-g}}$$

$$= \frac{-g}{\sqrt{-g}} \Delta_{abcd},$$

so that

$$\epsilon_{abcd} = \sqrt{-g}\, \Delta_{abcd}. \tag{4.18}$$

Here Δ_{abcd} comes from Δ^{abcd} on 'lowering of indices' by the metric η_{mn} of the Minkowski space. In particular we have $\Delta_{1234} = -1$.

Christoffel symbols From the transformation law (3.23) for the Christoffel symbols, that is, from

$$\Gamma^{m'}_{a'b'} = A^{m'}_m A^a_{a'} A^b_{b'} \Gamma^m_{ab} - A^{m'}_{a,b} A^a_{a'} A^b_{b'},$$

we deduce that they are geometrical objects, but not tensors.

Two-point tensors Two-point tensors are not geometrical objects in the strict sense. They appear in the description of physical processes in which a cause at a point P brings about an effect at the point $\bar{P}$. Their indices refer to the points $\bar{P}$ and P, and are written respectively with and without a bar over the index. Accordingly the transformation law for a two-point tensor, for example, reads

$$T_{\bar{a}'n'}(\bar{P}, P) = T_{an}(\bar{P}, P) A^{\bar{a}}_{\bar{a}'}(\bar{P}) A^n_{n'}(P). \tag{4.19}$$

An example of a two-point scalar is the arc-length of a geodesic which connects the points P and $\bar{P}$. We shall meet an example of a tensor in section 5.3 (the parallel propagator).

4.3 Tensor algebra

In tensor algebra only those operations are permitted which from a tensor produce another tensor. Since the transformation matrices $A^{a'}_a$ are position-dependent, one may in general combine tensors with one another only at the same point. Scalars form an exception; the addition of the values of a scalar function at two points again gives a scalar. The following rules are valid solely for tensors which are defined at the same point. It will seldom be proved explicitly that the operations described do lead again to a tensor – filling the gaps is recommended as an exercise.

Addition One adds tensors of the same rank and same index-form by adding their components

$$T^{ab} + S^{ab} = R^{ab}. \tag{4.20}$$

Structures of the form $T^a + S^{an}$ or $T^a{}_b + S^{ab}$, for example, are forbidden.

Multiplication Multiplication of the components of an nth rank tensor by those of an mth rank tensor produces an $(n+m)$th rank tensor, for example,

$$S^a{}_b{}^c T^{np}{}_q = N^{acnp}{}_{bq}. \tag{4.21}$$

4 Tensor algebra

Contraction Summing over a covariant and a contravariant index of a tensor gives another tensor, whose rank is reduced by 2:

$$T^{ab}{}_{nm} \to T^{ab}{}_{am} = S^{b}{}_{m}. \tag{4.22}$$

Let us prove the tensor property. From

$$T^{a'b'}{}_{n'm'} = A^{a'}_a A^{b'}_b A^{n}_{n'} A^{m}_{m'} T^{ab}{}_{nm}$$

we have, upon using (3.19),

$$T^{a'b'}{}_{a'm'} = S^{b'}{}_{m'} = A^{a'}_a A^{b'}_b A^{n}_{a'} A^{m}_{m'} T^{ab}{}_{nm}$$
$$= \delta^{n}_a A^{b'}_b A^{m}_{m'} T^{ab}{}_{nm}$$
$$= A^{b'}_b A^{m}_{m'} S^{b}{}_{m}.$$

The simplest example of a contraction is the *trace* $T = T^{a}{}_{a}$ of a second-rank tensor.

Inner product, raising and lowering of indices, scalar product The multiplication of two tensors with simultaneous contraction over indices of the two factors is called taking the inner product:

$$T^{na}{}_p \cdot S^{rp} = N^{nar}. \tag{4.23}$$

An important example of this operation is the raising and lowering of indices by contraction with the metric tensor, by means of which one can interchange covariant and contravariant components:

$$T^n = g^{na} T_a, \qquad T^{nr}{}_{pq} = g_{pm} g^{rs} T^{n}{}_{s}{}^{m}{}_{q}. \tag{4.24}$$

Another example is the scalar product of two vectors:

$$a_n b^n = g_{rn} a^r b^n = g^{ik} a_i b_k, \tag{4.25}$$

with the special case

$$a^2 = a_n a^n. \tag{4.26}$$

The sign of the invariant a^2 can serve to characterize the vector in a way independent of the coordinate system. As in special relativity one calls vectors with $a^2 > 0$ spacelike, those with $a^2 < 0$ timelike, and those with $a^2 = 0$ lightlike (null).

The quotient law A structure $N^{ab\cdots}{}_{pq\cdots}$ is a tensor if, and only if, the product with every tensor $T^{pq\cdots}{}_{ab\cdots}$ is an invariant:

$$N^{ab\cdots}{}_{pq\cdots} T^{pq\cdots}{}_{ab\cdots} = \text{invariant}.$$

As an example, the proof of this rule for the differential form

$$B_n dx^n = \text{invariant} \tag{4.27}$$

will be exhibited.

If B_n is a vector, the invariance of the scalar product follows trivially. Conversely it has to be shown that the correct transformation law for B_n follows from (4.27). And indeed

$$(B_n)' dx^{n'} = B_n dx^n = B_m \delta_n^m dx^n = B_m A_{n'}^m A_n^{n'} dx^n = B_m A_{n'}^m dx^{n'}$$

can only hold for all possible directions $dx^{n'}$ if

$$(B_n)' = B_m A_{n'}^m.$$

Formulae for products of ϵ-tensors If during a calculation a product of two ϵ-tensors occurs, then it can be expressed in terms of the Kronecker symbol $\delta_b^a = g_b^a$ and of the metric tensor g_{ab} as follows:

$$\begin{aligned}
\epsilon_{abcd}\epsilon^{pqnm} = &-g_a^p g_b^q g_c^n g_d^m + g_a^q g_b^n g_c^m g_d^p - g_a^n g_b^m g_c^p g_d^q \\
&+ g_a^m g_b^p g_c^q g_d^n + g_a^q g_b^p g_c^n g_d^m - g_a^p g_b^n g_c^m g_d^q \\
&+ g_a^n g_b^m g_c^q g_d^p - g_a^m g_b^q g_c^p g_d^n + g_a^n g_b^q g_c^p g_d^m \\
&- g_a^q g_b^p g_c^m g_d^n + g_a^p g_b^m g_c^n g_d^q - g_a^m g_b^n g_c^q g_d^p \\
&+ g_a^m g_b^q g_c^n g_d^p - g_a^q g_b^n g_c^p g_d^m + g_a^n g_b^p g_c^m g_d^q - g_a^p g_b^m g_c^q g_d^n \\
&+ g_a^p g_b^n g_c^q g_d^m - g_a^n g_b^q g_c^m g_d^p + g_a^q g_b^m g_c^p g_d^n - g_a^m g_b^p g_c^n g_d^q \\
&+ g_a^p g_b^q g_c^m g_d^n - g_a^q g_b^m g_c^n g_d^p + g_a^m g_b^n g_c^p g_d^q - g_a^n g_b^p g_c^q g_d^m,
\end{aligned} \tag{4.28}$$

$$\begin{aligned}
\epsilon_{abcd}\epsilon^{aqnm} = &-g_b^q g_c^n g_d^m - g_b^n g_c^m g_d^q - g_b^m g_c^q g_d^n \\
&+ g_b^q g_c^m g_d^n + g_b^m g_c^n g_d^q + g_b^n g_c^q g_d^m,
\end{aligned} \tag{4.29}$$

$$\epsilon_{abcd}\epsilon^{abnm} = -2(g_c^n g_d^m + g_c^m g_d^n), \tag{4.30}$$

$$\epsilon_{abcd}\epsilon^{abcm} = -6g_d^m, \tag{4.31}$$

and

$$\epsilon_{abcd}\epsilon^{abcd} = -24. \tag{4.32}$$

Formula (4.28) follows from (4.17) and (4.18) and the symmetry properties of the Levi-Civita symbol. In particular, the components of this symbol only differ from zero when all four indices have different values. The product therefore fails to vanish only when the values of the indices coincide pairwise. The remaining formulae follow from (4.28) by contraction, noticing that $g_a^a = 4$.

4.4 Symmetries of tensors

A tensor is called symmetric with respect to two indices r, s which are either both contravariant or both covariant if its components do not alter under interchange of these indices:

$$T^{ab}{}_{rs} = T^{ab}{}_{sr}. \tag{4.33}$$

It is called antisymmetric with respect to a pair of indices a and b if its sign alters under interchange of the indices:

$$T^{ab}{}_{rs} = -T^{ba}{}_{rs}. \tag{4.34}$$

These symmetry properties remain preserved under coordinate transformations.

The symmetric part with respect to two indices a and b of an arbitrary tensor is the sum of the component and its permutation:

$$T_{(a|bc|m)} = \tfrac{1}{2}(T_{abcm} + T_{mbca}). \tag{4.35}$$

One obtains the antisymmetric part analogously:

$$T_{[a|bc|m]} = \tfrac{1}{2}(T_{abcm} - T_{mbca}). \tag{4.36}$$

Here we have used the convention of *Bach brackets:* round brackets denote symmetrization, square brackets antisymmetrization. Indices in the brackets not touched by the formula are to be set between vertical lines. This convention is especially useful when one symmetrizes or antisymmetrizes tensors with respect to several indices. (One symmetrizes by forming the sum of the tensor components with all permutations of the indices and dividing by the number of permutations; to antisymmetrize one adds even permutations and subtracts odd permutations and then divides by the number of permutations.) For example,

$$T_{(n_1 \ldots n_\nu)} = \frac{1}{\nu!} \{T_{n_1 n_2 \ldots n_\nu} + T_{n_2 n_1 \ldots n_\nu} + \cdots + T_{n_\nu n_1 n_2 \ldots} + \cdots\}, \tag{4.37}$$

and

$$T_{[abc]} = \frac{1}{3!} \{T_{abc} - T_{bac} + T_{bca} - T_{cba} + T_{cab} - T_{acb}\}. \tag{4.38}$$

One sees the advantage of this convention by applying it to formula (4.28), which can be written simply as

$$\epsilon_{abcd} \epsilon^{pqnm} = -24 g_a^{[p} g_b^q g_c^n g_d^{m]}. \tag{4.39}$$

Tensors which are symmetric or antisymmetric with respect to all indices are called completely symmetric or completely antisymmetric, respectively.

A completely antisymmetric tensor of the third rank T_{abc} has exactly four essentially different components, for example, T_{123}, T_{124}, T_{134} and T_{234}, and therefore precisely the same number as a vector. One can exploit this fact to map it to a pseudo-vector T^n with the aid of the ϵ-tensor:

$$\epsilon^{abcn} T_{abc} = T^n, \qquad T_{abc} = \frac{1}{3!} \epsilon_{nabc} T^n, \tag{4.40}$$

in analogy to the mapping of an antisymmetric tensor (for example, the vector product of two vectors) to a pseudo-vector in three-dimensional Euclidean space.

A completely antisymmetric tensor of the fourth rank has essentially only one component, and, with the aid of the ϵ-tensor, can be mapped onto a pseudo-scalar:

$$\epsilon_{abcd} T^{abcd} = T; \tag{4.41}$$

it is proportional to the ϵ-tensor.

In a four-dimensional space there are no completely antisymmetric tensors of rank higher than four.

4.5 Algebraic properties of second-rank tensors

An arbitrary second-rank tensor can be decomposed into its symmetric and antisymmetric parts, and the symmetric part can be decomposed further into a trace-free term and a term proportional to the metric tensor;

$$\begin{aligned} T_{ab} &= T_{[ab]} + T_{(ab)} \\ &= T_{[ab]} + \{T_{(ab)} - \tfrac{1}{4} T^n_n g_{ab}\} + \tfrac{1}{4} T^n_n g_{ab}. \end{aligned} \tag{4.42}$$

The physically important second-rank tensors often belong to one of the symmetry classes, or at least their constituent parts have different physical meanings. Thus, for example, the electromagnetic field tensor is antisymmetric, the energy-momentum tensor of the electromagnetic field is symmetric and trace-free and the general energy–momentum tensor is symmetric.

Because of the particular importance of symmetric and antisymmetric tensors of second rank we shall examine more closely their algebraic properties (eigenvectors, eigenvalues, normal forms).

The metric tensor The defining equation

$$g_{ab} w^b = \lambda w_a \tag{4.43}$$

4 Tensor algebra

for an eigenvector w_a of the metric tensor is (trivially) satisfied with $\lambda = 1$ for every vector w_a. Every vector is an eigenvector of the metric tensor; g_{ab} singles out no direction in the space.

Symmetric tensors From the eigenvector equation for a symmetric tensor T_{ab},

$$T_{ab} w^b = \lambda w_a, \quad (4.44)$$

or

$$(T_{ab} - \lambda g_{ab}) w^b = 0, \quad (4.45)$$

which can be regarded as a linear system of equations for the w^b, the condition for the existence of a solution follows immediately, viz. the secular equation,

$$|T_{ab} - \lambda g_{ab}| = 0. \quad (4.46)$$

The eigenvalues λ can be determined from this equation. Under a coordinate transformation the secular equation is only multiplied by the square of the determinant $|A^a_{a'}|$ (because of (4.8) and (4.11)), and so the eigenvalues are invariant, and with them also the coefficients α_A of the equation

$$\lambda^4 + \alpha_1 \lambda^3 + \alpha_2 \lambda^2 + \alpha_3 \lambda + \alpha_4 = 0, \quad (4.47)$$

which follows from (4.46). As the α_A are derived from the components of the tensor T_{ab} and of the metric tensor by algebraic operations, they are algebraic invariants of the tensor T_{ab}. One can show that all other algebraic invariants can be constructed out of the α_A. The invariance property is also recognizable directly in, for example,

$$\alpha_1 = T^n_n, \qquad \alpha_4 = |T_{ab}| g^{-1}.$$

Equation (4.47) gives in general four different eigenvalues λ from which the eigenvectors can be determined. We shall not go into the details and special cases here, but instead indicate an important property. Whilst in three-dimensional space one can always transform simultaneously a symmetric tensor and the metric tensor to principal axes, in a pseudo-Riemannian four-dimensional space this is not always possible; that is, there are tensors which cannot be brought to diagonal form in a local Minkowski system by a real transformation. This possibility is intimately connected with the occurrence of null vectors, as can be seen in the example

$$g_{ab} = \eta_{ab} = \begin{pmatrix} 1 & & & \\ & 1 & & \\ & & 1 & \\ & & & -1 \end{pmatrix}, \quad T_{ab} = 2k_a k_b = \begin{pmatrix} 0 & & & \\ & 0 & & \\ & & 1 & -1 \\ & & -1 & 1 \end{pmatrix},$$

$$k_a = 2^{-1/2}(0, 0, 1, 1) \tag{4.48}$$

in which the tensor T_{ab} is constructed from the null vector k_a alone and cannot be transformed to principal axes by Lorentz transformations ($g_{ab} = \eta_{ab}$ invariant!).

Antisymmetric tensors An antisymmetric tensor F_{ab} can, of course, never be brought to diagonal form, but nevertheless the question of eigenvalues and eigenvectors is again significant. If the tensor F_{ab} is antisymmetric, then the eigenvalue equation

$$F_{ab} w^b = \lambda w_a \tag{4.49}$$

implies, by contraction with w^a, the relation

$$\lambda w_a w^a = 0,$$

that is, either the eigenvalue is zero, or the eigenvector w^a is a null vector (or both).

The antisymmetry of F_{ab} also implies that

$$|F_{ab} - \lambda g_{ab}| = |-F_{ba} - \lambda g_{ba}| = |F_{ba} + \lambda g_{ba}|,$$

and since one can interchange rows and columns in the determinant it follows that

$$|F_{ab} - \lambda g_{ab}| = |F_{ab} + \lambda g_{ab}|.$$

The secular equation

$$|F_{ab} - \lambda g_{ab}| = 0 \tag{4.50}$$

therefore transforms into itself when λ is replaced by $(-\lambda)$, and hence contains only even powers of λ:

$$\lambda^4 + \beta_2 \lambda^2 + \beta_4 = 0. \tag{4.51}$$

It thus furnishes only two invariants, β_2 and β_4.

Every antisymmetric tensor F_{ab} can be dualized; that is, with the aid of the ϵ-tensor its associated dual (pseudo-) tensor can be constructed:

$$\tilde{F}^{ab} = \tfrac{1}{2} \epsilon^{abcd} F_{cd}. \tag{4.52}$$

Because of the property (4.30) of the ϵ-tensor, a double application of the duality operation yields the original tensor, apart from a sign:

4 Tensor algebra

$$\tilde{F}_{nm} = \tfrac{1}{2}\epsilon_{nmab}\tilde{F}^{ab} = \tfrac{1}{4}\epsilon_{nmab}\epsilon^{abcd}F_{cd} = -F_{nm}. \tag{4.53}$$

One can show that the two invariants β_2 and β_4 can be simply expressed in terms of F_{ab} and $\tilde{F}_{ab}$. In fact

$$\beta_2 = F_{ab}F^{ab}, \qquad \beta_4 = (F_{ab}\tilde{F}^{ab})^2. \tag{4.54}$$

$F_{ab}\tilde{F}^{ab}$ is a pseudo-invariant.

An antisymmetric tensor is also called a *bivector*. If it can be formed out of two vectors a^i and b^i according to

$$F_{nm} = a_n b_m - a_m b_n, \tag{4.55}$$

then it is a *simple* or *decomposable bivector*. Since there are four linearly independent vectors in a four-dimensional space, one can construct out of them $\binom{4}{2} = 6$ bivectors, and every antisymmetric tensor (with six independent components) can be built up from these six linearly independent bivectors. We shall pursue this idea further in chapter 18, where the classification of antisymmetric tensors will be developed.

4.6* Tetrad and spinor components of tensors

Tetrads At every point of the space one can introduce systems of four linearly independent vectors $h_a^{(r)}$, which are known as tetrads. The index in brackets is the tetrad index; it numbers the vectors from one to four. These four vectors can have arbitrary lengths and form arbitrary angles with one another (as long as they remain linearly independent). The matrix

$$g^{(r)(s)} = h_a^{(r)} h_b^{(s)} g^{ab} \tag{4.56}$$

is an arbitrary symmetric matrix with negative-definite determinant. Its inverse $g_{(s)(t)}$, which is defined by

$$g_{(s)(t)} g^{(t)(r)} = \delta_{(s)}^{(r)} = g_{(s)}^{(r)}, \tag{4.57}$$

can be used to define tetrad vectors with tetrad indices downstairs

$$h_{(r)a} = g_{(r)(s)} h_a^{(s)}, \tag{4.58}$$

and to solve (4.56) for g_{ab}:

$$g_{ab} = g_{(r)(s)} h_a^{(r)} h_b^{(s)}. \tag{4.59}$$

Tetrad components of tensors Just as one can write any arbitrary vector as a linear combination of the four tetrad vectors, so one can use them to describe any tensor

$$T^{ab\ldots}{}_{nm\ldots} = T^{(r)(s)\ldots}{}_{(p)(q)\ldots} h^a_{(r)} h^b_{(s)} h^{(p)}_n h^{(q)}_m \ldots . \qquad (4.60)$$

The quantities $T^{(r)(s)\ldots}{}_{(p)(q)\ldots}$ are called the tetrad components of the tensor. They are calculated according to

$$T^{(r)(s)\ldots}{}_{(p)(q)\ldots} = T^{ab\ldots}{}_{nm\ldots} h^{(r)}_a h^{(s)}_b h^n_{(p)} h^m_{(q)} \ldots , \qquad (4.61)$$

which is consistent with (4.56) and (4.58). Tetrad indices are raised and lowered with $g^{(r)(s)}$ and $g_{(r)(s)}$, respectively.

Coordinate and tetrad transformations The advantages offered in many cases by the use of the tetrad components, which at first look very complicated, become clear when one examines their transformation properties and when one introduces tetrads which are appropriate to the particular problem being investigated.

As one can see from a glance at the defining equation (4.61), the tetrad components behave like scalars under coordinate transformations; clearly the labelling of the tetrad vectors, that is, their tetrad indices, does not change under a coordinate transformation. One has therefore a good way of investigating the algebraic properties of tensors and can simplify tensor components (that is, tetrad components) in a coordinate-independent fashion by the choice of the tetrads.

Besides the coordinate transformations – and completely independently of them – one can introduce a new tetrad system through a linear (position-dependent) transformation of the tetrad vectors $h^{(r)}_a$ at every point in the space:

$$h^{(r)'}_a = A^{(r)'}_{(r)} h^{(r)}_a, \quad h_{(r)'a} = A^{(r)}_{(r)'} h_{(r)a}, \quad A^{(r)'}_{(r)} A^{(r)}_{(s)'} = \delta^{(r)'}_{(s)'}. \qquad (4.62)$$

Under such transformations, of course, the tetrad components of tensors alter; indeed they will be transformed with the matrices $A^{(r)'}_{(r)}$ and $A^{(r)}_{(r)'}$, respectively, for example,

$$g_{(s)'(t)'} = g_{(s)(t)} A^{(s)}_{(s)'} A^{(t)}_{(t)'}. \qquad (4.63)$$

Special tetrad systems We can choose the tetrads in such a way that the four vectors at each point are in the directions of the coordinate axes; that is, parallel to the four coordinate differentials dx^a:

$$h^a_{(r)} = \delta^a_{(r)}, \qquad g_{rs} = g_{(r)(s)}. \qquad (4.64)$$

This choice has the consequence that tetrad and tensor components coincide. But on the other hand, given an arbitrary tetrad system in the space, it is not always possible to transform the coordinates so that the tetrads become tangent vectors to the coordinate lines.

4 Tensor algebra

A second important possibility is the identification of the tetrad vectors with the base vectors of a Cartesian coordinate system in the local Minkowski system of the point concerned:

$$g_{(r)(s)} = h^a_{(r)} h^b_{(s)} g_{ab} = \eta_{(r)(s)} = \begin{pmatrix} 1 & & & \\ & 1 & & \\ & & 1 & \\ & & & -1 \end{pmatrix}. \quad (4.65)$$

The four tetrad vectors, which we shall call z_a, w_a, v_a and u_a/c, form an orthonormal system of one timelike and three spacelike vectors. From (4.59) and (4.65), it follows that the metric tensor can be written as

$$g_{ab} = z_a z_b + w_a w_b + v_a v_b - u_a u_b/c^2. \quad (4.66)$$

A third special case is the use of null vectors as tetrad vectors. In our pseudo-Riemannian space of signature two there are two linearly independent null vectors, which we call k_a and l_a, and which can be constructed from the orthonormal system introduced above by

$$k_a = \frac{1}{\sqrt{2}}(u_a/c + v_a), \quad l_a = \frac{1}{\sqrt{2}}(u_a/c - v_a). \quad (4.67)$$

One can introduce two more null vectors t^a and $\bar{t}^a$ (which are complex conjugates of each other) by

$$t_a = \frac{1}{\sqrt{2}}(z_a - iw_a), \quad \bar{t}_a = \frac{1}{\sqrt{2}}(z_a + iw_a). \quad (4.68)$$

From (4.56) and (4.57) we obtain

$$g_{(r)(s)} = \begin{pmatrix} 0 & 1 & & \\ 1 & 0 & & \\ & & 0 & -1 \\ & & -1 & 0 \end{pmatrix}; \quad (4.69)$$

that is, of all possible scalar products of the basis vectors the only ones which differ from zero are

$$t_a \bar{t}^a = 1, \quad k_a l^a = -1. \quad (4.70)$$

According to (4.59), in this tetrad system the metric tensor has the form

$$g_{ab} = t_a \bar{t}_b + \bar{t}_a t_b + k_a l_b + l_a k_b. \quad (4.71)$$

Using this system, complex tetrad components can arise, although we have always allowed only real coordinates, transformations and tensors.

Spinors First-rank spinors are elements of a two-dimensional, complex vector-space, in which an alternative scalar product

$$[\varphi, \psi] = -[\psi, \varphi] \tag{4.72}$$

is defined. A spinor φ can be represented either by its contravariant components φ^A or by its covariant components φ_A. The scalar product of two spinors can be formed from these components with the help of the *metric spinor* ϵ_{AB},

$$[\varphi, \psi] = \epsilon_{AB} \varphi^A \psi^B = -\epsilon_{AB} \psi^A \varphi^B, \quad A, B = 1, 2. \tag{4.73}$$

The metric spinor is antisymmetric:

$$\epsilon_{AB} = -\epsilon_{BA}. \tag{4.74}$$

Together with its inverse, defined by

$$\epsilon_{AB} \epsilon^{CB} = \delta_A^C, \tag{4.75}$$

it can be used to shift indices:

$$\varphi^A = \epsilon^{AB} \varphi_B, \qquad \varphi_B = \varphi^A \epsilon_{AB}. \tag{4.76}$$

The scalar products (4.72) and (4.73) do not change if one carries out a unimodular transformation

$$\left.\begin{array}{c} \varphi^{A'} = \Lambda_A^{A'} \varphi^A, \qquad \varphi_{A'} = \Lambda_{A'}^{A} \varphi_A, \\ |\Lambda_A^{A'}| = 1, \qquad \Lambda_A^{A'} \Lambda_{B'}^{A} = \delta_{B'}^{A'}. \end{array}\right\} \tag{4.77}$$

The connection between the group of the unimodular transformations and the group of Lorentz transformations isomorphic to it plays a great rôle in special-relativistic field theory.

We denote quantities which transform with the complex matrix $\overline{(\Lambda_A^{A'})} = \Lambda_{\dot{A}}^{\dot{A}'}$ by a dot‡ over the index $\varphi_{\dot{A}}, \psi^{\dot{B}}, \ldots$. They obey

$$\varphi^{\dot{A}'} = \Lambda_{\dot{A}}^{\dot{A}'} \varphi^{\dot{A}}. \tag{4.78}$$

Scalar products $\varphi_{\dot{A}} \psi^{\dot{A}} = \epsilon_{\dot{A}\dot{B}} \varphi^{\dot{A}} \psi^{\dot{B}}$ remain invariant under such transformations. According to this convention one forms the conjugate complex of a spinor by dotting the index (with $\dot{A} \times A$, naturally):

$$\overline{(\varphi_A)} = \varphi_{\dot{A}}. \tag{4.79}$$

Spinors $\chi_{A\dot{B}...}^{M\dot{N}...}$ of higher rank are structures which behave with respect to unimodular transformations of each index like the corresponding

---

‡ *Editor's note:* Although this was the original convention it has been largely superseded by a dash, $\varphi_{A'}$, which is more legible, especially in handwritten formulae.

4 Tensor algebra

first-rank spinor. The rules for handling these spinors follow from the properties of first-rank spinors sketched above. Notice that upon multiplication and contraction, only summation over a contravariant and a covariant index of the same type (that is, dotted *or* undotted) yields a spinor again.

A spinor is *Hermitian* if it obeys the condition

$$\varphi_{A\dot{B}} = \overline{\varphi_{B\dot{A}}}. \tag{4.80}$$

Spinor components and tensors With the aid of the metric spin-tensors $\sigma_m{}^{A\dot{B}} = \sigma_m{}^{\dot{B}A}$, which are generalizations of the Pauli spin-matrices, one can map the four complex components $\varphi_{A\dot{B}}$ of an arbitrary second-rank spinor onto the four (now also complex in general) components of a four-vector:

$$\left. \begin{array}{l} T^a = \dfrac{1}{\sqrt{2}} \sigma^a{}_{A\dot{B}} \varphi^{A\dot{B}}, \\[2mm] T^a T_a = -\varphi_{A\dot{B}} \varphi^{A\dot{B}}, \\[2mm] \varphi_{A\dot{B}} = -\dfrac{1}{\sqrt{2}} \sigma_{aA\dot{B}} T^a. \end{array} \right\} \tag{4.81}$$

Here the four two-by-two matrices $\sigma_a{}^{A\dot{B}}$ satisfy the equations

$$\left. \begin{array}{l} \sigma^a{}_{A\dot{B}} \sigma_{aC\dot{D}} = -2\epsilon_{AC} \epsilon_{\dot{B}\dot{D}}, \\[2mm] \sigma_m{}^{A\dot{B}} \sigma_{nA\dot{B}} = -2g_{mn}. \end{array} \right\} \tag{4.82}$$

In analogous fashion one can map every nth-rank tensor to a spinor of rank $2n$.

From two basis spinors χ_A and μ_A, which satisfy the relations

$$\chi_A \mu^A = -\mu_A \chi^A = 1, \qquad \chi_A \chi^A = \mu_A \mu^A = 0, \tag{4.83}$$

one can form four second-rank spinors

$$\left. \begin{array}{ll} \chi_A \mu_{\dot{B}} = t_{A\dot{B}}, & \chi_A \chi_{\dot{B}} = k_{A\dot{B}}, \\[2mm] \mu_A \chi_{\dot{B}} = \bar{t}_{A\dot{B}}, & \mu_A \mu_{\dot{B}} = l_{A\dot{B}}. \end{array} \right\} \tag{4.84}$$

The vectors t_a, $\bar{t}_a$, k_a and l_a associated with them according to the prescription (4.81) satisfy the relations (4.69) and (4.70) of the null-tetrad system. Thus there exists a close relation between the representation of a tensor by its spinor components and its representation by components related to a null tetrad.

Bibliography to section 4

Textbooks   Eisenhart (1949), Hicks (1971), Penrose and Rindler (1984), Schmutzer (1968), Schouten (1954), Schutz (1980), Spivak (1979)

Review and research articles   Newman and Penrose (1962)

5 The covariant derivative and parallel transport

5.1 *Partial and covariant derivatives*

Physical laws are usually written down in mathematical form as differential equations. In order to guarantee that the laws are independent of the coordinate system, they should moreover have the form of tensor equations. We must therefore examine whether tensors can be differentiated in such a way that the result is again a tensor, and if so, how this can be done.

The partial derivative   We denote the usual partial derivative of a position-dependent tensor by a comma:

$$\frac{\partial T^{ab\ldots}_{c\ldots}}{\partial x^i} = T^{ab\ldots}_{c\ldots,i}. \tag{5.1}$$

However the components $T^{ab\ldots}_{c\ldots,i}$ are not the components of a tensor, as we can show from the example of the derivative of a vector. For we have

$$\begin{aligned}(T^n_{,i})' = \left(\frac{\partial T^n}{\partial x^i}\right)' &= \frac{\partial}{\partial x^{i'}}(A^{n'}_n T^n) = \frac{\partial x^i}{\partial x^{i'}}\frac{\partial}{\partial x^i}(A^{n'}_n T^n) \\ &= A^i_{i'} A^{n'}_n T^n_{,i} + A^i_{i'} A^{n'}_{n,i} T^n;\end{aligned} \tag{5.2}$$

that is, the $T^n_{,i}$ transform like the components of a tensor if, and only if, the transformation matrices $A^{n'}_n$ are independent of position (this is true, for example, for Lorentz transformations of Minkowski space).

The only exception is the generalized gradient $\varphi_{,a} = \partial \varphi / \partial x^a$ of a scalar φ; its components are those of a covariant vector. From $\varphi' = \varphi$, and hence $d\varphi' = d\varphi$, we have

$$\varphi_{,a} dx^a = (\varphi_{,a})' dx^{a'}, \tag{5.3}$$

and the quotient law ensures the vector property of $\varphi_{,a}$.

One can see why the partial derivatives of a tensor do not form a tensor if one describes a constant vector field in the plane by polar coordinates

5 The covariant derivative and parallel transport 43

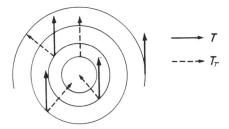

Fig. 5.1. Components of a constant vector field in polar coordinates.

(Fig. 5.1). The vector components of this constant vector field become position-dependent, because the directions of the coordinate lines change from point to point; the partial derivative of the vector components is a measure of the actual position-dependence of the vector only in Cartesian coordinate systems.

Definition of the covariant derivative The above considerations suggest that a covariant derivative (which produces tensors from tensors) can be constructed from the partial derivative by defining and making use of locally geodesic coordinates as follows.

The covariant derivative $T^{ab...}_{mn...;i}$ of a tensor $T^{ab...}_{mn...}$ is again a tensor, which coincides with the partial derivative in the locally geodesic coordinate system:

$$T^{ab...}_{mn...;i} = T^{ab...}_{mn...,i} \quad \text{for} \quad g_{mn} = \eta_{mn}, \quad \Gamma^a_{bc} = 0. \tag{5.4}$$

This definition of the covariant derivative is unique. It ensures the tensor property and facilitates the translation of physical laws to the Riemannian space, if these laws are already known in the Minkowski space (using partial derivatives).

Properties of, and rules for handling, the covariant derivatives Since the partial derivatives of η_{mn} and Δ^{abcd} are zero, we have the equations

$$g_{ab;m} = 0, \quad \epsilon^{abcd}_{;n} = 0. \tag{5.5}$$

The metric tensor and the ϵ-tensor are thus covariantly constant.

Because the product rule and the sum rule hold for partial derivatives, these rules also hold for covariant derivatives:

$$(T^{ab} + S^{ab})_{;n} = T^{ab}_{;n} + S^{ab}_{;n} \tag{5.6}$$

and

$$(T^{ab}S_c)_{;n} = T^{ab}_{;n}S_c + T^{ab}S_{c;n}. \tag{5.7}$$

44 Foundations of Riemannian geometry

Contraction, raising and lowering of indices, and taking the dual depend upon multiplication with the metric tensor or the ϵ-tensor. These operations therefore commute with covariant differentiation. For example,

$$(T^a{}_a)_{;n} = T^{ab}{}_{;n} g_{ab} = (T^a{}_a)_{,n}. \tag{5.8}$$

The covariant derivative of a scalar is equal to its partial derivative, as was shown above.

For practical calculations we naturally need also a formula for determining the covariant derivative in a given coordinate system – we certainly do not want to transform every time first to the local geodesic system, calculate the partial derivatives, and then transform back. We shall first give this formula without deriving it, and then demonstrate its agreement with the definition of the covariant derivative.

The covariant derivatives of the contravariant and covariant components of a vector are calculated according to the formulae

$$\left. \begin{array}{l} T^a{}_{;n} = T^a{}_{,n} + \Gamma^a{}_{nm} T^m \\ T_{a;n} = T_{a,n} - \Gamma^m{}_{an} T_m \end{array} \right\}, \tag{5.9}$$

and

respectively, and the covariant derivative of an arbitrary tensor is calculated by applying the prescription (5.9) to every contravariant and covariant index. For example,

$$T^a{}_{bc;d} = T^a{}_{bc,d} + \Gamma^a{}_{dm} T^m{}_{bc} - \Gamma^m{}_{bd} T^a{}_{mc} - \Gamma^m{}_{cd} T^a{}_{bm}. \tag{5.10}$$

In a locally geodesic system all the Christoffel symbols disappear, so that, there, covariant and partial derivatives coincide. The formulae (5.9) really do produce a tensor, although the two terms on the right-hand side are not separately tensors. For it follows from (5.2) and (5.3) that

$$(T^{a'}{}_{;n})' = \frac{\partial T^{a'}}{\partial x^{n'}} + \Gamma^{a'}{}_{n'm'} T^{m'}$$

$$= A^{a'}_a A^n_{n'} T^a{}_{,n} + A^n_{n'} A^{a'}_{a,n} T^a + A^{a'}_a A^n_{n'} A^m_{m'} \Gamma^a{}_{nm} A^{m'}_b T^b$$

$$- A^{a'}_{n,m} A^n_{n'} A^m_{m'} A^{m'}_b T^b$$

$$= A^{a'}_a A^n_{n'} (T^a{}_{,n} + \Gamma^a{}_{nm} T^m) + A^n_{n'} (A^{a'}_{a,n} - A^{a'}_{n,a}) T^a$$

and, because

$$A^{a'}_{a,n} = \frac{\partial^2 x^{a'}}{\partial x^n \partial x^a} = \frac{\partial^2 x^{a'}}{\partial x^a \partial x^n} = A^{a'}_{n,a}, \tag{5.11}$$

we obtain finally, as we claimed, the transformation law of a tensor:

5 The covariant derivative and parallel transport

$$(T^a{}_{;n})' = A^{a'}_a A^n_{n'} T^a{}_{;n}. \tag{5.12}$$

Equation (5.10) can be verified analogously.

Although the covariant derivative always produces a covariant index, one often writes it as a contravariant index; for example $T^{ab;n}$ is an abbreviation for

$$T^{ab;n} = g^{ni} T^{ab}{}_{;i}. \tag{5.13}$$

5.2 The covariant differential and local parallelism

There is an obvious geometric meaning to the covariant derivative which we shall describe in the following. One can visualize the covariant derivative – like the partial derivative – as the limiting value of a difference quotient. In this context one does not, however, simply form the difference in the value of the tensor components at the points x^i and $x^i + dx^i$:

$$dT^a = T^a(x^i + dx^i) - T^a(x^i) = T^a{}_{,i} dx^i \tag{5.14}$$

(this would correspond to the partial derivative), but rather uses

$$DT^a = dT^a + \Gamma^a_{nm} T^n dx^m = (T^a{}_{,m} + \Gamma^a_{nm} T^n) dx^m. \tag{5.15}$$

The deeper reason for this more complicated formula lies in the fact that tensors at two different points x^i and $x^i + dx^i$ obey different transformation laws, and hence their difference is not a tensor. Before forming the difference, the tensor at the point $x^i + dx^i$ must therefore be transported in a suitable manner (preserving the tensor property) to the point x^i, without of course changing it during this process. In our usual three-dimensional space we would translate 'without changing it' as 'keeping it parallel to itself.' We shall take over this way of speaking about the problem, but we must keep clearly in mind that the meaning of 'parallelism at different points' and 'parallel transport' is not at all self-evident in a non-Euclidean space.

Three simple examples may illustrate this. Referring to Fig. 5.2, we ask the following questions: (*a*) Are two vectors in a plane section still parallel after bending of the plane? (*b*) Are both the vectors, which are parallel in three-dimensional space, also parallel in the curved surface? (Obviously not, for vectors *in* the surface can have only two components, whereas the vector (2) juts out of the surface and has three components – but should one perhaps take the projection onto the surface?) (*c*) Which of the two vectors at the point *Q* of the sphere of Fig. 5.2c is parallel to that at the point *P*? Clearly *both* were parallel transported, the one along the equator

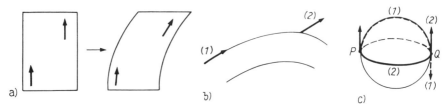

Fig. 5.2. Parallelism of vectors on surfaces.

always perpendicularly, and the other over the poles, always parallel to the curve joining P and Q!

What we should realize from these examples is that in a curved space one must *define* what one means by parallelism and parallel displacement. The definition used in the construction of the covariant derivative obviously reads: two vectors at infinitesimally close points are parallel if, and only if, we have

$$DT^a = dT^a + \Gamma^a_{nm} T^n dx^m = 0; \qquad (5.16)$$

that is to say, their covariant differential disappears. A vector field is parallel in the (infinitesimal) neighbourhood of a pint if its covariant derivative is zero there:

$$T^a{}_{;n} = T^a{}_{,n} + \Gamma^a_{nm} T^m = 0. \qquad (5.17)$$

If in a general affine (hence possibly even non-Riemannian) curved space one were also to use this definition, then the Γ^a_{mn} would in that case be arbitrary functions. A Riemannian space is distinguished by the fact that Γ^a_{mn} are precisely the Christoffel symbols formed from the metric tensor. The definitions (5.16) and (5.17) are of course so constructed that in the local geodesic system they lead to the usual parallel displacement in Minkowski space.

5.3 Parallel displacement along a curve and the parallel propagator

Let an arbitrary curve in our Riemannian space be given parametrically by $x^n = x^n(\lambda)$. It is then always possible to construct a parallel vector field along this curve from the requirement that the covariant differential of a vector along the curve vanishes; that is, from

$$\frac{DT^a}{D\lambda} \equiv \frac{T^a{}_{;n} dx^n}{d\lambda} = \frac{dT^a}{d\lambda} + \Gamma^a_{nm} T^m \frac{dx^n}{d\lambda} = 0. \qquad (5.18)$$

One can in fact specify arbitrarily the value of the vector components T^a at some initial point $\lambda = \bar{\lambda}$ and uniquely determine the vector at some other arbitrary point λ of the curve from the system of differential equations (5.18).

The geodesic equation is obviously an example of such an equation, which expresses the parallel transport of a vector:

$$\frac{d^2 x^a}{d\lambda^2} + \Gamma^a_{nm} \frac{dx^n}{d\lambda} \frac{dx^m}{d\lambda} = \frac{D}{D\lambda} \frac{dx^a}{d\lambda} = 0. \tag{5.19}$$

It says that the tangent vector $t^a = dx^a/d\lambda$ of a geodesic remains parallel to itself. The geodesic is thus not only the shortest curve between two points, but also the straightest. The straight line in Euclidean space also has these two properties.

There is precisely one geodesic between two points if one excludes the occurrence of conjugate points. (Such points of intersection of geodesics which originate in one point occur, for example, on a sphere: all great circles originating at the north pole intersect one another at the south pole.) The result of parallel transporting a vector (or a tensor) from the point $\bar{P}$ to the point P *along a geodesic* is therefore uniquely determined, while in general it certainly depends upon the choice of route (cf. section 6.2). Since the differential equation (5.18) to be integrated is linear in the components of the vector T^a to be transported, the vector components at the point P are linear functions of the components at the point $\bar{P}$:

$$T_a(P) = g_{a\bar{b}}(P, \bar{P}) T^{\bar{b}}(\bar{P}). \tag{5.20}$$

For tensor components we have analogously

$$T_{a\ n...}^{\ b} = g_{a\bar{a}} g^{b\bar{b}} g_{n\bar{n}}... T^{\bar{a}\ \bar{n}...}_{\ \bar{b}}. \tag{5.21}$$

The quantities $g_{a\bar{b}}$ are the components of the *parallel propagator*. It is a two-point tensor (cf. (4.19)); the indices (barred or not) of such a tensor also specify the coordinates of which point it depends upon. As the length of a vector remains constant during parallel transport, the parallel propagator is symmetrical,

$$g_{a\bar{b}} = g_{\bar{b}a}, \tag{5.22}$$

and we must obviously have

$$g_{a\bar{b}}(P, P) = g_{ab}(P). \tag{5.23}$$

5.4 Fermi–Walker transport

The parallel displacement of a vector appears to be the most natural way of comparing vectors at two different points of the space with one another or of transporting one to the other point. There are, however, physically

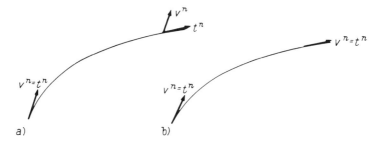

Fig. 5.3. Parallel transport (*a*) and Fermi–Walker transport (*b*) of the vector v^n.

important cases in which another kind of transport is more useful for the formulation of physical laws.

An observer who moves along an arbitrary timelike curve $x^n(\tau)$ under the action of forces will regard as natural, and use, a (local) coordinate system in which he himself is at rest and his spatial axes do not rotate. He will therefore carry along with him a tetrad system whose timelike vector is always parallel to the tangent vector $t^n = dx^n/d\tau$ of his path, for only then does the four-velocity of the observer possess no spatial components (he really is at rest), and he will regard as constant a vector whose components do not change with respect to this coordinate system.

The fact that the tangent vector to his own path does not change for the observer cannot, however, be expressed by saying that it is parallel transported along the path. Indeed the observer does not in general move along a geodesic, and therefore under parallel transport a vector pointing initially in the direction of the motion will later make an angle with the world line (cf. Fig. 5.3).

If, however, for every vector T^n one uses *Fermi–Walker transport,* defined by the vanishing of the *Fermi derivative,* that is, by

$$\frac{DT^n}{D\tau} - T_a \frac{1}{c^2} \left(\frac{dx^n}{d\tau} \frac{D^2 x^a}{D\tau^2} - \frac{dx^a}{d\tau} \frac{D^2 x^n}{D\tau^2} \right) = 0, \qquad (5.24)$$

then one can establish that the tangent vector t^n to an arbitrary timelike curve in the space is indeed Fermi–Walker-transported, since for $T^n = t^n$ (5.24) is satisfied identically as a consequence of the relation

$$\frac{D}{D\tau} \left(\frac{dx_n}{d\tau} \frac{dx^n}{d\tau} \right) = 0 = 2 \frac{dx_n}{d\tau} \frac{D^2 x^n}{D\tau^2}, \qquad (5.25)$$

which follows from $dx^n dx_n = -c^2 d\tau^2$. If the observer moves on a geodesic $D^2 x^n/D\tau^2 = 0$, then parallel transport and Fermi–Walker transport coincide.

5 The covariant derivative and parallel transport 49

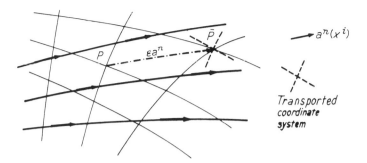

Fig. 5.4. How the Lie derivative is defined.

The equation (5.24) is to be interpreted in the following way. For a given curve $x^n(\tau)$ through the space it provides a definition of how the change of a vector T^n under advance along the curve is to be calculated from the initial values of the vector. The reader can confirm that the scalar product of vectors does not change under this type of transport, and therefore that lengths and angles remain constant.

A Fermi–Walker-transport tetrad-system is the best approximation to the coordinate system of an observer who employs locally a nonrotating inertial system in the sense of Newtonian mechanics (cf. section 8.2).

5.5 The Lie derivative

If in a space a family of world lines (curves) is available which covers the space smoothly and continuously, one speaks of a congruence of world lines. Such curves can be the world lines of particles of a fluid, for example. With every such congruence is associated a vector field $a^n(x^i)$, which at any given time has the direction of the tangent to the curve going through the point in question.

Let a vector field $T^n(x^i)$ also be given. One can now ask the question, how can the change of the vector T^n under motion of the observer in the direction of the vector field a^n be defined in an invariant (coordinate-system-independent) manner? Of course one will immediately think of the components $T^n{}_{;i}a^i$ of the covariant derivative of T^n in the direction of a^i. There is, however, yet a second kind of directional derivative, independent of the covariant derivative, namely the Lie derivative.

This derivative corresponds to the change determined by an observer who goes from the point P (coordinates x^i) in the direction of a^i to the infinitesimally neighbouring point $\bar{P}$ (coordinates $\bar{x}^i = x^i + \epsilon a^i(x^n)$) and takes his coordinate system with him (cf. Fig. 5.4).

If, however, at the point $\bar{P}$ he uses the coordinate system appropriate for P, then this corresponds to a coordinate transformation which associates with the point $\bar{P}$ the coordinate values of point P; that is, the transformation

$$\left.\begin{array}{l} x^{n'} = x^n - \epsilon a^n(x^i), \\ A_i^{n'} = \delta_i^n - \epsilon a^n{}_{,i}. \end{array}\right\} \quad (5.26)$$

He will therefore regard as components of the vector T^n at the point $\bar{P}$ the quantities

$$T^{n'}(\bar{P}) = A_i^{n'} T^i(x^k + \epsilon a^k) = (\delta_i^n - \epsilon a^n{}_{,i})[T^i(P) + \epsilon T^i{}_{,k}(P)a^k]$$
$$= T^n(P) + \epsilon T^n{}_{,k}(P)a^k - \epsilon a^n{}_{,k} T^k(P), \quad (5.27)$$

(ignoring terms in ϵ^2) and compare them with $T^n(P)$.

This consideration leads us to define the Lie derivative in the direction of the vector field a^n as the limiting value,

$$\underset{a}{\pounds} T^n = \lim_{\epsilon \to 0} \frac{1}{\epsilon} [T^{n'}(\bar{P}) - T^n(P)], \quad (5.28)$$

or the expression, which is equivalent because of (5.27),

$$\underset{a}{\pounds} T^n = T^n{}_{,k} a^k - T^k a^n{}_{,k}. \quad (5.29)$$

The Lie derivative of the covariant components T_n follows analogously as

$$\underset{a}{\pounds} T_n = T_{n,i} a^i + T_i a^i{}_{,n}. \quad (5.30)$$

One forms the Lie derivative of a tensor of higher rank by carrying over (5.29) or (5.30) to every contravariant or covariant index, respectively; thus, for example,

$$\underset{a}{\pounds} g_{mn} = g_{mn,i} a^i + g_{in} a^i{}_{,m} + g_{mi} a^i{}_{,n}. \quad (5.31)$$

The Christoffel symbols are not used in the calculation of the Lie derivative. One can, however, in (5.29)–(5.31) replace the partial derivatives by covariant derivatives, according to (5.9), obtaining

$$\left.\begin{array}{l} \underset{a}{\pounds} T^n = T^n{}_{;i} a^i - T^i a^n{}_{;i}, \\[4pt] \underset{a}{\pounds} T_n = T_{n;i} a^i + T_i a^i{}_{;n}, \\[4pt] \underset{a}{\pounds} g_{mn} = a_{m;n} + a_{n;m}. \end{array}\right\} \quad (5.32)$$

This result shows explicitly that the Lie derivative of a tensor is again a tensor, although only partial derivatives were used in its definition.

The Lie derivative of tensors has the following properties, which we list here without proof:

(a) It satisfies the Leibniz product rule.

(b) It commutes with the operation of contraction (although the Lie derivative of the metric tensor does not vanish).

(c) It can be applied to arbitrary, linear geometrical objects, to Christoffel symbols, for example.

(d) It commutes with the partial derivative.

(e) If the partial derivatives $a^i{}_{,n}$ are zero, for example, in a comoving coordinate system $a^i = (0, 0, 0, 1)$, then the Lie derivative and the usual directional derivative coincide.

The Lie derivative plays an important rôle in the investigation of symmetries of Riemannian spaces (cf. section 19.5).

Bibliography to section 5

Textbooks   Eisenhart (1949), Hicks (1971), Schmutzer (1968), Schouten (1954), Schutz (1980), Spivak (1979), Synge (1960)

Monographs and collected works   Yano (1955)

6 The curvature tensor

6.1 Intrinsic geometry and curvature

In the previous chapters of this book we have already frequently used the concept 'Riemannian space' or 'curved space.' Except in section 1.4 on the geodesic deviation, where we were dealing only with a Minkowski space with complicated curvilinear coordinates, or with a genuine curved space, it has not yet played any rôle. We shall now turn to the question of how to obtain a measure for the deviation of the space from a Minkowski space.

If one uses the word 'curvature' for this deviation, one most often has in mind the picture of a two-dimensional surface in a three-dimensional space; that is, one judges the properties of a two-dimensional space (the surface) from the standpoint of a flat space of higher dimensionality. This way of looking at things is certainly possible mathematically for a four-dimensional Riemannian space as well – one could regard it as a hypersurface in a ten-dimensional flat space. But this higher-dimensional space has no physical meaning and is also no more easy to represent or comprehend than the four-dimensional Riemannian space. Rather, we shall describe the properties of our space-time by four-dimensional concepts alone – we shall study 'intrinsic geometry.' In the picture of the two-dimensional surface we must therefore behave like two-dimensional beings, for whom the third dimension is inaccessible both practically and theoretically, and who can base predictions about the geometry of their surface through measurements on the surface alone.

The surface of, for example, a cylinder or a cone, which in fact can be constructed from a plane section without distortion, could not be distinguished locally from a plane by such beings (that is, without their going right around the cylinder or the cone and returning to their starting point). But they would be able to establish the difference between a plane and a sphere, because on the surface of the sphere:

(a) The parallel displacement of a vector depends upon the route. (Along route 1 in Fig. 5.2c the vector is a tangent vector of a geodesic; along route 2 it is always perpendicular to the tangent vector.)

(b) The sum of the angles of triangles bounded by 'straight lines' (geodesics) deviates from 180°; it can amount to 270°, for example.

(c) The circumference of a circle (produced by drawing out geodesics from a point and marking off a constant distance on them as radius) deviates from π multiplied by the diameter.

(d) The separation between neighbouring great circles is not proportional to the distance covered (cf. geodesic deviation).

As a detailed mathematical analysis shows, these four possibilities carried over to a four-dimensional space, all lead to the concept of curvature and to that of the curvature tensor. In the following sections we shall become more familiar with this tensor, and we begin with an investigation of the parallel transport of vectors.

6.2 The curvature tensor and global parallelism of vectors

The covariant derivative enables us to give, through (5.18),

$$T_{a;n}\frac{dx^n}{d\lambda} = \frac{dT_a}{d\lambda} - \Gamma^m_{an}T_m\frac{dx^n}{d\lambda} = 0, \qquad (6.1)$$

a unique formula for the parallel displacement of a vector along a fixed curve. When the result of the parallel displacement between two points

6 The curvature tensor

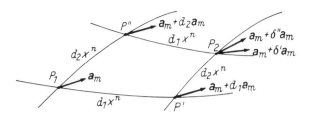

Fig. 6.1. Parallel displacement of a vector.

is *independent* of the choice of the curve, one speaks of global parallelism. A necessary condition for its existence is evidently that the parallel displacement should be independent of the route for infinitesimal displacements.

Upon applying (6.1) to the parallel displacement of the vector a_m along the sides of the infinitesimal parallelogram of Fig. 6.1 we obtain:

from P_1 to P':
$$d_1 a_m = \Gamma^i_{mn} a_i d_1 x^n,$$

from P_1 to P'':
$$d_2 a_m = \Gamma^i_{mn} a_i d_2 x^n,$$

from P_1 over P' to P_2 (Christoffel symbols are to be taken at the point P', therefore $\Gamma^r_{mq} + \Gamma^r_{mq,s} d_1 x^s$):

$$\delta' a_m = (\Gamma^r_{mq} + \Gamma^r_{mq,s} d_1 x^s)(a_r + d_1 a_r) d_2 x^q + d_1 a_m$$
$$\approx \Gamma^r_{mq} \Gamma^i_{rn} d_2 x^q d_1 x^n a_i + \Gamma^r_{mq,s} a_r d_1 x^s d_2 x^q + \Gamma^i_{mn} a_i (d_1 x^n + d_2 x^n),$$

from P_1 over P'' to P_2 (Christoffel symbols are to be taken at the point P''):

$$\delta'' a_m = (\Gamma^r_{mq} + \Gamma^r_{mq,s} d_2 x^s)(a_r + d_2 a_r) d_1 x^q + d_2 a_m$$
$$\approx \Gamma^r_{mq} \Gamma^i_{rn} d_1 x^q d_2 x^n a_i + \Gamma^r_{mq,s} a_r d_2 x^s d_1 x^q + \Gamma^i_{mn} a_i (d_1 x^n + d_2 x^n).$$

The vectors transported to P_2 by different routes thus differ by

$$\delta'' a_m - \delta' a_m = (-\Gamma^r_{mq,s} + \Gamma^r_{ms,q} + \Gamma^r_{nq} \Gamma^n_{ms} - \Gamma^r_{ns} \Gamma^n_{mq}) a_r d_1 x^s d_2 x^q. \tag{6.2}$$

The parallel transport is therefore independent of the route for all vectors a_r and all possible infinitesimal parallelograms ($d_1 x^n$ and $d_2 x^n$ arbitrary) if and only if the *Riemann curvature tensor (Riemann-Christoffel tensor)*, defined by

$$R^r_{msq} = \Gamma^r_{mq,s} - \Gamma^r_{ms,q} + \Gamma^r_{ns} \Gamma^n_{mq} - \Gamma^r_{nq} \Gamma^n_{ms}, \tag{6.3}$$

vanishes. If this condition is satisfied, then one can also define global parallelism for *finite* displacements; the parallel transport will be independent of path (as one can show by decomposing the surface enclosed by a curve into infinitesimal parallelograms).

Path independence of the parallel displacement is the pictorial interpretation of the commutation of the second covariant derivatives of a vector; in fact for every arbitrary vector a_m we have

$$a_{m;s;q} = a_{m;s,q} - \Gamma^r_{mq} a_{r;s} - \Gamma^r_{qs} a_{m;r}$$
$$= a_{m,s,q} - \Gamma^r_{ms,q} a_r - \Gamma^r_{ms} a_{r,q} - \Gamma^r_{qm} a_{r,s}$$
$$+ \Gamma^n_{mq} \Gamma^r_{ns} a_r - \Gamma^r_{qs} a_{m,r} + \Gamma^n_{qs} \Gamma^r_{mn} a_r,$$

and, after interchange of q and s bearing in mind (6.3), we obtain

$$a_{m;s;q} - a_{m;q;s} = R^r{}_{msq} a_r. \qquad (6.4)$$

Covariant derivatives commute if, and only if, the curvature tensor vanishes. One can also take (6.4) as the definition of the curvature tensor.

We can see the justification for the word *curvature* tensor in the fact that it disappears if, and only if, the space is flat, that is, when a Cartesian coordinate system can be introduced in the whole space. In Cartesian coordinates all the Christoffel symbols do indeed vanish, and with them the curvature tensor (6.3). Conversely if it does disappear, then one can create a Cartesian coordinate system throughout the space by (unique) parallel displacement of four vectors which are orthogonal at one point. That $R^r{}_{msq}$ really is a *tensor* can be most quickly realized from (6.4).

To summarize, we can thus make the following completely equivalent statements. The curvature tensor defined by (6.3) and (6.4) vanishes if, and only if: (*a*) the space is flat, that is, Cartesian coordinates with $g_{ab} = \eta_{ab}$ and $\Gamma^a_{bc} = 0$ can be introduced throughout the space; or (*b*) the parallel transport of vectors is independent of path; or (*c*) covariant derivatives commute; or (*d*) the geodesic deviation (the relative acceleration) of two arbitrary particles moving force-free vanishes (cf. section 1.4).

6.3 *The curvature tensor and second derivatives of the metric tensor*

The curvature tensor (6.3) contains Christoffel symbols and their derivatives, and hence the metric tensor and its first and second derivatives. We shall now examine more precisely the connection between the metric and the components of the curvature tensor.

6 The curvature tensor

To this end we carry out in a locally geodesic coordinate system,

$$\bar{g}_{ab} = \eta_{ab}, \quad \bar{\Gamma}^a_{bc} = 0 \quad \text{for } \bar{x}^n = 0, \tag{6.5}$$

a coordinate transformation

$$\left.\begin{aligned} x^n &= \bar{x}^n + \frac{1}{6} D^n{}_{pqr} \bar{x}^p \bar{x}^q \bar{x}^r, \\ \frac{\partial x^n}{\partial \bar{x}^i} &= A^n_i = \delta^n_i + \frac{1}{2} D^n{}_{pqi} \bar{x}^p \bar{x}^q, \end{aligned}\right\} \tag{6.6}$$

the constants $D^n{}_{pqi}$ being initially arbitrary, but symmetric in the lower indices. This transformation does not change the metric or the Christoffel symbols at the point $x^n = \bar{x}^n = 0$, but it can serve to simplify the derivatives of the Christoffel symbols. Because of the general transformation formula (3.23) we have

$$\Gamma^m_{ab,n} = (\bar{\Gamma}^r_{ik} A^m_r A^i_a A^k_b - A^m_{i,k} A^i_a A^k_b)_{,n}, \tag{6.7}$$

from which at the point $x^n = 0$ follows

$$\Gamma^m_{ab,n} = \bar{\Gamma}^m_{ab,n} - D^m{}_{abn} \tag{6.8}$$

because of (6.6).

Since the coefficients $D^m{}_{abn}$ of formula (6.6) are symmetric in the three lower indices, whereas the derivatives $\bar{\Gamma}^m_{ab,n}$ of the Christoffel symbols, which are to be regarded as specified, do not possess this symmetry property, not all the derivatives $\Gamma^m_{ab,n}$ can be made to vanish. Through the choice

$$D^m{}_{abn} = \tfrac{1}{3}(\bar{\Gamma}^m_{ab,n} + \bar{\Gamma}^m_{na,b} + \bar{\Gamma}^m_{bn,a}), \tag{6.9}$$

however, one can always ensure that

$$\Gamma^m_{ab,n} + \Gamma^m_{na,b} + \Gamma^m_{bn,a} = 0. \tag{6.10}$$

If (6.10) and (6.5) are satisfied at a point, one speaks of canonical coordinates.

In such a canonical coordinate system it follows from (6.3) that the components of the curvature tensor satisfy

$$R^r{}_{msq} = \Gamma^r_{mq,s} - \Gamma^r_{ms,q}, \tag{6.11}$$

and therefore, using also (6.10), that

$$R^r{}_{msq} + R^r{}_{smq} = -3\Gamma^r_{ms,q}. \tag{6.12}$$

From the definition of the Christoffel symbol (3.11), on the other hand, it follows that

$$g_{ia,bn} = g_{mi}\Gamma^m_{ab,n} + g_{ma}\Gamma^m_{ib,n}, \tag{6.13}$$

and (6.12) and (6.13) together yield finally

$$g_{ia,bn} = -\tfrac{1}{3}(R_{iabn} + R_{iban} + R_{aibn} + R_{abin}). \tag{6.14}$$

Assuming in advance the symmetry relations (6.18), this is equivalent to

$$g_{ia,bn} = -\tfrac{1}{3}(R_{iban} + R_{abin}). \tag{6.15}$$

The equations (6.14) and (6.15) lead to a conclusion which should be noted. At first sight they merely state that in canonical coordinates the second derivatives of the metric tensor can be constructed from the components of the curvature tensor. But because of (6.5) and (6.15), in canonical coordinates *all* tensors which can be formed out of the metric and its first and second derivatives can be expressed in terms of the curvature tensor and the metric tensor itself. This relation between tensors must be coordinate-independent, and so any tensor containing only the metric and its first and second derivatives can be expressed in terms of the curvature tensor and the metric tensor.

If one wants to apply this law to pseudo-tensors, then one must also admit as an additional building block the ϵ-tensor.

Canonical coordinates permit a simple geometrical interpretation. Their coordinate lines are pairwise orthogonal geodesics, and the coordinates of an arbitrary point are given by the product of the direction cosines of the geodesic to the point from the zero point with the displacement along this geodesic.

6.4 Properties of the curvature tensor

Symmetry properties The symmetry properties of the curvature tensor can, of course, immediately be picked out from the defining equation (6.3), or from

$$R_{amsq} = g_{ar}(\Gamma^r_{mq,s} - \Gamma^r_{ms,q} + \Gamma^r_{ns}\Gamma^n_{mq} - \Gamma^r_{nq}\Gamma^n_{ms}). \tag{6.16}$$

But, in the geodesic coordinate system, in which the Christoffel symbols vanish, and in which it follows from (6.16) that

$$R_{amsq} = (g_{ar}\Gamma^r_{mq})_{,s} - (g_{ar}\Gamma^r_{ms})_{,q},$$

and hence finally

6 The curvature tensor

$$R_{amsq} = \tfrac{1}{2}(g_{aq,ms} + g_{ms,aq} - g_{as,mq} - g_{mq,as}), \quad (6.17)$$

they are more quickly recognized. As one can immediately see from (6.17), the curvature tensor is antisymmetric under interchange of the first and second index, or of the third and fourth,

$$R_{amsq} = -R_{masq} = -R_{amqs} = R_{maqs}, \quad (6.18)$$

but it does not alter under exchange of the first and last pairs of indices,

$$R_{amsq} = R_{sqam}, \quad (6.19)$$

and, further, also satisfies the relation

$$3R_{a[msq]} = R_{amsq} + R_{asqm} + R_{aqms} = 0. \quad (6.20)$$

The equations (6.18) imply that under the relabelling $(12) \to 1$, $(23) \to 2$, $(34) \to 3$, $(41) \to 4$, $(13) \to 5$, $(24) \to 6$ the independent components of the curvature tensor can be mapped onto a 6×6 matrix R_{AB}. Because of (6.19) this matrix is symmetric, and therefore has at most $\binom{7}{2} = 21$ different components. The cyclic relation (6.20) is independent of (6.18) and (6.19) (that is, not trivially satisfied) if, and only if, all four indices of the curvature tensor are different, and (6.20) hence supplies only *one* additional equation. The result of this count is thus that in a four-dimensional space the Riemann curvature tensor has a maximum of twenty algebraically independent components. One can show that in an N-dimensional space there are precisely $N^2(N^2-1)/12$ independent components.

Ricci tensor, curvature tensor and Weyl tensor Because of the symmetry properties of the curvature tensor there is (apart from a sign) only one tensor that can be constructed from it by contraction, namely, the *Ricci tensor*:

$$R_{mq} = R^a{}_{maq} = -R^a{}_{mqa}. \quad (6.21)$$

It is symmetric, and has therefore 10 different components. Its trace

$$R = R^m{}_m \quad (6.22)$$

is called the *curvature scalar R*.

Just as (4.42) decomposes a symmetric tensor into a trace-free part and a term proportional to the metric tensor, the curvature tensor can be split into the *Weyl tensor* (or conformal curvature tensor) $C^{am}{}_{sq}$, and parts which involve only the Ricci tensor and the curvature scalar:

$$R^{am}{}_{sq} = C^{am}{}_{sq} + \tfrac{1}{2}(g^a_s R^m_q + g^m_q R^a_s - g^m_s R^a_q - g^a_q R^m_s)$$
$$- \tfrac{1}{6}(g^a_s g^m_q - g^a_q g^m_s)R. \quad (6.23)$$

The Weyl tensor defined by (6.23) is 'trace-free,'

$$C^{am}{}_{aq} = 0, \qquad (6.24)$$

and has all the symmetry properties of the full curvature tensor. The name 'conformal curvature tensor' or 'conformal tensor' relates to the fact that two different Riemannian spaces with the fundamental metric forms $d\bar{s}^2$ and ds^2 which are conformally related

$$d\bar{s}^2 = M^2(x^i)ds^2 \qquad (6.25)$$

(all lengths are multiplied by the position-dependent conformal factor M^2, independent of direction), have the same conformal curvature tensor, although their Riemann curvature tensors are different.

In summary we can therefore make the following statement. At every point of a four-dimensional Riemannian space of the 100 possible different second derivatives of the metric tensor only twenty cannot be eliminated by coordinate transformations: they correspond to the twenty algebraically independent components of the curvature tensor. These twenty components can always be expressed by the ten components of the Ricci tensor and the ten of the Weyl tensor, as (6.23) shows.

In three-dimensional space the curvature tensor has only six independent components, exactly as many as the Ricci tensor, and the curvature tensor can be expressed in terms of the Ricci tensor:

$$R^{\alpha\mu}{}_{\sigma\tau} = (g_\sigma^\alpha R_\tau^\mu + g_\tau^\mu R_\sigma^\alpha - g_\sigma^\mu R_\tau^\alpha - g_\tau^\alpha R_\sigma^\mu) - \tfrac{1}{2}R(g_\sigma^\alpha g_\tau^\mu - g_\tau^\alpha g_\sigma^\mu). \qquad (6.26)$$

On a two-dimensional surface the curvature tensor has essentially only one component, the curvature scalar R:

$$R^{AM}{}_{ST} = R(g_S^A g_T^M - g_T^A g_S^M) \qquad (6.27)$$

(cf. (1.38)).

Bianchi identities Until now we have always thought of the metric as given, and derived the curvature tensor from it. Conversely, one can also ask the question whether the curvature tensor (with the correct symmetry properties, of course) can be specified as an arbitrary function of position, and the metric belonging to it determined. The answer to this apparently abstract mathematical question will reveal a further property of the Riemann tensor, which is particularly important for gravitation.

The determination of the metric from a specified curvature tensor amounts, because of (6.17), to the solution of a system of twenty second-order differential equations for the ten metric functions g_{ab}. In general

6 The curvature tensor

such a system will possess no solutions; given a tensor with the algebraic properties of the curvature tensor there does not correspond a metric whose curvature tensor it is. Rather, additional integrability conditions must be satisfied. Although (6.15) holds only at one point, and therefore may not be differentiated, one can recognize the basis of the integrability condition in it. Since the third partial derivatives of the metric commute, there must be some relations among the derivatives of components of the curvature tensor.

To set up these relations we write down the covariant derivative of the curvature tensor

$$R_{amsq;i} = g_{ar}(\Gamma^r_{mq,s} - \Gamma^r_{ms,q} + \Gamma^r_{ns}\Gamma^n_{mq} - \Gamma^r_{nq}\Gamma^n_{ms})_{;i} \quad (6.28)$$

in locally geodesic coordinates. Since the Christoffel symbols vanish in these coordinates one can replace the covariant derivative by the partial and drop the products of Christoffel symbols:

$$R_{amsq;i} = g_{ar}(\Gamma^r_{mq,si} - \Gamma^r_{ms,qi}).$$

If we add to this equation the two produced on permuting indices

$$R_{amqi;s} = g_{ar}(\Gamma^r_{mi,sq} - \Gamma^r_{mq,is})$$

and

$$R_{amis;q} = g_{ar}(\Gamma^r_{ms,iq} - \Gamma^r_{mi,sq}),$$

then we obtain the *Bianchi identities:*

$$3R_{am[is;q]} = R_{amis;q} + R_{amsq;i} + R_{amqi;s} = 0. \quad (6.29)$$

Every curvature tensor must satisfy these equations; if they hold, then one can determine the metric for a given curvature tensor, and conversely, if one expresses the curvature tensor through the metric, then they are satisfied identically. Because of the symmetry properties of the curvature tensor exhibited in (6.18)–(6.20), many of the Bianchi identities are trivially satisfied, for example, if not all the indices i, q, s are different. In four-dimensional space-time the system (6.29) contains only twenty non-trivial independent solutions.

Upon contracting the Bianchi identities, we obtain identities for the Ricci tensor. We have

$$3g^{aq}R_{am[is;q]} = R^a{}_{mis;a} - R_{ms;i} + R_{mi;s},$$

and

$$3q^{ms}g^{aq}R_{am[is;q]} = R^a{}_{i;a} - R_{,i} + R^a{}_{i;a},$$

and therefore
$$(R^{ai} - \tfrac{1}{2}g^{ai}R)_{;i} = 0. \tag{6.30}$$

Finally in this section on the properties of the curvature tensor we should point out that various sign conventions occur in the literature. With respect to our definition, the Riemann tensor can have the opposite sign, and the Ricci tensor can be formed by contraction over a different pair of indices and hence again change its sign. When reading a book or an article it is recommended that the convention used there be written out in order to avoid mistakes arising from comparison with this book or with other publications.

Bibliography to section 6

Textbooks Eisenhart (1949), Hicks (1971), Schmutzer (1968), Schouten (1954), Schutz (1980), Spivak (1979), Synge (1960)

7 Differential operators, integrals and integral laws

7.1 The problem

In the formulation of physical laws in three-dimensional flat space one often uses the vector operators div, grad, curl and $\Delta = \text{div grad}$, which can also be applied to tensor components in Cartesian coordinates. They satisfy the identities

$$\text{div curl} = 0 \quad \text{and} \quad \text{curl grad} = 0, \tag{7.1}$$

and, because of the integral laws

$$\oint \text{div } A \, dV = \oint A \, dS \quad \text{(Gauss)} \tag{7.2}$$

and

$$\oint \text{curl } A \, dS = \oint A \, dr \quad \text{(Stokes)}, \tag{7.3}$$

they make an integral formulation of physical statements possible, for example in electrodynamics. The integral laws can also be applied to tensors of higher rank.

While the differential operators can be carried over relatively easily to a four-dimensional curved space, the generalization of integral laws leads to difficulties. One cause of the difficulties is that integrals can never be taken over tensor components, but only over scalars, if the result is to be

7 Differential operators, integrals and integral laws

a tensor. A second cause is the fact that the reverse of an integration is really a partial differentiation, whereas for tensor equations we have to choose the covariant derivative; for this reason we shall be especially interested in those differential operators which are covariant, and yet which can be expressed simply by partial derivatives.

The comprehensibility of the calculations is further obscured by the complicated way in which we write volume and surface elements in covariant form. The use of differential forms here can indeed produce some improvement, but for actual calculations the gain is small.

7.2 Some important differential operators

The covariant derivative is the generalized gradient; for a scalar, covariant and partial derivatives coincide:

$$\varphi_{;a} = \varphi_{,a}. \tag{7.4}$$

The generalized curl of a vector A_m is the antisymmetric part of the tensor $A_{n;m}$:

$$A_{n;m} - A_{m;n} = A_{n,m} - A_{m,n} - \Gamma^a_{nm} A_a + \Gamma^a_{mn} A_a.$$

Because of the symmetry of the Christoffel symbol in the lower indices, one can replace covariant derivatives by partial derivatives in this expression:

$$A_{n;m} - A_{m;n} = A_{n,m} - A_{m,n}. \tag{7.5}$$

One obtains the generalized divergence by contraction over the index with respect to which the covariant derivative has been taken. Because of the relation (3.14), $\Gamma^a_{ab} = (\ln \sqrt{-g})_{,b}$, we have for a vector

$$B^n{}_{;n} = B^n{}_{,n} + \Gamma^n_{na} B^a = B^n{}_{,n} + (\ln \sqrt{-g})_{,a} B^a,$$

and therefore

$$B^n{}_{;n} = \frac{1}{\sqrt{-g}} (B^n \sqrt{-g})_{,n}. \tag{7.6}$$

For an antisymmetric tensor $F_{ab} = -F_{ba}$ we have, because of the symmetry property of the Christoffel symbol,

$$F^{ab}{}_{;b} = F^{ab}{}_{,b} + \Gamma^a_{bm} F^{mb} + \Gamma^b_{bm} F^{am} = F^{ab}{}_{,b} + (\ln \sqrt{-g})_{,m} F^{am};$$

thus, just as for a vector, its divergence can also be expressed as a partial derivative:

$$F^{ab}{}_{;b} = \frac{1}{\sqrt{-g}} (\sqrt{-g} F^{ab})_{,b}. \tag{7.7}$$

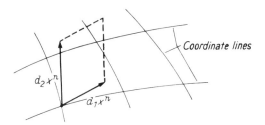

Fig. 7.1. A surface element.

Similarly, for every completely antisymmetric tensor we have

$$F^{[mn...ab]}{}_{;b} = \frac{1}{\sqrt{-g}} (\sqrt{-g} F^{[mn...ab]})_{,b}. \tag{7.8}$$

For the divergence of a symmetric tensor there is no comparable simple formula.

The generalized Δ-operator is formed from div and grad; from (7.4) and (7.6) we have

$$\Delta \varphi = \varphi^{,n}{}_{;n} = \frac{1}{\sqrt{-g}} (\sqrt{-g} \, g^{na} \varphi_{,a})_{,n}. \tag{7.9}$$

7.3 Volume, surface and line integrals

In an N-dimensional ($N \leq 4$) space, an s-dimensional hypersurface element ($s \leq N$) is spanned by s infinitesimal vectors $d_1 x^n, d_2 x^n, \ldots, d_s x^n$ which are linearly independent and do not necessarily have to point in the direction of the coordinate axes (cf. Fig. 7.1).

We shall need the generalized Kronecker symbol $\delta^{n_1...n_s}_{m_1...m_s}$, which is antisymmetric both in all upper and all lower indices, and for $n_i = m_i$ takes the value 1 (when these numbers are all different), so that

$$\delta^{n_1...n_s}_{m_1...m_s} = \delta^{[n_1...n_s]}_{m_1...m_s} = \delta^{n_1...n_s}_{[m_1...m_s]}, \quad \delta^{n_1...n_s}_{m_1...m_s} = 1 \quad \text{for} \quad n_i = m_i. \tag{7.10}$$

We next define the object

$$dV^{n_1...n_s} = \delta^{n_1...n_s}_{m_1...m_s} d_1 x^{m_1} \cdots d_s x^{m_s} \tag{7.11}$$

as a hypersurface (volume) element. As one can see, and can verify from examples, this is a tensor which is antisymmetric in all indices. Its components become particularly simple when the $d_i x^n$ point in the directions of the coordinate axes, for example, when

$$d_1 x^n = (dx^1, 0, 0, \ldots), \quad d_2 x^n = (0, dx^2, 0, \ldots), \quad \text{etc.} \tag{7.12}$$

7 Differential operators, integrals and integral laws

For $s=1$, (7.11) simply defines the line element

$$dV^n = dx^n. \tag{7.13}$$

For $s=2$, since

$$\delta^{n_1 n_2}_{m_1 m_2} = \delta^{n_1}_{m_1}\delta^{n_2}_{m_2} - \delta^{n_1}_{m_2}\delta^{n_2}_{m_1}, \tag{7.14}$$

the hypersurface element is associated in a simple manner with the surface element $d_1 r \times d_2 r$

$$dV^{n_1 n_2} = d_1 x^{n_1} d_2 x^{n_2} - d_1 x^{n_2} d_2 x^{n_1}, \tag{7.15}$$

and for $s=N$ it has, using the differentials (7.12), essentially one component

$$dV^{12\ldots N} = dx^1 dx^2 \cdots dx^N. \tag{7.16}$$

Since we can in principle integrate only over scalars, if we demand that the integral be a tensor, then we must always contract the hypersurface element with a tensor of the same rank. Thus only integrals of the form

$$\int_{G_s} T_{n_1 \ldots n_s} dV^{n_1 \ldots n_s} = I_s, \quad s \leq N, \tag{7.17}$$

are allowed.

In an N-dimensional space there are therefore precisely N different types of integral, each corresponding to the dimension s of the hypersurface being integrated over. We may suppose that the tensors $T_{n_1 \ldots n_s}$ are completely antisymmetric, because in contraction with $dV^{n_1 \ldots n_s}$ all symmetric parts would drop out anyway. G_s denotes the region over which the s-dimensional integration is to be carried out.

In four-dimensional space there are thus four types of integral contained in (7.17). When $s=1$ we have, for example, the simple line integral

$$I_1 = \int_{G_1} T_n \, dx^n. \tag{7.18}$$

For $s=3$, as we have shown in (4.4), one can map the tensor $T_{n_1 n_2 n_3}$ onto a vector, according to (4.40):

$$T_{n_1 n_2 n_3} = \frac{1}{3!} \epsilon_{a n_1 n_2 n_3} T^a = \frac{\sqrt{-g}}{3!} \Delta_{a n_1 n_2 n_3} T^a. \tag{7.19}$$

Here it is meaningful to introduce

$$df_a = \frac{1}{3!} \epsilon_{a n_1 n_2 n_3} dV^{n_1 n_2 n_3}, \tag{7.20}$$

the pseudo-vector df_a which is perpendicular to the hypersurface element (that is, perpendicular to the vectors $d_i x^n$), and whose length in a system (7.12) is just $\sqrt{-g}\,dx^1 dx^2 dx^3$. df_a is thus the generalized surface element. In this way we obtain the simpler form

$$I_3 = \int_{G_3} T^a df_a \left(= \int_{G_3} T^a \sqrt{-g}\, dx^1 dx^2 dx^3 \delta_a^4 \right) \tag{7.21}$$

of the hypersurface integral (the expression in parentheses is valid only in the system in which G_3 is the surface $x_4 =$ constant and in which (7.12) holds).

Finally, for $s = 4$, every antisymmetric tensor $T_{n_1 n_2 n_3 n_4}$ is proportional to the ϵ-tensor:

$$T_{n_1 n_2 n_3 n_4} = \frac{1}{4!} T \epsilon_{n_1 n_2 n_3 n_4}. \tag{7.22}$$

Because

$$\delta^{n_1 n_2 n_3 n_4}_{m_1 m_2 m_3 m_4} = -\epsilon^{n_1 n_2 n_3 n_4} \epsilon_{m_1 m_2 m_3 m_4} \tag{7.23}$$

(cf. (4.28)), it is appropriate to introduce the volume element dV by

$$dV = -\epsilon_{m_1 m_2 m_3 m_4} d_1 x^{m_1} d_2 x^{m_2} d_3 x^{m_3} d_4 x^{m_4}, \tag{7.24}$$

which in the preferred system (7.12) has the form

$$dV = \sqrt{-g}\, dx^1 dx^2 dx^3 dx^4. \tag{7.25}$$

Volume integrals thus always have the simple form

$$I_4 = \int_{G_4} T\, dV \left(= \int_{G_4} T \sqrt{-g}\, dx^1 dx^2 dx^3 dx^4 \right) \tag{7.26}$$

(the expression in parentheses is the form when (7.12) is valid).

7.4 Integral laws

Integral laws which are valid in a Riemannian space all have the form of a Stokes law, that is, they reduce the integral over a generalized curl

$$T_{[n_1 n_2 \ldots n_{s-1}; n_s]} = T_{[n_1 \ldots n_{s-1}, n_s]} \tag{7.27}$$

to an integral over the boundary G_{s-1} of the original (simply connected) region of integration G_s:

$$\int_{G_s} T_{n_1 \ldots n_{s-1}, n_s}\, dV^{n_s n_1 \ldots n_{s-1}} = \int_{G_{s-1}} T_{n_1 \ldots n_{s-1}}\, dV^{n_1 \ldots n_{s-1}}. \tag{7.28}$$

7 Differential operators, integrals and integral laws

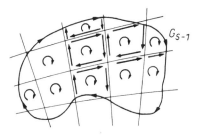

Fig. 7.2. The Stokes law for a surface G_s.

(Because of the antisymmetry of the volume element we are able to drop the antisymmetrizing brackets on the tensor field.) In spite of the partial derivatives, (7.28) is a tensor equation – one can in all cases replace the partial by the covariant derivative.

We shall not go through the proof of this law here, but merely indicate the idea on which it is based. Just as with the proof of the Stokes law for a two-dimensional surface, one decomposes the region G_s into infinitesimal elements, demonstrates the validity of the law for these elements, and sums up over all elements. In the summation the contributions from the 'internal' boundary surfaces cancel out, because in every case they are traversed twice, in opposite directions (cf. Fig. 7.2).

In three-dimensional flat space there are consequently three integral laws. For $s = 1$ we obtain from (7.28)

$$\int_{P_1}^{P_2} T_{,n}\,\mathrm{d}x^n = T(P_2) - T(P_1) \qquad (7.29)$$

(the boundary of a curve is represented by the two end points P_1 and P_2). The Stokes law proper corresponds to $s = 2$, and $s = 3$ yields the Gauss law.

In four-dimensional space, too, the Gauss law is a special case of the general Stokes law (7.28). Because of (7.19) and (7.20), for $s = N = 4$ we obtain from (7.28)

$$\int_{G_4} \frac{1}{3!} \epsilon_{an_1 n_2 n_3} T^a{}_{;n_4}\,\mathrm{d}V^{n_4 n_1 n_2 n_3} = \int_{G_3} T^a\,\mathrm{d}f_a.$$

We next substitute for $\mathrm{d}V^{n_4 n_1 n_2 n_3}$ from (7.11), (7.23) and (7.24):

$$-\int_{G_4} \frac{1}{3!} \epsilon_{an_1 n_2 n_3} T^a{}_{;n_4} \epsilon^{n_4 n_1 n_2 n_3}\,\mathrm{d}V = \int_{G_3} T^a\,\mathrm{d}f_a,$$

and, finally, taking into account the rule (4.31), we obtain the Gauss law

$$\int_{G_4} T^a{}_{;a} \, dV = \int_{G_3} T^a \, df_a. \tag{7.30}$$

When making calculations with integrals and integral laws one has to make sure that the orientation of the hypersurface element is correctly chosen and remains preserved; under interchange of coordinates the sign of the hypersurface element $dV^{n_1 \cdots n_2}$ changes. Such a fixing of the orientation occurs also, of course, in the case of the usual Stokes law in three dimensions, where the sense in which the boundary curve is traversed is related to the orientation of the surface.

7.5 Integral conservation laws

We want to describe in detail a particularly important physical application of the Gauss integral law. From special relativity one already knows that a mathematical statement of the structure

$$T^{a \cdots c}{}_{;a} = 0 \tag{7.31}$$

(the vanishing of the divergence of a tensor field) corresponds physically to a conservation law, establishing that some physical quantity does not change with time. In order to prove this connection, one uses the Gauss law, which is also valid for tensor components.

In a Riemannian space the number of possible integral conservation laws is already restricted by the fact that the Gauss law (7.30) can only be applied to the divergence of a vector. Now what conclusions can one draw from

$$T^a{}_{;a} = 0? \tag{7.32}$$

Let us imagine a vector field $T^a(x^i)$, which differs from zero only within a finite spatial region, and apply (7.30) to a four-dimensional 'cylindrical' region whose three-dimensional 'convex' surface $G_3^{(M)}$ lies outside this region of space (hatched in Fig. 7.3).

Since the contributions from the convex surface $G_3^{(M)}$ disappear, it follows from (7.30) and (7.32) that

$$\int_{G_3^{(1)}} T^a \, df_a + \int_{G_3^{(2)}} T^a \, df_a = 0. \tag{7.33}$$

If we now let the convex surfaces go to (spatial) infinity, then the regions of integration $G_3^{(1)}$ and $G_3^{(2)}$ cover the whole space $x^4 = \text{constant}$. And if we further notice the opposite orientations of df_a in the two regions, then it follows from (7.33) and (7.20) that

8 Fundamental laws of physics in Riemannian spaces

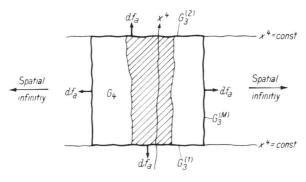

Fig. 7.3. The region of integration used in deriving the conservation law (7.34).

$$\int_{x^4=\text{constant}} T^a \, df_a = \int_{x^4=\text{constant}} T^4 \sqrt{-g} \, dx^1 dx^2 dx^3 = \text{constant}. \quad (7.34)$$

The integral (7.34) defines a quantity whose value does not depend upon the (arbitrary) time coordinate x^4; it defines a conserved quantity. We have derived this law under the supposition of a so-called isolated vector field T^a, that is, one restricted to a finite region of space. It is, however, also valid when there are no convex surfaces $G_3^{(M)}$, that is, when the space is closed (like a two-dimensional spherical surface), or when the integral over the convex surface tends to zero (T^a falls off sufficiently quickly when the convex surface is pushed to spatial infinity).

Bibliography to section 7

Textbooks Hicks (1971), Schmutzer (1968), Schouten (1954), Schutz (1980), Spivak (1979), Straumann (1984)

8 Fundamental laws of physics in Riemannian spaces

8.1 *How does one find the fundamental physical laws?*

Before turning in the next chapter to the laws governing the gravitational field, that is, to the question of how the matter existing in the Universe determines the structure of the Riemannian space, we shall enquire into the physical laws which hold in a *given* Riemannian space; that is to say, how a given gravitational field influences other physical processes. How can one transcribe a basic physical equation, formulated in Minkowski space without regard to the gravitational force, into the Riemannian space, and thereby take account of the gravitational force?

In this context the word 'transcribe' somewhat conceals the fact that it is really a matter of searching for entirely new physical laws, which are very similar to the old laws only because of the especially simple way in which the gravitational field acts. It is clear that we shall not be forced to the new form of the laws by logical or mathematical considerations, but that we can attain the answer only by observation and experiment. In searching for a transcription principle we therefore want our experience to be summarized in the simplest possible formulae.

In the history of relativity theory the principle of covariance plays a large rôle in this connection. There is no clear and unique formulation of this principle; the opinions of different authors diverge here. Rather, the principle of covariance expresses the fact that physical laws are to be written covariantly by the use of tensors, to ensure the equivalence, in principle, of all coordinate systems. Many criticisms have been raised against this principle, their aim being to assert that neither is it a *physical* principle, nor does it guarantee the correctness of the equations thus obtained. An example from special relativity will illustrate this. The potential equation

$$\Delta V = \eta^{\alpha\beta} V_{,\alpha,\beta} = 0, \quad \alpha, \beta = 1, 2, 3, \tag{8.1}$$

is certainly not Lorentz invariant. But we can make it so by introducing an auxiliary field u^n which in a special coordinate system (in which (8.1) holds) has the form $u^n = (0, 0, 0, c)$. The equation

$$\Delta V = (\eta^{ab} + u^a u^b / c^2) V_{,a,b} = 0, \quad a, b = 1, \ldots, 4, \tag{8.2}$$

thus obtained is certainly Lorentz invariant (covariant), but it is definitely false, because according to it effects always propagate with infinitely large velocity. Of course one has to criticize (8.2) on the grounds that a vector field u^a was intorduced *ad hoc* which singles out the three-dimensional coordinates used in (8.1) and thereby favours the rest system of an 'aether.'

It is constructive to compare this example with the transition from the Lorentz-invariant wave equation,

$$\Box V = \eta^{ab} V_{,a,b} = 0, \tag{8.3}$$

to the generally covariant equation

$$\Box V = g^{ab} V_{;a;b} = 0. \tag{8.4}$$

In place of the auxiliary quantity u^n, the auxiliary quantity g^{ab} has entered, which also singles out special coordinate systems (for example, locally

8 Fundamental laws of physics in Riemannian spaces

geodesic ones). How do we know whether (8.4) is correct? The fundamental difference between (8.2) and (8.4) consists in the fact that g^{ab}, in contrast to u^a, possesses a physical significance; the metric describes the influence of the gravitational field. One can therefore interpret the requirement that physical equations should be covariant, and that all the metrical quantities being introduced to guarantee covariance should correspond to properties of the gravitational field, as the physical basis of the principle of covariance.

A much more meaningful transcription formula follows from the principle of equivalence. Consistent with experience, we can generalize the identity of inertial and gravitational mass. *All* kinds of interactions between the constituent parts of a body (nuclear forces in the nuclei, electromagnetic forces in atoms and molecules) contribute to its mass. The principle of equivalence says that locally (in a region of space-time not too large) one cannot *in principle* distinguish between the action of a gravitational field and an acceleration. In other words, a freely falling observer in a gravitational field cannot detect the gravitational field by physical experiments in his immediate neighbourhood; for him *all* events occur as in an inertial system.

We have already encountered coordinate systems, local geodesic coordinates, in which the orbits of freely moving particles are described by $d^2x^a/d\tau^2 = 0$ as in an inertial system. Because of this coincidence we shall identify inertial systems and local geodesic coordinate systems. As we know, such a local geodesic system can only be introduced in the immediate neighbourhood of a point; it is only useful so long as derivatives of the Christoffel symbols, and hence the influence of the space curvature, can be ignored. Accordingly the freely falling observer too can establish the action of the gravitational force by examining larger regions of space-time; for him the planetary orbits are not straight lines, and upon bouncing on the Earth the freely falling box is no longer an inertial system.

The identification (by the freely falling observer) of inertial system and local geodesic coordinates and the definition (5.4) of the covariant derivative make plausible the following transcription principle: one formulates the physical laws in a Lorentz-invariant manner in an inertial system and substitutes covariant for partial derivatives. This prescription ensures simultaneously the covariance of the resulting equations and their validity upon using curvilinear coordinates in Minkowski space.

Two criticisms can at once be raised, pointing out that this prescription is neither unique nor logically provable. The first criticism concerns the order of higher derivatives. Partial derivatives commute, covariant ones do not. Practical examples nevertheless show that one can solve this problem simply in most cases. The second objection concerns the question of how we know that the curvature tensor and its derivatives do not also enter the basic physical laws. The resulting covariant equations would then *not* go over to the corresponding equations of Minkowski space in local-geodesic coordinates; the difference would certainly be small, however, and would be difficult to detect. Such a modification of our transcription formula cannot be excluded in principle. But up until now no experiments or other indications are known which make it necessary.

In the following sections we shall formulate the most important physical laws in Riemannian spaces, without referring every time to the transcription prescription 'partial → covariant derivative.' For the special-relativistic formulation of these laws we refer the reader to appropriate textbooks.

8.2 Particle mechanics

The momentum p^n of a particle is the product of the mass m_0 and the four-velocity u^n:

$$p^n = m_0 \frac{dx^n}{d\tau} = m_0 u^n, \tag{8.5}$$

in which τ is the proper time, defined by

$$ds^2 = g_{nm} dx^n dx^m = -c^2 d\tau^2. \tag{8.6}$$

For force-free motion in Minkowski space the momentum is constant. Accordingly, a particle upon which no force acts apart from the gravitational force moves on a geodesic of the Riemannian space,

$$\frac{Dp^n}{D\tau} = m_0 \frac{D^2 x^n}{D\tau^2} = m_0 u^n{}_{;i} u^i = m_0 \left(\frac{d^2 x^n}{d\tau^2} + \Gamma^n_{ab} \frac{dx^a}{d\tau} \frac{dx^b}{d\tau} \right) = 0. \tag{8.7}$$

External forces K^n cause a deviation from the geodesic equation:

$$\frac{Dp^n}{D\tau} = m_0 \frac{D^2 x^n}{D\tau^2} = K^n. \tag{8.8}$$

Since the magnitude $-c^2 = u_n u^n$ of the four-velocity is constant, we have

$$u_{n;i} u^n = 0; \tag{8.9}$$

that is, the four-velocity is perpendicular to the four-acceleration $\dot{u}^n = u^n{}_{;i} u^i$ and the force K^n,

$$u^n u_{n;i} u^i = u^n \dot{u}_n = 0 = K^n u_n. \tag{8.10}$$

8 Fundamental laws of physics in Riemannian spaces

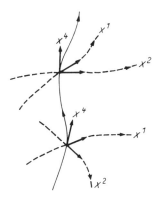

Fig. 8.1. Coordinate system of an arbitrarily moving observer.

The four equations of motion (8.8) are therefore not independent of one another (the energy law is a consequence of the momentum law).

In order to understand better the connection between Newtonian mechanics and mechanics in a Riemannian space we shall sketch how the guiding acceleration a and the Coriolis acceleration $2\omega \times \dot{r}$, which an accelerated observer moving in a rotating coordinate system would experience, are contained in the geodesic equation (8.7) which is valid for all coordinate systems.

We therefore imagine an observer who is moving along an arbitrary (timelike) world line and carries with him an orthogonal triad of vectors, whose directions he identifies with the directions of his spatial coordinate axes. For the description of processes in his immediate neighbourhood he will therefore prefer a coordinate system with the following properties: the observer is permanently at the origin of the spatial system; as time he uses his proper time; along his worldline $x^\alpha = 0$ he always uses a Minkowski metric (cf. Fig. 8.1). Summarizing, this gives up to terms quadratic in the x^n

$$ds^2 = \eta_{ab} dx^a dx^b + g_{ab,\nu}(O) x^\nu dx^a dx^b, \quad g_{ab,4}(O) = 0. \tag{8.11}$$

As spatial coordinate lines he will take the lines which arise from 'straight' extension of his triad axes (which are thus geodesics), marking off as coordinates along them the arc-length, and so completing this system that in his space $x^4 = $ constant *all* geodesics have locally the form of straight lines $x^\alpha = s\lambda^\alpha$ (s is the arc-length, λ^α the direction cosine of the geodesics). For these geodesics we have then for arbitrary constant λ^α

$$\frac{d^2 x^a}{ds^2} + \Gamma^a_{mn} \frac{dx^m}{ds} \frac{dx^n}{ds} = \Gamma^a_{\mu\nu} \lambda^\mu \lambda^\nu = 0; \tag{8.12}$$

that is, all Christoffel symbols $\Gamma^a_{\mu\nu}$ ($a=1,\ldots,4$; $\mu\nu=1,2,3$) vanish. Because of (3.11) the derivatives of the metric (8.11) therefore satisfy the conditions

$$g_{\alpha\beta,\nu} = 0; \qquad g_{4\beta,\nu} = -g_{4\nu,\beta}. \tag{8.13}$$

The equations (8.13) show that there are only three independent components of the derivatives $g_{4\beta,\nu}$; one can thus map these onto the components of a 'three-vector' ω^μ,

$$g_{4\beta,\nu} = -\epsilon_{4\beta\nu\mu}\frac{\omega^\mu}{c} = -\epsilon_{\beta\nu\mu}\frac{\omega^\mu}{c}. \tag{8.14}$$

The derivatives $g_{44,\nu}$ not yet taken into account in (8.11), (8.13) and (8.14) can be expressed through the acceleration a_ν of the observer, for whose world line $x^\alpha = 0$, $x^4 = ct = c\tau$ we have

$$a^\nu \equiv \frac{d^2 x^\nu}{dt^2} + \Gamma^\nu_{ab}\frac{dx^a}{dt}\frac{dx^b}{dt} = \Gamma^\nu_{44}c^2, \tag{8.15}$$

and hence

$$g_{44,\nu} = -2\frac{a_\nu}{c^2} = -2g_{\nu b}\frac{a^b}{c^2}. \tag{8.16}$$

To summarize, an observer, who carries with him his local Minkowski system and in whose position space all geodesics diverging from him are straight lines, uses in a neighbourhood of his world line $x^\alpha = 0$, $x^4 = c\tau$, the coordinate system

$$ds^2 = \eta_{\alpha\beta}dx^\alpha dx^\beta - 2\epsilon_{\beta\nu\mu}x^\nu\frac{\omega^\mu}{c}dx^4 dx^\beta - \left(1 + \frac{2a_\nu x^\nu}{c^2}\right)(dx^4)^2. \tag{8.17}$$

For him the only non-vanishing Christoffel symbols are

$$\Gamma^\alpha_{4\nu} = \epsilon_\nu{}^\alpha{}_\mu\frac{\omega^\mu}{c}, \qquad \Gamma^\alpha_{44} = \frac{a^\alpha}{c^2}, \qquad \Gamma^4_{4\nu} = \frac{a_\nu}{c^2}. \tag{8.18}$$

If the observer moves on a geodesic, then a_ν vanishes. In the coordinates (8.17) and (8.18) the equation (5.24) which defines Fermi–Walker transport has the form

$$\frac{dT^\mu}{d\tau} + \epsilon_\alpha{}^\mu{}_\nu\omega^\nu T^\alpha = 0, \qquad \frac{dT^4}{d\tau} = 0. \tag{8.19}$$

Hence for an observer who subjects his triad, formed out of vectors which he regards as constant, to a Fermi–Walker transport, the vector ω^μ must vanish. If a^ν disappears as well as ω^μ, then the coordinate system (8.17) is an inertial system along the entire world line of the observer.

8 Fundamental laws of physics in Riemannian spaces

To describe the motion of a particle the observer will naturally use his coordinate system (8.17), (8.18) and examine the acceleration d^2x^α/dt^2 of this particle in it. From the geodesic equation we have for the three spatial components of the acceleration the relation

$$\frac{d^2x^\alpha}{dt^2} = 2\epsilon^\alpha{}_{\nu\mu}\omega^\mu \frac{dx^\nu}{dt} - a^\alpha - \frac{d\lambda}{dt}\frac{d}{dt}\left(\frac{dt}{d\lambda}\right)\frac{dx^\alpha}{dt}. \tag{8.20}$$

We can understand the connection between λ and t from the time component of the geodesic equation

$$\frac{d^2t}{d\lambda^2} + 2\frac{a_\nu}{c^2}\frac{dt}{d\lambda}\frac{dx^\nu}{d\lambda} = 0. \tag{8.21}$$

Substitution of (8.21) into (8.20) yields

$$\frac{d^2x^\alpha}{dt^2} = -a^\alpha + 2\epsilon^\alpha{}_{\nu\mu}\omega^\mu \frac{dx^\nu}{dt} + \frac{2a_\nu}{c^2}\frac{dx^\nu}{dt}\frac{dx^\alpha}{dt}, \tag{8.22}$$

or, in vector form,

$$\ddot{r} = -a - 2\omega \times \dot{r} + \frac{2(a\dot{r})}{c^2}\dot{r}. \tag{8.23}$$

One recognizes the guiding acceleration a and its relativistic correction $2(a\dot{r})\dot{r}/c^2$ (both vanish if the observer is moving freely on a geodesic), and also the Coriolis acceleration $2\omega \times \dot{r}$, caused by the rotation of the triad carried by the observer relative to a Fermi–Walker-transported triad. The vanishing of the Coriolis term in the Fermi–Walker system justifies the statement that for an observer who is not falling freely ($a^\nu \neq 0$) a local coordinate system produced by Fermi–Walker transport of the spatial triad of vectors is the best possible realization of a non-rotating system.

8.3 Electrodynamics in vacuo

The field equations As in Minkowski space, the electromagnetic field is described by an antisymmetric field-tensor F_{mn}. Because it satisfies the equations

$$3F_{[mn;a]} = F_{mn;a} + F_{na;m} + F_{am;n} = F_{mn,a} + F_{na,m} + F_{am,n} = 0, \tag{8.24}$$

it can be represented as the curl of a four-potential A_n:

$$F_{mn} = A_{n;m} - A_{m;n} = A_{n,m} - A_{m,n}. \tag{8.25}$$

This potential is determined only up to a four-dimensional gradient. The field is produced by the four-current j^m:

$$F^{mn}{}_{;n} = \frac{1}{\sqrt{-g}} (\sqrt{-g}\, F^{mn})_{,n} = \frac{1}{c} j^m. \tag{8.26}$$

Because of the antisymmetry of F_{mn}, (8.26) is only integrable (self-consistent) if the continuity equation

$$j^m{}_{;m} = \frac{1}{\sqrt{-g}} (\sqrt{-g}\, j^m)_{,m} = 0 \tag{8.27}$$

is satisfied. For an isolated charge distribution the conservation law for the total charge Q follows from it (cf. section 7.5):

$$\int_{x^4 = \text{constant}} j^a \, df_a = \text{constant} = Q. \tag{8.28}$$

By substituting (8.25) into (8.26) one can derive the generalized inhomogeneous wave equation for the potential. Using the expressions written with covariant derivatives, one obtains

$$A^{n;m}{}_{;n} - A^{m;n}{}_{;n} = A^n{}_{;n}{}^{;m} + R_n{}^m A^n - A^{m;n}{}_{;n} = \frac{1}{c} j^m. \tag{8.29}$$

If, on the other hand, one sets out directly from the special-relativistic equation,

$$(A^n{}_{,n})^{,m} - A^{m,n}{}_{,n} = \frac{1}{c} j^m, \tag{8.30}$$

and in it replaces the partial by covariant derivatives, then one obtains (8.29) *without* the term in the Ricci tensor $R_n{}^m$ (which arises by interchange of covariant derivatives). One clearly sees here that the transcription formula 'partial → covariant derivative' is not unique when applied to the potential. Potentials, however, are not directly measurable, and for the physically important field strengths and their derivatives the prescription which we gave above is unique. Since (8.29) follows directly from this prescription it is the correct generalization of the inhomogeneous wave equation.

The Lorentz gauge,

$$A^n{}_{;n} = 0, \tag{8.31}$$

does not possess the same significance in Riemannian spaces that it has in Minkowski space. There need not exist coordinates for which the Lorentz gauge ensures the decoupling of the four equations (8.29).

8 Fundamental laws of physics in Riemannian spaces

For practical calculations it is often convenient to use partial derivatives; (8.25) and (8.29) give

$$[\sqrt{-g}\, g^{ma} g^{nb}(A_{b,a} - A_{a,b})]_{,n} = \sqrt{-g}\, \frac{1}{c} j^m. \tag{8.32}$$

Lagrangian and energy–momentum tensor The Maxwell equations can be derived from the action principle

$$W = \int \mathcal{L}\, d^4x = \int \left[\frac{1}{c} j^a A_a - \frac{1}{4}(A_{n,m} - A_{m,n})(A^{n,m} - A^{m,n})\right] \sqrt{-g}\, d^4x$$

$$= \text{extremum}, \tag{8.33}$$

where the components of the potential are varied as the independent field quantities.

The symmetric energy–momentum tensor,

$$T^{mn} = F^{am} F_a{}^n - \tfrac{1}{4} g^{mn} F_{ab} F^{ab}, \tag{8.34}$$

is trace-free, $T^n{}_n = 0$. Its divergence is, up to a sign, equal to the Lorentz force-density,

$$T^{mn}{}_{;n} = -\frac{1}{c} F^{mn} j_n. \tag{8.35}$$

Description of the solutions in terms of the sources In a Riemannian space it is still possible to express the solution to the differential equation (8.29) in the form of an integral

$$A^m(x^i) = \int G^{m\bar{n}}(x^i, \bar{x}^i) j_{\bar{n}}(\bar{x}^i) \sqrt{-\bar{g}}\, d^4\bar{x}. \tag{8.36}$$

The two-point tensor $G^{m\bar{n}}$, the generalization of the Green function, is now in general a very complicated function. We want to point out (without proof) a notable difference in the way in which effects propagate (for example, the propagation of light) in a Riemannian space in comparison with that in a Minkowski space. While in Minkowski space the propagation of effects *in vacuo* takes place exactly on the light cone, that is, a flash of light at the point $\bar{P}$ of Fig. 8.2a reaches the observer at precisely the point P, in Riemannian space the wave can also propagate inside the future light cone, a (weak) flash of light being noticeable also at points later than P (for example, P'). The reason for this deviation, which one can also interpret as a deviation from Huygens' principle, can be thought

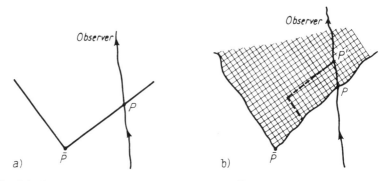

Fig. 8.2. Propagation of effects between source $\bar{P}$ and observer: (*a*) Minkowski space: on the light cone, (*b*) Riemannian space: within the entire (hatched) interior of the light cone. (Dashed line = possible light path ('dispersion').)

of as a kind of scattering of the light wave by the space curvature. In particularly simple Riemannian spaces this effect does not occur; for example, the Robertson–Walker metrics belong to this class (cf. section 25).

Special properties of source-free fields   Since one can convert (8.24) into the system

$$\tilde{F}^{ab}{}_{;b} = 0$$

by use of the dual field tensor,

$$\tilde{F}_{ab} = \tfrac{1}{2} \epsilon_{abmn} F^{mn} \qquad (\tilde{E} = -B, \tilde{B} = E), \tag{8.37}$$

then for $j^m = 0$ the Maxwell equations are equivalent to the equations

$$\Phi^{ab}{}_{;b} = (F^{ab} + i\tilde{F}^{ab})_{;b} = 0 \tag{8.38}$$

for the complex field tensor Φ^{ab}. A solution Φ^{ab} remains a solution after multiplication by a complex number $e^{i\alpha}$ (a 'duality rotation'). The energy-momentum tensor

$$T^{ab} = \tfrac{1}{2} \Phi^{ac} \bar{\Phi}^b{}_c \tag{8.39}$$

does not change under such a duality rotation. In three-dimensional form in flat space a duality rotation corresponds to a transformation

$$\hat{E} = \cos \alpha E + \sin \alpha B, \qquad \hat{B} = \cos \alpha B - \sin \alpha E. \tag{8.40}$$

The source-free Maxwell equations are 'conformally invariant.' By this one means that a potential A_a, which is a solution of the Maxwell equations in a Riemannian space with metric ds^2, also satisfies them in a conformally

8 *Fundamental laws of physics in Riemannian spaces* 77

equivalent space with metric $d\bar{s}^2 = M^2 ds^2$. The proof can be obtained directly from (8.32) with $j^m = 0$, remembering the relations

$$\left.\begin{array}{ll} d\bar{s}^2 = M^2 ds^2, & \bar{g}_{ab} = M^2 g_{ab}, \quad \bar{A}_a = A_a, \\ & \bar{g}^{ab} = M^{-2} g^{ab}, \quad \bar{A}_{a,b} = A_{a,b}, \\ & \bar{g} = M^8 g. \end{array}\right\} \quad (8.41)$$

Null electromagnetic fields The electromagnetic field tensor possesses two invariants, namely,

$$I_1 = F_{ab} F^{ab}, \qquad I_2 = F_{ab} \tilde{F}^{ab} \quad (8.42)$$

(cf. (4.54)), which in Minkowski space have the form

$$I_1 = -2(E^2 - B^2), \qquad I_2 = 4EB. \quad (8.43)$$

Null electromagnetic fields are fields for which both invariants vanish. They are therefore generalizations of plane waves in flat space. One can show that the conditions

$$\left.\begin{array}{l} F_{mn} = \lambda(p_m k_n - k_m p_n), \\ k^n k_n = 0 = p_n k^n, \quad p_n p^n = 1 \end{array}\right\} \quad (8.44)$$

on the field tensor are necessary and sufficient for the vanishing of both invariants (for example, by introduction of a local Minkowski system and choice of coordinate axes so that $E = (E_x, 0, 0)$ and $B = (B_x, B_y, 0)$ hold). F_{mn} is therefore a simple bivector, in the sense of (4.55), with the null vector k_n as eigenvector. The associated energy-momentum tensor has the simple structure

$$T_{mn} = \lambda^2 k_m k_n. \quad (8.45)$$

Incoherent radiation fields In superposing incoherent electromagnetic fields one has to add (and average) the energy-momentum tensors and not the field strengths. A field tensor can no longer be associated with this superposition.

If the fields being superposed single out locally no spatial direction in the rest system $u^m = (0, 0, 0, c)$ of an observer, then the resulting energy-momentum tensor has the form

$$T_{mn} = p g_{mn} + (p/c^2 + \mu) u_m u_n, \quad (8.46)$$

and hence in a local Minkowski frame with $u_m = (0, 0, 0, -c)$

$$T_{mn} = \begin{pmatrix} p & & & \\ & p & & \\ & & p & \\ & & & \mu c^2 \end{pmatrix}. \qquad (8.47)$$

Under the superposition the properties of vanishing trace, $T^n{}_n = 0$, and vanishing divergence, $T^{mn}{}_{;n} = 0$, are of course preserved. Consequently the radiation pressure p and the energy density μc^2 are related by

$$3p = \mu c^2. \qquad (8.48)$$

8.4 Geometrical optics

The transition from wave solutions of the source-free Maxwell equations to geometrical optics can be accomplished by substituting into the field equations

$$(A^{n;m} - A^{m;n})_{;n} = \frac{1}{\sqrt{-g}}[\sqrt{-g}\, g^{ma} g^{nb}(A_{b,a} - A_{a,b})]_{,n} = 0 \qquad (8.49)$$

the ansatz

$$A_a = \hat{A}_a(x^n) e^{i\omega S(x^n)} \quad (\hat{A}_a \text{ complex, } S \text{ real}) \qquad (8.50)$$

and setting the coefficients of ω^2 and ω separately to zero. As in flat space, this splitting into amplitude $\hat{A}_a$ and eikonal (phase) S is meaningful only in certain finite regions of space and represents a good approximation only for large ω.

Substitution of (8.50) into (8.49) gives, on taking into account only the terms in ω^2,

$$S^{,m}(\hat{A}^n S_{,n}) - \hat{A}^m(S_{,n} S^{,n}) = 0. \qquad (8.51)$$

Since the part of the field tensor proportional to ω is

$$F_{mn} = (\hat{A}_n S_{,m} - \hat{A}_m S_{,n}) i\omega e^{i\omega S} = i\omega(A_n S_{,m} - A_m S_{,n}), \qquad (8.52)$$

this part vanishes if $\hat{A}_m$ is parallel to $S^{,m}$. We are therefore interested only in the solution

$$S_{,n} S^{,n} = 0, \qquad A_n S^{,n} = 0 \qquad (8.53)$$

of (8.51). The gradient $S_{,n}$ of the surfaces of equal phase is therefore a null vector and the field tensor (8.52) has the structure (8.44) of the field tensor of a null field, with k_m proportional to $S_{,m}$ and p_n proportional to $\text{Re}\, A_n$ (note that for comparison purposes one must take the real part of

8 Fundamental laws of physics in Riemannian spaces

the complex field quantities used here). In this approximation the field consequently behaves locally like a plane wave.

Differentiating (8.53) gives

$$S^{,n} S_{,n;m} = 0.$$

Since the curl of a gradient vanishes ($S_{,n;m} = S_{,m;n}$), this equation implies that

$$S_{,m;n} S^{,n} = 0. \tag{8.54}$$

This equation says that the curves $x^n(\lambda)$, whose tangent vector is $S^{,n}$, are geodesics,

$$\frac{\mathrm{d}x^n}{\mathrm{d}\lambda} = S^{,n}, \tag{8.55}$$

and because $S_{,n} S^{,n} = 0$ they are null geodesics.

If we characterize the wave not by $\hat{A}_a$ and the surfaces of constant phase $S = \text{constant}$, but by the curves $x^n(\lambda)$ orthogonal to them (which we call light rays), then we have accomplished the transition from wave optics to geometrical optics. In words, (8.54) then says that light rays are null geodesics.

We shall take the approximation one step further, investigating the terms in the Maxwell equations proportional to ω, and hence obtaining statements about how the intensity and polarization of the wave change along a light ray.

From (8.49) and (8.53) one obtains immediately

$$-\mathrm{i}\omega[2\hat{A}^m{}_{;n} S^{,n} + \bar{A}^m S^{,n}{}_{;n} - \hat{A}^n{}_{;n} S^{,m}] = 0. \tag{8.56}$$

If one contracts this equation with the vector $\bar{A}_m$, which is the complex conjugate of $\hat{A}_m$, and takes note of (8.53), then the result can be written in the form

$$(\bar{A}_m \hat{A}^m)_{;n} S^{,n} + (\bar{A}_m \hat{A}^m) S^{,n}{}_{;n} = 0, \tag{8.57}$$

or in the equivalent form

$$(\bar{A}_m \hat{A}^m S^{,n})_{;n} = 0. \tag{8.58}$$

Because the intensity of the wave is proportional to $\bar{A}_m \hat{A}^m$, (8.57) can be read as a statement about the change in intensity of the light ray in the direction $S^{,n}$. However, (8.58) suggests the following picture, which is clearer. If one interprets

$$J^n = \bar{A}_m \hat{A}^m S^{,n} \tag{8.59}$$

as a photon current, then this current is source-free (conservation of photon number) and in the direction of the light rays.

We obtain a further physical consequence from (8.56) if we decompose the vector $\hat{A}_m$ into its magnitude a and the unit vector P_m:

$$\hat{A}_m = aP_m. \tag{8.60}$$

Then (8.57) is equivalent to

$$a_{,n}S^{,n} = -\tfrac{1}{2}aS^{,n}{}_{;n}, \tag{8.61}$$

and from (8.65) we have

$$P^m{}_{;b}S^{,b} = \frac{1}{2}\left(\frac{a_{,n}}{a}P^n + P^n{}_{;n}\right)S^m. \tag{8.62}$$

This means, however, that the tensor f_{mn} associated with the field tensor (8.52),

$$f_{mn} = P_n S_{,m} - P_m S_{,n}, \tag{8.63}$$

which contains the characteristic directions of the wave (direction of propagation $S^{,n}$ and polarization P^m), is parallelly transported along the rays; we have

$$f_{mn;i}S^{,i} = 0. \tag{8.64}$$

8.5 Thermodynamics

Thermodynamical systems can be extraordinarily complicated; for example, a great number of processes can be going on in a star simultaneously. We want to try to explain the basic general ideas, restricting ourselves to the simplest systems.

During thermodynamical processes certain elements of matter, with their properties, remain conserved, for example, in non-relativistic thermodynamics molecules or atoms and their masses. In the course of transformations in stars and during nuclear processes the baryons with their rest mass are conserved instead. We shall therefore relate all quantities to these baryons. If, for example, we choose a volume element of the system, then we shall take as four-velocity u^i of this element the average baryon velocity. The flow (motion) of the system will therefore be characterized by a four-velocity field

$$u^i = u^i(x^n), \quad u^i u_i = -c^2. \tag{8.65}$$

To set up the basic thermodynamical equations one first goes to the local rest system,

$$u^i = (0, 0, 0, c), \tag{8.66}$$

8 Fundamental laws of physics in Riemannian spaces

of the volume element under consideration and regards this volume element as a system existing in equilibrium (of course it interacts with its surroundings, so that the whole system is not necessarily in equilibrium); that is, one introduces for this volume element the fundamental thermodynamic state variables, for example,

$$\left.\begin{array}{ll} n \text{ baryon number density,} & s \text{ entropy per baryon mass,} \\ \rho \text{ baryon mass density,} & p \text{ isotropic pressure,} \\ T \text{ temperature,} & \bar{\mu} \text{ chemical potential,} \\ u \text{ internal energy per unit mass,} & f \text{ free energy per unit mass.} \end{array}\right\} \quad (8.67)$$

'Density' here always means 'per three-dimensional volume in the local rest system'; the entropy density, for example, would be given by $s\rho$. There exist relationships between these state variables which in the simplest case express the fact that only two of them are really independent, and, from knowledge of the entropy as a function of the energy and the density, or of the specific volume $v = 1/\rho$,

$$s = s(u, v), \quad (8.68)$$

one can calculate the other quantities, for example,

$$\frac{\partial s}{\partial u} = \frac{1}{T}, \quad \frac{\partial s}{\partial v} = \frac{p}{T}, \quad f = u - Ts, \quad \text{etc.} \quad (8.69)$$

For the interaction of the volume element with its surroundings we have balance equations. These are the law of conservation of baryon number,

$$(\rho u^n)_{;n} = 0 \quad (8.70)$$

(generalized mass conservation), the balance equations for energy and momentum, formulated as the vanishing of the divergence of the energy-momentum tensor T^{mn},

$$T^{mn}_{;n} = 0 \quad (8.71)$$

(generalized first main law), and the balance equation for the entropy,

$$s^n_{;n} = \sigma \geq 0, \quad (8.72)$$

which says that the density of entropy production σ is always positive or zero (generalized second law of thermodynamics). Of course these equations take on a physical meaning only if the entropy current density s^n and the energy-momentum tensor T^{mn} are tied up with one another and with the thermodynamic quantities (8.67).

This can be done as follows. One uses the projection tensor,

$$h_{ab} = g_{ab} + u_a u_b/c^2, \quad (8.73)$$

to decompose the energy–momentum tensor into components parallel and perpendicular to the four-velocity,

$$T_{ab} = \mu u_a u_b + p h_{ab} + (u_a q_b + q_a u_b)/c^2 + \pi_{ab},$$
$$q_a u^a = 0, \qquad \pi_{ab} u^a = 0, \qquad \pi^a{}_a = 0, \qquad (8.74)$$

and links the quantities which then occur to the thermodynamic state variables and to the entropy current vector. The coefficient of h_{ab} is the isotropic pressure p, the internal energy per unit mass u is coupled to the mass density μ in the rest system of the matter by

$$\mu = \rho(1 + u/c^2), \qquad (8.75)$$

and the heat current q^i (momentum current density in the rest system) goes into the entropy current:

$$s^i = \rho s u^i + q^i/T. \qquad (8.76)$$

Equation (8.76) says that the entropy flows in such a way that it is carried along convectively with the mass (first term) or transported by the flow of heat (generalization of $dS = dQ/T$).

We now want to obtain an explicit expression for the entropy production density σ. Upon using (8.70) and the equation

$$s_{,n} u^n = \frac{1}{T} \left(\frac{\mu c^2}{\rho} \right)_{,n} u^n + \frac{p}{T} \left(\frac{1}{\rho} \right)_{,n} u^n = \frac{1}{\rho T} [(p + \mu c^2) u^i{}_{;i} + \mu_{,n} u^n c^2], \qquad (8.77)$$

which follows from (8.68), (8.69) and (8.75), we obtain

$$\sigma = s^n{}_{;n} = \frac{1}{T} [(p + \mu c^2) u^i{}_{;i} + \mu_{,n} u^n c^2] + \left(\frac{q^n}{T} \right)_{;n}. \qquad (8.78)$$

Since the terms in square brackets can be written in the form

$$(p + \mu c^2) u^i{}_{;i} + \mu_{,n} u^n c^2 = (\mu u_a u_b + p h_{ab})^{;b} u^a, \qquad (8.79)$$

and the divergence of the energy–momentum tensor vanishes, (8.78) implies the relation

$$\sigma = [T^{mn} - \mu u^m u^n - p h^{mn}]_{;m} \frac{u_n}{T} + \left(\frac{q^n}{T} \right)_{;n}, \qquad (8.80)$$

which, bearing in mind the definition (8.74) of q^n, and using (5.32), can be cast finally into the form

8 Fundamental laws of physics in Riemannian spaces

$$\sigma = -(T^{mn} - \mu u^m u^n - p h^{mn})\left(\frac{u^m}{T}\right)_{;n}$$

$$= -\frac{1}{2}(T^{mn} - \mu u^m u^n - p h^{mn}) \underset{u_a/T}{\pounds} g_{mn}. \tag{8.81}$$

In irreversible thermodynamics one can satisfy the requirement that $\sigma \geq 0$ in many cases by writing the right-hand side of (8.81) as a positive-definite quadratic form, that is, by making an assumption of linear phenomenological equations. For the particular case $\pi^{mn} = 0$, when (8.81) reduces, because of (8.74), to

$$\sigma = -q_a \frac{c^2}{T^2}\left(\frac{T}{c^2}\dot{u}^a + T^{,a}\right), \quad \dot{u}^a \equiv u^a{}_{;n}u^n, \tag{8.82}$$

this ansatz means that

$$q_a = -\bar{\kappa}(T_{,n} + \dot{u}_n T/c^2)h_a{}^n, \tag{8.83}$$

which represents the relativistic generalization of the linear relation between heat current and temperature gradient.

In many cases one can ignore irreversible processes. If the system is determined by only two state quantities in the sense of (8.68), this means because of (8.81) that complete, exact reversibility ($\sigma = 0$) is possible either only for certain metrics (whose Lie derivatives vanish), or for especially simple media, whose energy–momentum tensor has the form

$$T^{mn} = \mu u^m u^n + p h^{mn} = (\mu + p/c^2)u^m u^n + p g^{mn}. \tag{8.84}$$

Such a medium is called a *perfect fluid* or, for $p = 0$, *dust*.

8.6 Perfect fluids and dust

According to the definition given in section 8.5, a perfect fluid is characterized by having an energy–momentum tensor of the form (8.84).

The equation of motion of this flow reads, using the notation $\dot{\mu} = \mu_{,n}u^n$, $\dot{u}^n = u^n{}_{;i}u^i$, etc.,

$$T^{mn}{}_{;n} = (\mu + p/c^2)u^n{}_{;n}u^m + (\mu + p/c^2)\dot{u}^m + (\mu + p/c^2)\dot{u}^m + p^{,m} = 0. \tag{8.85}$$

Contraction with u_m/c^2 gives the energy balance

$$\dot{\mu} + (\mu + p/c^2)u^n{}_{;n} = 0, \tag{8.86}$$

and contraction with the projection tensor, $h^i_m = g^i_m + u^i u_m/c^2$, the momentum balance

$$(\mu + p/c^2)\dot{u}^i + h^{im}p_{,m} = 0. \tag{8.87}$$

Equation (8.87) shows that the pressure too contributes to the inertia of the matter elements, the classical analogue of this equation being of course

$$\rho \frac{dv}{dt} = -\operatorname{grad} p. \tag{8.88}$$

The equations of motion (8.86) and (8.87) must in each case be completed by the specification of an equation of state. One can regard as the simplest equation of state that of dust, $p = 0$. From this and the equations of motion follow

$$\dot{u}^m = Du^m/D\tau = 0 \quad \text{and} \quad (\mu u^n)_{;n} = 0; \tag{8.89}$$

that is, the stream-lines of the matter are geodesics, and the rest mass μ is conserved.

8.7 Other fundamental physical laws

Just as with the examples of particle mechanics, electrodynamics, thermodynamics and mechanics of continua which have been described in detail, so also one can carry over to Riemannian spaces other classical theories, for example, those of the Dirac equation, of the Weyl equation for the neutrino field, and for the Klein–Gordon equation. Although the foundations of these theories have been thoroughly worked out, convincing examples and applications within the theory of gravitation are still lacking, and we shall therefore not go into them further. For the Einstein gravitation theory only the following property of closed systems, that is, systems upon which act no forces whose origins lie outside the systems, will be important; namely, that their energy–momentum tensor T_{mn} is symmetric (expressing the law of conservation of angular momentum in special relativity) and its divergence vanishes (generalization of the law of conservation of energy–momentum in special relativity):

$$T_{mn} = T_{nm}, \qquad T^{mn}{}_{;n} = 0. \tag{8.90}$$

With the fundamental laws of quantum mechanics and quantum field theory things are rather different. Here indeed some work has been done, addressed to particular questions, but one cannot yet speak of a real synthesis between quantum theory and gravitation theory (cf. section 30).

8 Fundamental laws of physics in Riemannian spaces

Bibliography to section 8

Textbooks   Schmutzer (1968), Synge (1960)

Monographs and collected works   Neugebauer (1980), Tolman (1934)

Review and research articles   de Witt and Brehme (1960), Ehlers (1961, 1966, 1971), Ellis and Sciama (1972), Israel (1972), Israel and Stewart (1980), Neugebauer (1974)

Foundations of Einstein's theory of gravitation

9 The fundamental equations of Einstein's theory of gravitation

9.1 The Einstein field equations

As we have already indicated more than once, the basic idea of Einstein's theory of gravitation consists in geometrizing the gravitational force, that is, mapping all the properties of the gravitational force and its influence upon physical processes onto the properties of a Riemannian space. While up until the present we have concerned ourselves only with the mathematical structure of such a space and the influence of a given Riemannian space upon physical laws, we want now to turn to the essential physical question. Gravitational fields are produced by masses, and so we may ask, how are the properties of the Riemannian space calculated from the distribution of matter? Here 'matter,' in the context of general relativity, means everything that can produce a gravitational field (i.e. that contributes to the energy–momentum tensor), for example, not only atomic nuclei and electrons, but also the electromagnetic field.

Of course one cannot derive logically the required new fundamental physical law from the laws already known; however, one can set up several very plausible requirements. We shall do this in the following and discover, surprisingly, that once one accepts the Riemannian space, the Einstein field equations follow almost directly.

The following requirements appear reasonable:

(a) The field equations should be tensor equations (independence of coordinate systems of the laws of nature).
(b) Like all other field equations of physics they should be partial differential equations of at most second order for the functions to be determined (the components of the metric tensor g_{mn}), which are linear in the highest derivatives.
(c) They should (in the appropriate limit) go over to the Poisson (potential) equation:

$$\Delta U = 4\pi f \mu \tag{9.1}$$

9 Fundamental equations of Einstein's theory of gravitation

of Newtonian gravitation theory (here U is the potential, f is the Newtonian gravitational constant, and μ is the mass density).

(d) Since the energy–momentum tensor T^{mn} is the special relativistic analogue of the mass density, it should be the cause (source) of the gravitational field.

(e) If the space is flat, T^{mn} should vanish.

We now want to see where these requirements lead us. Plainly we need a tensor (requirement (a)) that contains only derivatives of the metric up to second order (requirement (b)); as building blocks for this *Einstein tensor* G_{mn}, only the curvature tensor, the metric tensor and the ϵ-tensor are available, as we have already shown in section 6.3. Requirement (d) means that the field equations have the structure

$$G_{mn} = \kappa T_{mn}, \qquad (9.2)$$

with a constant of nature κ which is still to be determined; this is consistent with the symmetry and vanishing divergence (8.90) of the energy–momentum tensor only if

$$G^{mn}{}_{;n} = 0 \quad \text{and} \quad G_{mn} = G_{nm}. \qquad (9.3)$$

There is now, as one can show, only one tensor which is linear (requirement (b)) in the components of the curvature tensor and which satisfies (9.3); namely, $R_{mn} - \tfrac{1}{2} g_{mn} R$, which we have already met in (6.30) during the discussion of the Bianchi identities. Since the metric tensor itself also satisfies (9.3), G^{mn} has the form

$$G^{mn} = R^{mn} - \tfrac{1}{2} g^{mn} R + \Lambda g^{mn}. \qquad (9.4)$$

The natural constant Λ is the *cosmological constant*, discussed by Einstein. If it does not vanish, a completely matter-free space ($T^{mn} = 0$) would always be curved, in contradiction to requirement (e), since because of (9.2) and (9.4) the Ricci tensor R^{mn} cannot vanish. This requirement (e) is, however, difficult to prove. It is only possible to distinguish the cases $\Lambda = 0$ and $\Lambda \neq 0$ by making observations and relating them to cosmological models. We shall assume that $\Lambda = 0$, but we could also discuss for a series of examples the influence of the term Λg^{mn} by bringing it into the right-hand side of (9.2) and formally regarding it as part of the energy–momentum tensor.

The Einstein tensor thus has the form

$$G_{mn} = R_{mn} - \tfrac{1}{2} R g_{mn}. \qquad (9.5)$$

For actual calculations, use of the explicit representation in terms of the components of the curvature tensor

88 Foundations of Einstein's theory of gravitation

$$\left.\begin{array}{l}G^1_1 = -(R^{23}{}_{23}+R^{24}{}_{24}+R^{34}{}_{34}),\\ G^2_2 = -(R^{13}{}_{13}+R^{14}{}_{14}+R^{34}{}_{34}),\\ G^1_2 = R^{31}{}_{32}+R^{41}{}_{42},\\ G^2_3 = R^{12}{}_{13}+R^{42}{}_{43}, \quad \text{etc.,}\end{array}\right\} \tag{9.6}$$

is often useful.

Our demands have led us in a rather unambiguous manner to the *Einstein field equations*

$$G_{mn} = R_{mn} - \tfrac{1}{2} R g_{mn} = \kappa T_{mn}. \tag{9.7}$$

Einstein himself derived them in the year 1915, after about ten years of research. The natural law (9.7) shows how the space curvature (represented by the Ricci tensor R_{mn}) is brought about by the matter distribution (represented by the energy–momentum tensor T_{mn}).‡

The field equations (9.7) constitute a system of 10 differential equations to determine the 10 metric functions g_{mn}. But even for fixed initial conditions this system has no unique solution; it must still always be possible to carry out arbitrary coordinate transformations. In fact precisely this under-determinacy in the system of field equations is guaranteed by the existence of the contracted Bianchi identities

$$G^{mn}{}_{;n} = \kappa T^{mn}{}_{;n} = 0. \tag{9.8}$$

They of course express the fact that the 10 field equations (9.7) are not all independent of one another.

The equations (9.8) permit yet another interpretation and conclusion of great physical significance. Since the divergence of the Einstein tensor G^{mn} vanishes identically, the Einstein field equations are integrable and free of internal contradiction only if $T^{mn}{}_{;n} = 0$. The covariant derivative in this condition is, however, to be calculated with respect to the metric g_{mn}, which should be first determined from this very energy–momentum tensor! It is therefore in principle impossible first to specify the space-time distribution of the matter (the matter and its motion) and from this to calculate the space structure. Space structure (curvature) and motion of the matter in this space constitute a dynamical system whose elements are so closely coupled with one another that they can only be solved simultaneously. The space is not the stage for the physical event, but rather an aspect of the interaction and motion of the matter.

---

‡ *Editor's note:* In general things are not quite as simple as this. In many cases T_{mn}, which is being regarded as the source of the curvature, actually contains g_{mn}, e.g. the energy–momentum tensor (8.46).

Sometimes one can assume to good approximation that the space structure is determined by a part of the energy–momentum tensor (for example, by the masses of the stars) and that the remainder (for example, the starlight) no longer alters the curvature. One then speaks of test fields. These are fields which do not cause gravitational fields, but are only influenced by the gravitational fields already existing and hence serve to demonstrate the properties of these fields; they do not appear on the right-hand side of the field equations (9.7).

9.2 The Newtonian limit

In every new physical theory the previous one is contained as a limiting case. This experience is confirmed also in the theory of gravitation. The purpose of this section is to bring out the connection between the Einstein equations (9.7) and the Newtonian theory of gravitation and thereby to clarify the physical meaning of the natural constant κ introduced in (9.7). First of all we must define what we mean by 'Newtonian limit.' In the Newtonian theory of gravitation the mass density μ is the only source of the field. In the applications in which its predictions have been verified, such as planetary motion, all velocities in the rest system of the centre of gravity of the field-producing masses, for example, the Sun, are small compared to the velocity of light. Therefore the following characterization of the Newtonian limit is appropriate:

(i) There exists a coordinate system in which the energy density

$$T_{44} = \mu c^2 \tag{9.9}$$

is the effective source of the gravitational field and all other components of the energy–momentum tensor are ignorable.

(ii) The fields vary only slowly; derivatives with respect to $x^4 = ct$, which of course contain the factor c^{-1}, are to be ignored.

(iii) The metric deviates only slightly from that of a Minkowski space:

$$g_{mn} = \eta_{nm} + f_{nm}, \quad \eta_{nm} = \begin{pmatrix} 1 & & & \\ & 1 & & \\ & & 1 & \\ & & & -1 \end{pmatrix}. \tag{9.10}$$

Terms which are quadratic in f_{mn} and its derivatives are ignored; the Einstein field equations are linearized (cf. section 13.2).

We have now to incorporate these three ideas into the field equations (9.7).

By contraction we have quite generally from (9.7) the relation

90   *Foundations of Einstein's theory of gravitation*

$$-R = \kappa T^a{}_a = \kappa T, \tag{9.11}$$

so that one can also write the Einstein equations in the form

$$R_{mn} = \kappa(T_{mn} - \tfrac{1}{2} g_{mn} T). \tag{9.12}$$

Of these ten equations only

$$R_{44} = \kappa\left(T_{44} - \frac{1}{2}\eta_{44} T\right) = \kappa\left(\mu c^2 - \frac{1}{2}\mu c^2\right) = \frac{\kappa}{2}\mu c^2 \tag{9.13}$$

is of interest in the Newtonian approximation (cf. section 13.1). In order to calculate R_{44} from the metric (9.10) we start from the defining equation (6.3) for the curvature tensor and ignore terms which are quadratic in the Christoffel symbols, that is, we use

$$R^a{}_{mbn} = \Gamma^a{}_{mn,b} - \Gamma^a{}_{mb,n} = \tfrac{1}{2}\eta^{as}(f_{sn,mb} + f_{mb,sn} - f_{mn,bs} - f_{bs,mn}). \tag{9.14}$$

Then we have

$$R_{44} = R^a{}_{4a4} = \tfrac{1}{2}\eta^{as}(f_{s4,a4} + f_{a4,s4} - f_{44,as} - f_{as,44}),$$

or, on ignoring all time derivatives,

$$R_{44} = -\tfrac{1}{2}\eta^{as} f_{44,as} = -\tfrac{1}{2}\eta^{\alpha\sigma} f_{44,\alpha\sigma} = -\tfrac{1}{2}\Delta f_{44}, \tag{9.15}$$

and the field equation (9.13) simplifies to

$$\Delta f_{44} = -\kappa\mu c^2. \tag{9.16}$$

This equation has indeed the structure of a Poisson equation – but not every quantity which satisfies a Poisson equation necessarily coincides with the Newtonian gravitational potential! In order not to make a mistake in the physical interpretation of (9.16) we need one additional piece of information, which is furnished by the geodesic equation

$$\frac{d^2 x^n}{d\tau^2} = -\Gamma^n_{ab} \frac{dx^a}{d\tau} \frac{dx^b}{d\tau}. \tag{9.17}$$

For slowly moving particles (for example, planets) proper time almost coincides with coordinate time $t = x^4/c$, and the four-velocity on the right-hand side of (9.17) can be replaced by $u^a = (0, 0, 0, c)$:

$$\frac{d^2 x^\nu}{dt^2} = -\Gamma^\nu_{44} c^2 = \frac{1}{2}\eta^{\nu\mu} g_{44,\mu} c^2 = \frac{1}{2}\eta^{\nu\mu} f_{44,\mu} c^2. \tag{9.18}$$

If we compare this equation of motion with that for a particle in the gravitational potential U, that is, with

9 *Fundamental equations of Einstein's theory of gravitation* 91

$$\frac{d^2\mathbf{r}}{dt^2} = -\text{grad } U, \tag{9.19}$$

then we see that the Newtonian gravitational potential U is related to the metric by the relation

$$U = -c^2 f_{44}/2, \quad g_{44} = -(1 + 2U/c^2), \tag{9.20}$$

and that because of (9.1), (9.16) and (9.20) we have the relation

$$8\pi f/c^4 = \kappa = 2.07 \times 10^{-48} \text{g}^{-1}\text{cm}^{-1}\text{s}^2 \tag{9.21}$$

between the Newtonian constant of gravitation f and the Einstein natural constant κ. This establishes the required connection between Newtonian and Einsteinian gravitational theories.

The relation (9.20) between g_{44} and the potential U is in agreement with equation (8.17), since for small spatial regions we certainly have $U(x^\nu) = U_{,\nu} x^\nu = a_\nu x^\nu$ ($a_\nu = +U_{,\nu}$, because we are dealing with components of the acceleration seen from a freely falling inertial system).

9.3 The equations of motion of test particles

Monopole particle It is one of its particular merits that, in the Einstein theory, the equations of motion are a consequence of the field equations. If we take, for example, the Maxwell theory, then charge conservation is of course a consequence of the field equations, but the motion of the sources and the distribution of the charges are arbitrarily specifiable. Also the field of two point charges at rest a finite distance apart is an exact solution of the Maxwell equations – although the charges exert forces upon one another and therefore would be immediately accelerated into motion.

Even after the Einstein field equations had been set up it was thought that one had to demand in addition that the geodesic equation be the equation of motion of a test particle, but eventually it was realized that this can be deduced from the relation

$$T^{mn}{}_{;n} = 0 \tag{9.22}$$

which is always valid in the Einstein theory, and is thus a consequence of the local energy–momentum conservation.

In order to show this we first of all need the energy–momentum tensor for a pointlike particle of constant rest mass m. We obtain it by starting from the tensor $T^{mn} = \mu u^m u^n$ of dust and using the four-dimensional δ-function defined by

$$\begin{aligned}\int F(x^i) \frac{\delta^4[x^n - a^n]}{\sqrt{-g}} \sqrt{-g}\, d^4x &= F(a^n), \\ \delta^4(x^n) &= \delta(x^1)\delta(x^2)\delta(x^3)\delta(x^4), \\ \int f(\lambda) \frac{d\delta(\lambda)}{d\lambda} d\lambda &\equiv -\int \frac{df}{d\lambda} \delta(\lambda)\, d\lambda, \end{aligned} \quad (9.23)$$

to obtain the transition to pointlike particles:

$$T^{ik}(y^n) = mc \int \frac{\delta^4[y^n - x^n(\tau)]}{\sqrt{-g(x^a)}} \frac{dx^i}{d\tau} \frac{dx^k}{d\tau} d\tau. \quad (9.24)$$

In the local Minkowski rest-system ($\sqrt{-g} = 1$, $t = \tau$, $x^\nu = 0$) it corresponds precisely to the transition

$$\mu \to m\delta(x)\delta(y)\delta(z) \quad (9.25)$$

of a continuous distribution of matter to a point mass.

We now insert the energy-momentum tensor (9.24) into (9.22). Using (9.23) and (3.14) we can rewrite the partial derivatives as:

$$T^{ik}{}_{,k} = \frac{\partial T^{ik}}{\partial y^k} = mc \int \frac{\frac{\partial}{\partial y^k}\delta^4[y^n - x^n(\tau)]}{\sqrt{-g(x^a)}} \frac{dx^i}{d\tau} \frac{dx^k}{d\tau} d\tau$$

$$= -mc \int \frac{\frac{\partial}{\partial x^k}\delta^4[y^n - x^n(\tau)]}{\sqrt{-g(x^a)}} \frac{dx^k}{d\tau} \frac{dx^i}{d\tau} d\tau$$

$$= mc \int \delta^4[y^n - x^n(\tau)] \frac{d}{d\tau}\left(\frac{dx^i}{d\tau} \frac{1}{\sqrt{-g}}\right) d\tau$$

$$= mc \int \frac{\delta^4[y^n - x^n(\tau)]}{\sqrt{-g}} \left(\frac{d^2x^i}{d\tau^2} - \Gamma^a_{ab} \frac{dx^i}{d\tau} \frac{dx^b}{d\tau}\right) d\tau.$$

From (9.22) we thus obtain

$$0 = T^{ik}{}_{;k} = T^{ik}{}_{,k} + \Gamma^i_{nk} T^{nk} + \Gamma^k_{nk} T^{in}$$

$$= mc \int \frac{\delta^4[y^n - x^n(\tau)]}{\sqrt{-g}} \left(\frac{d^2x^i}{d\tau^2} - \Gamma^a_{ak} \frac{dx^i}{d\tau} \frac{dx^k}{d\tau}\right.$$

$$\left. + \Gamma^i_{nk} \frac{dx^n}{d\tau} \frac{dx^k}{d\tau} + \Gamma^a_{ak} \frac{dx^i}{d\tau} \frac{dx^k}{d\tau}\right) d\tau,$$

9 Fundamental equations of Einstein's theory of gravitation

and hence

$$0 = mc \int \frac{\delta^4[y^n - x^n(\tau)]}{\sqrt{-g}} \left(\frac{D^2 x^i}{D\tau^2}\right) d\tau. \tag{9.26}$$

On the world line $y^n = x^n(\tau)$ of the particle this equation can only be satisfied if

$$\frac{D^2 x^i}{D\tau^2} = 0, \tag{9.27}$$

and so the particle must move on a geodesic.

At first sight it is perhaps not apparent where it has in fact been assumed in this rather formal derivation that we are dealing with a *test* particle. The gravitational field produced by a pointlike particle will certainly not be regular at the position of the particle (the electrical field at the position of a point-charge is also singular), so that the metric and Christoffel symbols do not exist there at all.

Spinning particle An extended body, for example, a planet, will in general not move exactly along a geodesic. This is due not so much to the gravitational field caused by the body itself as to the action of 'tidal forces.' Because of the space curvature, the distance between neighbouring geodesics is not constant (cf. section 1.4); that is, the gravitational forces (which try to move every point of the body along a geodesic) deform the body, change its state of rotation and thereby lead to a complicated orbit. We can take account of one part of this effect by starting off from the model of a pointlike body, but associating with it higher moments (dipole moment, spin) in addition to the mass. Mathematically we can do this by using for its description not just δ-functions, but also their derivatives.

As we shall be interested later on in the action of the gravitational field upon a top, the equations of motion of a spinning (monopole–dipole) particle will be described in brief; for details and proofs we refer to the extensive literature on this problem.

An extended body can be approximately described by its mass $m(\tau)$, the four-velocity $u^a(\tau)$ of a suitably chosen point, and the antisymmetric spin-tensor $S^{ab}(\tau)$. From the vanishing of the divergence of the energy-momentum tensor

$$T^{ik}(y^n) = c \int \frac{1}{\sqrt{-g}} \left[\delta^4[y^n - x^n(\tau)]\left(mu^i u^k + \frac{1}{2c^2}\dot{S}^{ni} u_n u^k + \frac{1}{2c^2}\dot{S}^{nk} u_n u^i\right) \right.$$
$$\left. - \left\{\delta^4[y^n - x^n(\tau)]\frac{1}{2}(S^{mi} u^k + S^{mk} u^i)\right\}_{;m} \right] d\tau \tag{9.28}$$

follow the equations of motion

$$\frac{D}{D\tau}S^{ab} \equiv \dot{S}^{ab} = \frac{1}{c^2}(u_n \dot{S}^{na} u^b - u_n \dot{S}^{nb} u^a) \tag{9.29}$$

and

$$\frac{D}{D\tau}\left(mu^a + \frac{1}{c^2}u_b \dot{S}^{ba}\right) = \frac{1}{2}R^a{}_{bcd}S^{dc}u^b. \tag{9.30}$$

Of these equations only seven are independent, as contraction with u^a shows. Thus they do not suffice for the determination of the ten unknown functions (m, S^{ab}, and three components of u^a). The physical reason for this is that S^{ab}, like angular momentum or dipole moment in Newtonian mechanics, depends upon the reference point, and we have not yet fixed this point and its world line. We now define the reference world line $x^a(\tau)$ by the requirement that in the instantaneous rest frame of an observer moving on the world line, the dipole moment of the body is zero. Since the total mass is positive, such a line always exists. One possible version of this condition is, as one can show,

$$S^{ab}u_b = 0. \tag{9.31}$$

Because of this subsidiary condition the antisymmetric tensor S^{ab} has only three independent components, which can be mapped uniquely onto the *spin vector (intrinsic angular-momentum vector)* S_a according to

$$S_a = \frac{1}{2}\epsilon_{abmn}u^b S^{mn}, \quad S^{mn} = \frac{1}{c^2}\epsilon_{aq}{}^{nm}S^a u^q. \tag{9.32}$$

Substitution of (9.31) and (9.32) into the equations of motion (9.29) to (9.30) yields

$$\frac{DS_c}{D\tau} = \frac{1}{c^2}u_a S_n \frac{Du^n}{D\tau}, \tag{9.33}$$

$$\frac{D}{D\tau}m = 0, \tag{9.34}$$

and

$$m\frac{D}{D\tau}u^a = -\frac{1}{c^2}\epsilon^{arq}{}_p S_r u_q \frac{D^2 u^p}{D\tau^2} + \frac{1}{2c^2}R^a{}_{bcd}\epsilon^{cdpq}S_p u_q u^b. \tag{9.35}$$

The first of these equations says that the spin vector S_a is Fermi-Walker transported along the orbit (cf. (5.24), with $S^n u_n = 0$); its magnitude then remains constant:

9 *Fundamental equations of Einstein's theory of gravitation* 95

$$S^a S_a = \text{constant.} \qquad (9.36)$$

From the third equation (9.35) we see that the point defined by (9.31) does not move on a geodesic; this effect will in general be ignorable.

The equation of motion for spin, (9.33), is also valid when additional non-gravitational forces act, provided only that these forces exert no couple on the body. An observer can therefore realize his Fermi–Walker transported triad in the directions of the axes of three tops which are suspended freely.

9.4 A variational principle for Einstein's theory

All known fundamental, physically significant equations of classical fields can be derived from a variational principle, including the Einstein field equations. What demands must one make regarding the Lagrangian density $\mathcal{L}$ in order that precisely the field equations (9.7) follow from

$$\delta \int \mathcal{L} \, d^4 x = 0? \qquad (9.37)$$

Of course the quantity in (9.37) must be an invariant; that is, it must be the product of a scalar L and $\sqrt{-g}$. But for the pure gravitational field there is only one unique scalar which is quadratic in first derivatives of the metric and linear in the second derivatives, namely, the scalar curvature R (there is no scalar which contains only first derivatives). Since the matter must also be represented in L, we couple it – as usual in field theory – by simply adding a part $\kappa \overset{M}{L}$ arising from the matter distribution (for example, from an electromagnetic field). Our variational principle reads, upon appropriate choice of numerical factors, thus,

$$\delta W = \delta \int \left(\frac{R}{2} + \kappa \overset{M}{L} \right) \sqrt{-g} \, d^4 x = 0 \qquad (9.38)$$

(Hilbert, 1915). We shall now show that the Einstein field equations (9.7) really do follow from this ansatz.

As fundamental quantities of the gravitational field, which are to be varied independently of one another, we shall naturally take the components of the metric tensor g_{mn}. (If in (9.38) one varies the non-metrical field quantities contained in $\overset{M}{L}$, one obtains the corresponding field equations, for example, the Maxwell equations.) As usual with action integrals containing second derivatives, the variations δg_{mn} of the basic quantities and the variations of their first derivatives (combined into the variations $\delta \Gamma^n{}_{ab}$ of the Christoffel symbols) will be restricted so as to vanish on the bounding surfaces of the four-dimensional region of integration. Our first goal is to express the variations occurring in the equation

$$\delta W = \tfrac{1}{2} \int [R\delta\sqrt{-g} + \sqrt{-g}\, R_{mn}\delta g^{mn} + \sqrt{-g}\, g^{mn}\delta R_{mn} + \delta(2\kappa \overset{M}{L}\sqrt{-g})]\, d^4x$$
$$= 0 \qquad (9.39)$$

in terms of δg_{mn}.

From the properties of the metric tensor and its determinant described in section 3.2 we obtain immediately

$$\delta\sqrt{-g} = \frac{\partial\sqrt{-g}}{\partial g_{mn}}\delta g_{mn} = \frac{1}{2}\sqrt{-g}\, g^{mn}\delta g_{mn} \qquad (9.40)$$

and

$$\delta g^{mn} = -g^{ma}g^{nb}\delta g_{ab}. \qquad (9.41)$$

The defining equations (6.3) and (6.21) for the curvature tensor and the Ricci tensor, respectively, lead to

$$\delta R_{mn} = -(\delta\Gamma^a_{ma})_{,n} + (\delta\Gamma^a_{mn})_{,a} - \delta(\Gamma^a_{rn}\Gamma^r_{ma} - \Gamma^a_{ra}\Gamma^r_{mn}). \qquad (9.42)$$

The evaluation of the variational principle will be seen later to depend only upon the structure of the term containing δR_{mn}, which is not easily found by direct calculation from (9.42). We therefore write down the result

$$g^{mn}\delta R_{mn} = \frac{1}{\sqrt{-g}}[\sqrt{-g}\,(g^{mn}\delta\Gamma^a_{mn} - g^{ma}\delta\Gamma^n_{mn})]_{,a}$$
$$\equiv \frac{1}{\sqrt{-g}}(\sqrt{-g}\,F^a)_{,a} = F^a_{;a} \qquad (9.43)$$

without calculation and prove the correctness of this equation by showing that it is a tensor equation and that it is satisfied in a particular coordinate system.

The tensor property of equation (9.43) follows from the fact that not only the δR_{mn}, but also the difference $\delta\Gamma^a_{mn}$ of Christoffel symbols (the disturbing terms in (3.23) cancel when the difference is formed), are tensors. The equation (9.43) is clearly correct in locally geodesic coordinates; for $\sqrt{-g} = 1$, $g^{mn}{}_{,a} = 0$, $\Gamma^a_{bc} = 0$, and (9.42) and (9.43) lead to the same equation, namely, to

$$g^{mn}\delta R_{mn} = (g^{mn}\delta\Gamma^a_{mn} - g^{ma}\delta\Gamma^n_{mn})_{,a} = g^{mn}(\delta\Gamma^a_{mn,a} - \delta\Gamma^a_{ma,n}). \qquad (9.44)$$

We can only really work out the last term in (9.39) with exact knowledge of the Lagrangian $\overset{M}{L}$ of the matter. In order to obtain $\overset{M}{L}$, we shall

9 Fundamental equations of Einstein's theory of gravitation

invoke the aid of the usual transcription formula. One starts from the corresponding special-relativistic Lagrangian, replaces the partial by covariant derivatives and now forms scalar products with g_{mn} instead of with η_{mn}. $\overset{M}{L}$ will thus certainly contain Christoffel symbols, but no second derivatives of the metric, since it is in general constructed from the field quantities and their first derivatives only. We can therefore write generally

$$\delta(\sqrt{-g}\,\overset{M}{L}) = \frac{\delta(\sqrt{-g}\,\overset{M}{L})}{\delta g_{mn}}\delta g_{mn} + \left(\sqrt{-g}\,\frac{\partial \overset{M}{L}}{\partial g_{mn,a}}\delta g_{mn}\right)_{,a}, \quad (9.45)$$

in which we have used the usual abbreviation

$$\frac{\delta(\sqrt{-g}\,\overset{M}{L})}{\delta g_{mn}} = \frac{\partial(\sqrt{-g}\,\overset{M}{L})}{\partial g_{mn}} - \left(\frac{\partial(\sqrt{-g}\,\overset{M}{L})}{\partial g_{mn,a}}\right)_{,a} \quad (9.46)$$

for the so-called variational derivative.

If we now substitute (9.40), (9.41), (9.43) and (9.45) into the variational principle (9.39) then we obtain

$$\delta W = \frac{1}{2}\int\left[\left(\frac{1}{2}g^{mn}R - R^{mn} + \kappa\frac{2}{\sqrt{-g}}\frac{\delta(\sqrt{-g}\,\overset{M}{L})}{\delta g_{mn}}\right)\delta g_{mn} + F^{a}{}_{;a}\right.$$
$$\left.+ \frac{2\kappa}{\sqrt{-g}}\left(\sqrt{-g}\,\frac{\partial \overset{M}{L}}{\partial g_{mn,a}}\delta g_{mn}\right)_{,a}\right]\sqrt{-g}\,d^{4}x = 0. \quad (9.47)$$

With the help of the Gauss law (7.40) we can reduce the last two terms of the sum to a surface integral; this surface integral vanishes, however, because we have demanded that $\delta g_{mn}=0$ and $\delta\Gamma^{a}_{bc}=0$ on the boundary. Hence (9.47) simplifies to

$$\delta W = \frac{1}{2}\int\left(\frac{1}{2}g^{mn}R - R^{mn} + \kappa\frac{2}{\sqrt{-g}}\frac{\delta(\sqrt{-g}\,\overset{M}{L})}{\delta g_{mn}}\right)\delta g_{mn}\sqrt{-g}\,d^{4}x = 0. \quad (9.48)$$

Because of the independence of the variations δg_{mn} the sum of the terms contained in the parentheses must vanish identically, so that from our variational principle we obtain precisely the Einstein equations

$$R^{mn} - \tfrac{1}{2}g^{mn}R = \kappa T^{mn}, \quad (9.49)$$

if we identify the energy–momentum tensor with the variational derivative according to

$$T^{mn} = \frac{2}{\sqrt{-g}}\frac{\delta(\sqrt{-g}\,\overset{M}{L})}{\delta g_{mn}}. \quad (9.50)$$

How can one justify this identification? Two standpoints are possible. On the one hand one can regard (9.50) as defining the energy–momentum tensor or, put more exactly, *the* energy–momentum tensor out of the many possible energy–momentum tensors of a classical field theory, which must stand on the right-hand side of the Einstein field equations. In this sense (9.50) is the construction principle for the symmetric energy–momentum tensor, which is remarkably complicated to find in many field theories. Although this procedure is quite natural in the Einstein theory, one can of course also corroborate it by comparison with the energy–momentum tensor T^{mn}, which is already known. We will do this for the example of the Maxwell field. Since its Lagrangian

$$\overset{M}{L} = -\tfrac{1}{4} F_{ab} F^{ab} = -\tfrac{1}{4}(A_{b,a} - A_{a,b})(A_{s,r} - A_{r,s}) g^{ar} g^{bs} \qquad (9.51)$$

does not depend at all upon derivatives of the metric, we have, because of (9.45) and (9.41),

$$\frac{\delta(\sqrt{-g}\,\overset{M}{L})}{\delta g_{mn}} = \frac{\partial(\sqrt{-g}\,\overset{M}{L})}{\partial g_{mn}} = \frac{1}{2}\sqrt{-g}\,\overset{M}{L} g^{mn} + \frac{1}{2}\sqrt{-g}\,F^{nb} F^{m}{}_{b}, \qquad (9.52)$$

and from the definition (9.50)

$$T^{mn} = F^{nb} F^{m}{}_{b} - \tfrac{1}{4} g^{mn} F_{ab} F^{ab}, \qquad (9.53)$$

which is indeed the correct energy–momentum tensor of electrodynamics.

Finally we want to draw attention to a peculiar property of the action function of the gravitational field. Although the Lagrangian contains derivatives of second order, it does not give rise to differential equations of fourth order for the field equations – as we would normally have expected. This has to do with the fact that the curvature scalar R contains derivatives of second order in precisely such a combination that a four-dimensional divergence can be formed from them,

$$R = \frac{1}{\sqrt{-g}} [\sqrt{-g}\,(g^{ma} \Gamma^{n}_{ma} - g_{mn} \Gamma^{a}_{ma})]_{,n} + F(g_{mn}, \Gamma^{a}_{mn}), \qquad (9.54)$$

which, by means of the Gauss law, can be turned into a surface integral and hence supplies no contribution to the variation. This is also the deeper reason why one part of the variation of the integrand appears as a divergence (cf. (9.43)). The non-covariant decomposition (9.54) of the Lagrangian of the gravitational field into a divergence and a remainder containing only first derivatives plays an important rôle in the attempt to define an energy–momentum tensor (energy–momentum complex) of the gravitational field in the context of the Lagrangian function.

Bibliography to section 9

Textbooks   Schmutzer (1968)

Monographs and collected works   International Conference on Relativistic Theories of Gravitation (1965)

Review and research articles   Ehlers (1971), Mehra (1974), Taub (1965), Westpfahl (1967)

10 The Schwarzschild solution

10.1 *The field equations*

The gravitational fields which are most important in our daily life, namely, that of the Earth and that of the Sun, are produced by slowly rotating, nearly spherical mass distributions; they are approximately spherically symmetric. Since, however, we may hope that spherically symmetric gravitational fields are especially simple, we discuss, as a first application of the Einstein field equations, the problem of obtaining exact spherically symmetric solutions.

The line element   We shall naturally try to introduce coordinates appropriate to the problem. Since a choice of coordinates always leads to requirements on the metric functions, we must proceed carefully in order not to lose solutions by making the restrictions too strong. Spherical symmetry evidently signifies that in the three-dimensional space, $T=$ constant, all radial directions are equivalent and no perpendicular direction is singled out; in spherical coordinates R, ϑ, φ we have

$$d^{(3)}s^2 = g_{11}(R, cT)dR^2 + f(R, cT)[d\vartheta^2 + \sin^2\vartheta d\varphi^2].$$

The angular coordinates at different times can be so chosen that $g_{T\vartheta}$ and $g_{T\varphi}$ do not appear in the metric (they would single out tangential directions). Our ansatz thus reads

$$ds^2 = g_{11}(R, cT) dR^2 + f(R, cT)[d\vartheta^2 + \sin^2 \vartheta d\varphi^2]$$
$$+ 2g_{14}(R, cT) dR\, dcT + g_{44}(R, cT) c^2 dT^2. \qquad (10.1)$$

For many calculations it is expedient to simplify ds^2 further. By the coordinate transformation

$$r^2 = f(R, cT),$$

(f is positive, for if not ϑ, φ would be additional timelike coordinates) we bring the line element into the form

$$ds^2 = h^2(r, T) dr^2 - 2a(r, T) b(r, T) c\, dT\, dr - b^2(r, T) c^2 dT^2$$
$$+ r^2(d\vartheta^2 + \sin^2 \vartheta\, d\varphi^2)$$

which already contains the usual two-dimensional spherical surface element. Here we have assumed implicitly that r is a spacelike and t a timelike coordinate. A further transformation

$$e^{\nu/2} d(ct) = b\, d(cT) + a\, dr$$

($e^{\nu/2}$ plays the rôle of an integrating factor) eliminates the undesired non-orthogonal term. Thus we arrive at the Schwarzschild form,

$$ds^2 = e^{\lambda(r, t)} dr^2 + r^2(d\vartheta^2 + \sin^2 \vartheta\, d\varphi^2) - e^{\nu(r, t)} d(ct)^2, \qquad (10.2)$$

of the line element of a spherically symmetric metric.

The Christoffel symbols The Christoffel symbols associated with a metric are constructed most quickly by comparing the Euler–Lagrange equations,

$$\frac{d}{d\tau} \frac{\partial L}{\partial \left(\frac{dx^i}{d\tau}\right)} - \frac{\partial L}{\partial x^i} = 0, \qquad (10.3)$$

for the Lagrangian,

$$L = \frac{1}{2} \left[e^\lambda \left(\frac{dr}{d\tau}\right)^2 + r^2 \left(\frac{d\vartheta}{d\tau}\right)^2 + r^2 \sin^2 \vartheta \left(\frac{d\varphi}{d\tau}\right)^2 - e^\nu \left(\frac{dx^4}{d\tau}\right)^2 \right], \qquad (10.4)$$

with the geodesic equation

$$\frac{d^2 x^i}{d\tau^2} + \Gamma^i_{mn} \frac{dx^m}{d\tau} \frac{dx^n}{d\tau} = 0. \qquad (10.5)$$

The Christoffel symbols can then easily be read off. With the abbreviations $\dot{} \equiv \partial/\partial ct$ and $' \equiv \partial/\partial r$ equations (10.3) become

10 The Schwarzschild solution

$$e^\lambda \left[\frac{dr^2}{d\tau^2} + \frac{1}{2}\left(\frac{dr}{d\tau}\right)^2 \lambda' + \frac{dr}{d\tau}\frac{dx^4}{d\tau}\dot\lambda \right] - r\left(\frac{d\vartheta}{d\tau}\right)^2 + r\sin^2\vartheta\left(\frac{d\varphi}{d\tau}\right)^2$$
$$+ \frac{1}{2}\left(\frac{dx^4}{d\tau}\right)^2 e^\nu \nu' = 0,$$

$$r^2 \frac{d^2\vartheta}{d\tau^2} + 2r\frac{dr}{d\tau}\frac{d\vartheta}{d\tau} - r^2 \sin\vartheta\cos\vartheta\left(\frac{d\varphi}{d\tau}\right)^2 = 0,$$

$$r^2 \sin^2\vartheta \frac{d^2\varphi}{d\tau^2} + 2r\sin^2\vartheta\frac{dr}{d\tau}\frac{d\varphi}{d\tau} + 2r^2 \sin\vartheta\cos\vartheta\frac{d\varphi}{d\tau}\frac{d\vartheta}{d\tau} = 0,$$

and

$$e^\nu \left[\frac{d^2 x^4}{d\tau^2} + \frac{\dot\nu}{2}\left(\frac{dx^4}{d\tau}\right)^2 + \nu' \frac{dx^4}{d\tau}\frac{dr}{d\tau} \right] + \frac{1}{2} e^\lambda \dot\lambda \left(\frac{dr}{d\tau}\right)^2 = 0.$$

(10.6)

Of the total of forty independent Christoffel symbols only the following twelve are non-zero (here $x^1 = r$, $x^2 = \vartheta$, $x^3 = \varphi$ and $x^4 = ct$):

$$\Gamma^1_{11} = \frac{\lambda'}{2}, \qquad \Gamma^1_{14} = \frac{\dot\lambda}{2}, \qquad \Gamma^1_{22} = -re^{-\lambda},$$

$$\Gamma^1_{33} = -r\sin^2\vartheta\, e^{-\lambda}, \qquad \Gamma^1_{44} = \frac{1}{2}e^{\nu-\lambda}\nu', \qquad \Gamma^2_{12} = \frac{1}{r},$$

$$\Gamma^2_{33} = -\sin\vartheta\cos\vartheta, \qquad \Gamma^3_{13} = \frac{1}{r}, \qquad \Gamma^3_{23} = \cot\vartheta,$$

$$\Gamma^4_{11} = \frac{\dot\lambda}{2}e^{\lambda-\nu}, \qquad \Gamma^4_{14} = \frac{\nu'}{2}, \qquad \Gamma^4_{44} = \frac{\dot\nu}{2}.$$

(10.7)

The Ricci tensor From the general defining equation

$$R^a{}_{mbn} = \Gamma^a_{mn,b} - \Gamma^a_{mb,n} + \Gamma^a_{rb}\Gamma^r_{mn} - \Gamma^a_{rn}\Gamma^r_{mb} \tag{10.8}$$

we obtain, bearing in mind (10.7),

$$R^1{}_{m1n} = \Gamma^1_{mn,1} - \Gamma^1_{1m,n} + \Gamma^1_{11}\Gamma^1_{mn} + \Gamma^1_{14}\Gamma^4_{mn} - \Gamma^1_{rn}\Gamma^r_{m1},$$

$$R^2{}_{m2n} = \Gamma^2_{mn,2} - \Gamma^2_{2m,n} + \Gamma^2_{12}\Gamma^1_{mn} - \Gamma^2_{rn}\Gamma^r_{m2},$$

$$R^3{}_{m3n} = -\Gamma^3_{3m,n} + \Gamma^3_{13}\Gamma^1_{mn} + \Gamma^3_{23}\Gamma^2_{mn} - \Gamma^3_{rn}\Gamma^r_{m3},$$

and

$$R^4{}_{m4n} = \Gamma^4_{mn,4} - \Gamma^4_{4m,n} + \Gamma^4_{14}\Gamma^1_{mn} + \Gamma^4_{44}\Gamma^4_{mn} - \Gamma^4_{rn}\Gamma^r_{m4}.$$

(10.9)

Unless $m = n$ or $(m, n) = (1, 4)$ these components vanish. Also $R_{1234} = 0$.

Thus only the following components of the Ricci tensor differ from zero:

$$\begin{aligned}
R_{11} &= -\frac{\nu''}{2} - \frac{\nu'^2}{4} + \frac{\nu'\lambda'}{4} + \frac{\lambda'}{r} + e^{\lambda-\nu}\left[\frac{\ddot{\lambda}}{2} + \frac{\dot{\lambda}^2}{4} - \frac{\dot{\lambda}\dot{\nu}}{4}\right], \\
R_{44} &= e^{\nu-\lambda}\left[\frac{\nu''}{2} + \frac{\nu'^2}{4} - \frac{\nu'\lambda'}{4} + \frac{\nu'}{r}\right] - \frac{\ddot{\lambda}}{2} - \frac{\dot{\lambda}^2}{4} + \frac{\dot{\lambda}\dot{\nu}}{4}, \\
R_{14} &= \frac{\dot{\lambda}}{r}, \\
R_{22} &= -e^{-\lambda}\left[1 + \frac{r}{2}(\nu' - \lambda')\right] + 1, \\
R_{33} &= \sin^2\vartheta\, R_{22}.
\end{aligned} \qquad (10.10)$$

The vacuum field equations Outside the field-producing masses the energy–momentum tensor vanishes, and, since it follows immediately by taking the trace of

$$R^{mn} - \tfrac{1}{2}g^{mn}R = 0$$

that $R = 0$, the field equations for the vacuum are simply

$$R_{mn} = 0; \qquad (10.11)$$

that is, all the components of the Ricci tensor written down in (10.10) must vanish.

10.2 The solution of the vacuum field equations

Birkhoff's theorem From $R_{14} = 0$ we have immediately that $\dot{\lambda} = 0$; thus λ depends only upon the radial coordinate r. $R_{22} = 0$ can then only be satisfied if ν' is also independent of time,

$$\nu = \nu(r) + f(t). \qquad (10.12)$$

Since ν occurs in the line element in the combination $e^{\nu(r)}e^{f(t)}d(ct)^2$ one can always make the term $f(t)$ in (10.12) vanish by a coordinate transformation

$$dt' = e^{f/2}dt, \qquad (10.13)$$

so that in the new coordinates we have

$$\lambda = \lambda(r), \quad \nu = \nu(r); \qquad (10.14)$$

that is, the metric no longer depends upon time. And thus we have proved *Birkhoff's theorem: every spherically symmetric vacuum solution is independent of t.* (The assumptions made in section 10.1 may fail, so that t is not a timelike coordinate and r is not a spacelike coordinate, for example, in a black hole (see section 22.3). However the theorem still holds, although one would no longer describe the solution as static.)

If one considers the vacuum gravitational field produced by a spherically symmetric star, then this field remains static even if the material inside the star experiences a spherically symmetric radial displacement (explosion). Thus Birkhoff's theorem is the analogue of the statement in electrodynamics that a spherically symmetric distribution of charges and currents does not radiate, that there are no spherically symmetric electromagnetic waves.

The Schwarzschild solution For static vacuum fields the field equations (10.11) simplify to

and
$$\left.\begin{array}{c} \dfrac{\nu''}{2} + \dfrac{\nu'^2}{4} - \dfrac{\nu'\lambda'}{4} - \dfrac{\lambda'}{r} = 0, \\ (\nu'+\lambda')\dfrac{1}{r} = 0, \\ e^{-\lambda}(1-r\lambda')-1 = 0. \end{array}\right\} \quad (10.15)$$

The second of these equations is equivalent to

$$\lambda(r) = -\nu(r), \quad (10.16)$$

since the possible additive constant in (10.16) can be made to vanish by a coordinate transformation (time transformation) $dt' = \text{constant} \times dt$ (a special case of (10.13)).

Under the substitution $\alpha = e^{-\lambda}$, the third equation transforms into the differential equation

$$\alpha' + \alpha/r = 1/r, \quad (10.17)$$

whose general solution is

$$\alpha = e^{-\lambda} = e^{\nu} = 1 - 2M/r, \quad (10.18)$$

with $2M$ as a freely adjustable constant of integration. The spherically symmetric vacuum solution, found first in 1916 by Schwarzschild, therefore has the line element

$$ds^2 = \dfrac{dr^2}{1-2M/r} + r^2(d\vartheta^2 + \sin^2\vartheta\, d\varphi^2) - \left(1 - \dfrac{2M}{r}\right)c^2 dt^2. \quad (10.19)$$

One can verify by direct substitution that the first of the field equations (10.15) is also satisfied and furnishes no further conditions. We shall discuss the physical meaning of the constant of integration M in the following section.

10.3 General discussion of the Schwarzschild solution

In order to understand the physical properties of the Schwarzschild line element (10.19) we have first to clarify the physical significance of the integration parameter M. This is best done through a comparison with Newtonian theory. For large values of the coordinate r, (10.19) deviates only a little from the metric of a flat space, and, from the relation (9.20), which is valid in this limit and links the Newtonian gravitational potential U to the metric, we have

$$U = -\frac{c^2}{2}(1 + g_{44}) = -\frac{Mc^2}{r}. \tag{10.20}$$

We have thus to interpret the Schwarzschild solution as the gravitational field present outside a spherically symmetric mass distribution whose (Newtonian) mass is

$$m = \frac{Mc^2}{f} = \frac{8\pi M}{\kappa c^2} > 0. \tag{10.21}$$

According to (10.21), the (positive constant of integration $2M$ is a measure of the total mass; since it has the dimension of a length, one also calls $r_G = 2M$ the *Schwarzschild radius* or *gravitational radius* of the source. For normal stars or planets r_G is very small in relation to the geometrical radius. The Schwarzschild radius of the Sun, for example, has the value $r_G = 2.96$ km, that of the Earth $r_G = 8.8$ mm. Since the Schwarzschild metric describes only the gravitational field outside the matter distribution (we shall discuss the interior field in chapter 11), whilst the Schwarzschild radius mostly lies far in the interior, we shall initially suppose that $r \gg 2M$ always. (However see also section 22, where we shall investigate the Schwarzschild metric again, and in more detail.)

In the discussion of physical properties of the Schwarzschild metric (10.19) one must always remember that r and t in particular are only coordinates and have no immediate physical significance. We therefore call t the *coordinate time,* to distinguish it, for example, from the proper time τ of an observer at rest in the gravitational field; in the Schwarzschild field these two quantities are related by

10 The Schwarzschild solution

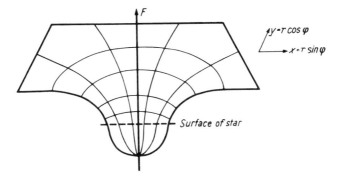

Fig. 10.1. Illustration of the section $t = $ constant, $\vartheta = \pi/2$ of the Schwarzschild metric; $F = \sqrt{8M(r-2M)}$.

$$d\tau = \sqrt{1 - 2M/r}\, dt. \tag{10.22}$$

The radial coordinate r is so defined that the surface area of a sphere $r = $ constant, $t = $ constant has the value $4\pi r^2$. The infinitesimal displacement in the radial direction ($d\vartheta = d\varphi = dt = 0$) is given, however, by

$$ds = dR = \frac{dr}{\sqrt{1 - 2M/r}}, \tag{10.23}$$

and is therefore always greater than the difference of the radial coordinates. One can illustrate the metrical relations in the surface $t = $ constant, $\vartheta = \pi/2$ by means of a surface of revolution $F = F(r)$, which for $r \to \infty$ goes over to a plane and for small r has a bulge out of this plane (cf. Fig. 10.1). When discussing paths of motion (t variable) one must always remember that g_{44} is also dependent upon position.

10.4 The motion of the planets and perihelion precession

Figure 10.1 gives a qualitative idea of the planetary orbits if one imagines the planets as spheres which roll about on the surface under the influence of a downwardly directed gravitational field. According to the Newtonian gravitational theory the orbits of planets are ellipses (in the xy-plane of Fig. 10.1). Does the Einstein theory in any way change this well verified result?

To answer this question properly we must integrate the Lagrange equations of the second kind already set up in (10.6). One can always satisfy the initial conditions $\vartheta = \pi/2$ and $d\vartheta/d\tau = 0$ by a suitable rotation of the coordinate system, and (10.6) then implies that $d^2\vartheta/d\tau^2$ also vanishes, so that the orbit remains permanently in the plane $\vartheta = \pi/2$. As in Newtonian theory, the orbit of a planet runs in a 'plane' which passes through the middle of the Sun. We can therefore proceed from the simplified Lagrangian

$$L = \frac{1}{2}\left[\frac{1}{1-2M/r}\left(\frac{dr}{d\tau}\right)^2 + r^2\left(\frac{d\varphi}{d\tau}\right)^2 - \left(1 - \frac{2M}{r}\right)\left(\frac{dx^4}{d\tau}\right)^2\right] \qquad (10.24)$$

which results from substitution of the Schwarzschild metric (10.19) and $\vartheta = \pi/2$ into (10.3).

Since φ and x^4 are cyclic coordinates, two conservation equations hold, namely, that of angular momentum,

$$r^2 \frac{d\varphi}{d\tau} = B, \qquad (10.25)$$

and that of energy,

$$\left(1 - \frac{2M}{r}\right)\frac{dct}{d\tau} = A. \qquad (10.26)$$

In place of a third equation of motion, we use the defining equation

$$\frac{1}{1-2M/r}\left(\frac{dr}{d\tau}\right)^2 + r^2\left(\frac{d\varphi}{d\tau}\right)^2 - \left(1 - \frac{2M}{r}\right)\left(\frac{dct}{d\tau}\right)^2 = -c^2 \qquad (10.27)$$

for the proper time τ, which like the energy law and the momentum law has the form of a first integral of the equations of motion.

From now on the procedure is analogous to that of Newtonian mechanics. In order to obtain the orbits $r = r(\varphi)$ we replace the variable τ by φ, with the aid of the angular-momentum law, and simplify the equation of motion by the substitution $u = r^{-1}$. Putting

$$r = \frac{1}{u}, \quad \frac{d\varphi}{d\tau} = Bu^2, \quad \frac{dct}{d\tau} = \frac{A}{1-2Mu}, \quad \frac{dr}{d\tau} = -B\frac{du}{d\varphi} \qquad (10.28)$$

into (10.27), we have

$$B^2 u'^2 + B^2 u^2 (1 - 2Mu) - A^2 = -c^2(1 - 2Mu), \quad u' \equiv du/d\varphi. \qquad (10.29)$$

This equation can in fact be integrated immediately, but it leads to elliptic integrals, which are awkward to handle. We therefore differentiate (10.29) and obtain the equation

$$u'' + u = Mc^2/B^2 + 3Mu^2, \qquad (10.30)$$

which is easier to discuss. The term $3Mu^2$ is absent in the Newtonian theory, where we have

10 The Schwarzschild solution

$$u_0'' + u_0 = Mc^2/B^2. \tag{10.31}$$

The solutions of this latter differential equation are, as is well known, the conics

$$u_0 = \frac{Mc^2}{B^2}(1 + \epsilon \cos \varphi). \tag{10.32}$$

We can obtain an approximate solution u_1 to the exact orbit equation (10.30) valid for $M/r \ll 1$, if we substitute the Newtonian solution (10.32) into the term quadratic in u, that is, we solve

$$u_1'' + u_1 = \frac{Mc^2}{B^2} + 3Mu_0^2 = \frac{Mc^2}{B^2} + \frac{3M^3c^4}{B^4}(1 + 2\epsilon \cos \varphi + \epsilon^2 \cos^2 \varphi). \tag{10.33}$$

This differential equation is of the type due to a forced oscillation. As one can confirm by substitution, the first approximation sought for is

$$u_1 = u_0 + \frac{3M^3c^4}{B^4}\left[1 + \epsilon\varphi \sin \varphi + \epsilon^2\left(\frac{1}{2} - \frac{1}{6}\cos 2\varphi\right)\right]. \tag{10.34}$$

The most important term on the right-hand side is the term linear in $\epsilon\varphi$, because it is the only one which in the course of time (with many revolutions of the planet) becomes larger and larger. We therefore ignore the other corrections to u_0 and obtain (after substituting for u_0)

$$u_1 = \frac{Mc^2}{B^2}\left[1 + \epsilon \cos \varphi + \epsilon\frac{3M^2c^2}{B^2}\varphi \sin \varphi\right], \tag{10.35}$$

or, since $r_0 = u_0^{-1}$ is large compared with M ($M^2c^2/B^2 \ll 1$),

$$u_1 = \frac{1}{r} = \frac{Mc^2}{B^2}\left[1 + \epsilon \cos\left(1 - \frac{3M^2c^2}{B^2}\right)\varphi\right]. \tag{10.36}$$

The orbit of the planet is thus only approximately an ellipse (cf. Fig. 10.2). The solution (10.36) is indeed still a periodic function, but no longer, however, with the period 2π. The point at which the orbit is closest to the Sun is reached again only after an additional rotation through the angle

$$\Delta\varphi_P = 6\pi M^2 c^2/B^2. \tag{10.37}$$

This effect is the famous perihelion precession. If, using the equation of the ellipse (10.32), we express the factor Mc^2/B^2 in (10.37) in terms of the semi-major axis a of the ellipse and of ϵ, so that

$$\Delta\varphi_P = 6\pi M/a(1 - \epsilon^2), \tag{10.38}$$

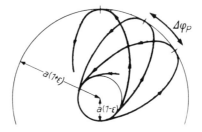

Fig. 10.2. Rosette motion of a planet due to perihelion precession ($\Delta\varphi_P$ exaggerated in magnitude).

then we see that the precession of the perihelion is greatest for a large central mass M and an elongated ellipse ($\epsilon \approx 1$) with a small (motion close to the centre). For circular orbits it disappears.

10.5 The propagation of light in the Schwarzschild field

Light rays and deflection of light Light rays are null geodesics, that is, geodesics with $ds^2 = 0$. We can compute the corresponding geodesic equation as in the previous section and hence immediately take over a part of the results found there. However, we must use the affine parameter λ in place of the proper time τ and the relation

$$\frac{1}{1-2M/r}\left(\frac{dr}{d\lambda}\right)^2 + r^2\left(\frac{d\varphi}{d\lambda}\right)^2 - \left(1 - \frac{2M}{r}\right)\left(\frac{dct}{d\lambda}\right)^2 = 0 \quad (10.39)$$

in place of (10.27) (cf. (3.9)). We then arrive at the statement that for suitable choice of coordinates light rays in the Schwarzschild metric travel in the surface $\vartheta = \text{constant} = \pi/2$ and satisfy the differential equation

$$\frac{d^2u}{d\varphi^2} + u = 3Mu^2, \quad u \equiv 1/r, \quad (10.40)$$

which is analogous to (10.30).

In flat space ($M = 0$) the light rays are of course straight lines. With our choice of coordinates these straight lines are represented by

$$u_0 = \frac{1}{r} = \frac{1}{D}\sin(\varphi - \varphi_0). \quad (10.41)$$

They run in the directions $\varphi = \varphi_0$ and $\varphi = \varphi_0 + \pi$ to infinity ($u = 0$) and have displacement D from the centre ($r = 0$).

10 The Schwarzschild solution

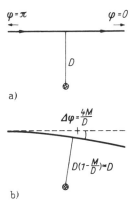

Fig. 10.3. Deflection of light in the Schwarzschild metric: (*a*) flat space $u = u_0$, (*b*) Schwarzschild metric $u = u_1$.

To obtain an approximation solution u_1 to (10.40) we put the Newtonian value (10.41) into the term quadratic in u and solve

$$\frac{d^2 u_1}{d\varphi^2} + u_1 = \frac{3M}{D^2} \sin^2(\varphi - \varphi_0). \tag{10.42}$$

As one can verify by substitution, for suitable choice of φ_0,

$$u_1 = \frac{1}{r} = \pm \frac{\sin \varphi}{D} + \frac{M(1 + \cos \varphi)^2}{D^2} \tag{10.43}$$

is a family of solutions. We are thus dealing with curves which come in parallelly from infinity (from the direction $\varphi = \pi$). The sign in (10.43) is always to be chosen so that $u_1 = 1/r$ is positive. Since a curve always leaves the field in the direction in which u_1 becomes zero (r infinite), its total deflection relative to a straight line is (ignoring terms quadratic in M)

$$\Delta \varphi = \frac{4M}{D} \tag{10.44}$$

(cf. Fig. 10.3).

This effect is the familiar deflection of light in a gravitational field, one of the most important predictions of the Einstein theory. The deflection is inversely proportion to the (Newtonian) displacement from the centre; since $M/D \ll 1$ always holds in the Solar System (this was presupposed in the derivation) the effect is very small. In very strong gravitational fields (10.44) is no longer applicable (cf. section 22).

110   Foundations of Einstein's theory of gravitation

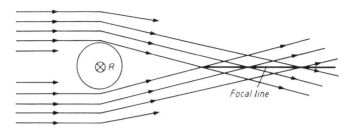

Fig. 10.4. The spherically symmetric gravitational field as a gravitational lens. (R is the radius of the gravitational source.)

The family of curves (10.43) can obviously be interpreted as the family of light rays arriving from a distant point source. As can be seen in Fig. 10.4, they converge; a (spherically symmetric) gravitational field behaves like a lens. A gravitational lens is far from ideal, possessing two closely related peculiarities; it produces double images, and incoming parallel rays are focussed onto a *focal line* rather than a point.

An observer at P will see two images of the source, with differing intensities, corresponding to the two rays shown arriving at P. The details of these double images depend critically on interference effects caused by the coherence properties of the source.

However, geometrical (ray) optics cannot treat correctly the superposition of infinitely many incoming rays on and near the focal line. Taking into account the wave nature of light it appears that the double image appears not as a bounding ring, as suggested by geometrical optics, but as an image point of increased intensity. For a source of wavelength λ the intensity on the focal line is increased by a factor $4\pi M/\lambda$. For a solar mass lens and visible light this factor is of the order 3×10^{10}.

In interpreting Figs. 10.3(b) and 10.4 it should be noted that the $r\varphi$-section of the Schwarzschild metric has been drawn as an $r\varphi$-plane. Thus they will give a good description only for large r.

Red shift   During the propagation of the light in a gravitational field, changes occur not only in the direction of the light rays, but also in the frequency of the light. Since the corresponding formulae can be derived for arbitrary static fields at no extra effort, we shall carry out this generalization and only substitute again the special case of the Schwarzschild metric in the final result.

A more general relationship, valid for arbitrary gravitational fields, will be described in section 25.2.

10 The Schwarzschild solution

In a static gravitational field, that is, in a metric g_{mn} which is independent of time and which satisfies the condition $g_{4\alpha}=0$, it is possible making the ansatz

$$A_\mu(x^n) = a_\mu(x^\alpha) e^{i\omega t} \tag{10.45}$$

in the gauge

$$A_4 = 0, \quad [\sqrt{-g}\, g^{44} A^\beta]_{,\beta} = 0, \tag{10.46}$$

to separate the Maxwell equations with respect to time and reduce them to the time-independent wave-equation

$$\frac{1}{\sqrt{-g}} \left[\sqrt{-g}\, g^{\mu\alpha} g^{\nu\beta} (a_{\beta,\alpha} - a_{\alpha,\beta}) \right]_{,\nu} + \frac{\omega^2}{c^2} g^{44} a^\mu = 0. \tag{10.47}$$

An observer who is at rest at the location P_1 of the transmitter will naturally use not the coordinate time t but the proper time τ_1 to measure the frequency ν_1 of the wave. Because of the general relation

$$\tau_P = \sqrt{-g_{44}(P)}\, t \tag{10.48}$$

between proper time and coordinate time, he will therefore associate the frequency

$$\nu_1 = \frac{\omega t}{2\pi \tau_1} = \frac{\omega}{2\pi \sqrt{-g_{44}(1)}} = \frac{\omega \sqrt{-g^{44}(1)}}{2\pi} \tag{10.49}$$

with the monochromatic wave (10.45), in agreement with the interpretation of the factor $-g^{44}\omega^2/c^2$ in (10.47) as the square of a position-dependent wave number. Analogously, an observer at rest at the location P_2 of the receiver measures the frequency

$$\nu_2 = \frac{\omega}{2\pi \sqrt{-g_{44}(2)}} \tag{10.50}$$

in his local Minkowski system. The frequencies ν_1 and ν_2 measured by observers at rest at the points P_1 and P_2 are thus related by

$$\frac{\nu_1}{\nu_2} = \sqrt{\frac{g_{44}(2)}{g_{44}(1)}} = 1 + z. \tag{10.51}$$

Although ν_2 can just as well be larger as smaller than ν_1, in general one speaks of the effect of the red shift in the gravitational field and calls the quantity z defined by (10.51) the *red shift* at the position P_2.

For the Schwarzschild metric we have from (10.51) the relation

$$\frac{\nu_1}{\nu_2} = \sqrt{\frac{1 - 2M/r_2}{1 - 2M/r_1}} \approx 1 + M\left(\frac{1}{r} - \frac{1}{r_2}\right). \tag{10.52}$$

Light reaching the Earth (P_2) from the Sun (P_1) ($r_1 < r_2$) is shifted to the red-wavelength region.

If we express g_{44} in terms of the Newtonian gravitational potential U according to (9.20) then we obtain to first approximation

$$\frac{\nu_1}{\nu_2} = 1 + \frac{U_2 - U_1}{c^2}. \tag{10.53}$$

In the picture of light as particles (photons) the gravitational red-shift corresponds to a change in the kinetic energy $E = h\nu$ by the gain or loss of potential energy $m\Delta U = \Delta U \times E/c^2$, in accordance with (10.53).

Light-travel times and the Fermat principle Here too we again generalize the problem and permit arbitrary static gravitational fields ($g_{ab,4} = 0$, $g_{\alpha 4} = 0$). In all these fields then the equation

$$\frac{d^2 x^a}{d\lambda^2} + \Gamma^a_{mn} \frac{dx^m}{d\lambda} \frac{dx^n}{d\lambda} = 0 \tag{10.54}$$

of a null geodesic can be brought into a form which allows a particularly simple physical interpretation. In place of the parameter λ we introduce the coordinate time t and substitute for the Christoffel symbols the explicit expressions

$$\left.\begin{array}{l} \Gamma^\alpha_{\mu\nu} = \tfrac{1}{2} g^{\alpha\beta}(g_{\beta\mu,\nu} + g_{\beta\nu,\mu} - g_{\mu\nu,\beta}), \\[4pt] \Gamma^\alpha_{4\nu} = 0, \quad \Gamma^4_{\mu\nu} = 0, \quad \Gamma^4_{\alpha 4} = \tfrac{1}{2} g^{44} g_{44,\alpha} = \dfrac{g_{44,\alpha}}{2 g_{44}}, \\[4pt] \Gamma^\alpha_{44} = -\tfrac{1}{2} g^{\alpha\beta} g_{44,\beta}, \quad \Gamma^4_{44} = 0. \end{array}\right\} \tag{10.55}$$

The fourth of equations (10.54),

$$\frac{d^2 t}{d\lambda^2} + \frac{g_{44,\alpha}}{g_{44}} \frac{dx^\alpha}{dt} \left(\frac{dt}{d\lambda}\right)^2 = 0, \tag{10.56}$$

enables us to eliminate λ from the three spatial equations

$$\frac{d^2 x^\alpha}{dt^2} + \Gamma^\alpha_{\mu\nu} \frac{dx^\mu}{dt} \frac{dx^\nu}{dt} + \Gamma^\alpha_{44} c^2 + \frac{d^2 t}{d\lambda^2} \left(\frac{d\lambda}{dt}\right)^2 \frac{dx^\alpha}{dt} = 0, \tag{10.57}$$

finding

$$\frac{d^2 x^\alpha}{dt^2} + g^{\alpha\beta} \tfrac{1}{2}(g_{\beta\mu,\nu} + g_{\beta\nu,\mu} - g_{\mu\nu,\beta}) \frac{dx^\mu}{dt} \frac{dx^\nu}{dt} - \frac{c^2}{2} g^{\alpha\beta} g_{44,\beta}$$

$$- \frac{g_{44,\nu}}{g_{44}} \frac{dx^\nu}{dt} \frac{dx^\alpha}{dt} = 0. \tag{10.58}$$

10 The Schwarzschild solution

If we remember the property of null geodesics that $ds = 0$, that is,

$$c^2 dt^2 = \frac{g_{\alpha\beta}}{-g_{44}} dx^\alpha dx^\beta \equiv \gamma_{\alpha\beta} dx^\alpha dx^\beta \equiv \frac{dl^2}{-g_{44}}, \qquad (10.59)$$

then (10.58) can be reduced to

$$\left.\begin{array}{c} \dfrac{d^2 x^\alpha}{dt^2} + \hat{\Gamma}^\alpha_{\mu\nu} \dfrac{dx^\mu}{dt} \dfrac{dx^\nu}{dt} = 0, \\[6pt] \hat{\Gamma}^\alpha_{\mu\nu} \equiv \tfrac{1}{2} \gamma^{\alpha\beta}(\gamma_{\beta\mu,\nu} + \gamma_{\beta\nu,\mu} - \gamma_{\mu\nu,\beta}), \quad \gamma^{\alpha\beta}\gamma_{\beta\nu} = \delta^\alpha_\nu. \end{array}\right\} \qquad (10.60)$$

The curves described by this equation are, however, just the extremals which follow from the variational principle

$$\int dt = \frac{1}{c} \int \sqrt{\gamma_{\alpha\beta} \frac{dx^\alpha}{dt} \frac{dx^\beta}{dt}}\, dt = \frac{1}{c} \int \frac{dl}{\sqrt{-g_{44}}} = \text{extremal} \qquad (10.61)$$

(cf. section 1.3).

The variational principle (10.61) is the generalization of the Fermat principle, that light propagates in a three-dimensional space in such a way that the light-travel time t is an extremum.

The variational principle (10.61) can also be interpreted as saying that the three-dimensional space (metric $g_{\alpha\beta}$) has a refractive index $n = (-g_{44})^{-1/2}$, which is caused by the gravitational force (and which produces the deflection of light), and that the velocity of light v in the gravitational field is decreased according to $c = nv$. But this latter interpretation is only to be used with the proviso that v is the velocity of light with respect to the coordinate time t and therefore has, like t itself, no immediate physical significance. Predictions about the numerical value of the velocity of light have little value in general relativity; the only essential thing is that light propagates along null geodesics (and that in local inertial systems one can give the velocity of light the value c through choice of the unit of time).

10.6 Further aspects of the Schwarzschild metric

Isotropic coordinates The Schwarzschild coordinates in which we have until now described the spherically symmetric gravitational field go over to spherical coordinates at a great distance from the centre (for $r \to \infty$). For many calculations or considerations it is more convenient to use coordinates which are related to Cartesian coordinates. We introduce them by the transformation

$$r = \bar{r}(1+M/2\bar{r})^2, \quad \begin{aligned} \bar{x} &= \bar{r}\cos\varphi\sin\vartheta, \\ \bar{y} &= \bar{r}\sin\varphi\sin\vartheta, \\ \bar{z} &= \bar{r}\cos\vartheta, \end{aligned} \quad (10.62)$$

which turns (10.19) into

$$ds^2 = (1+M/2\bar{r})^4(d\bar{x}^2+d\bar{y}^2+d\bar{z}^2) - \left(\frac{1-M/2\bar{r}}{1+M/2\bar{r}}\right)^2 c^2 dt^2. \quad (10.63)$$

Since in this form of the line element the three spatial directions enter on an equal footing, one speaks of *isotropic coordinates*.

Harmonic coordinates Coordinates which are restricted by

$$\Box x^a = \frac{1}{\sqrt{-g}}[\sqrt{-g}\,g^{nm}x^a{}_{,m}]_{,n} = \frac{1}{\sqrt{-g}}(\sqrt{-g}\,g^{an})_{,n} = 0 \quad (10.64)$$

are called *harmonic coordinates* (of course (10.64) is not a covariant equation, rather it serves to pick out a coordinate system). Such coordinates are useful in approximation procedures for the solution of Einstein's equations (cf. section 13.2). In such coordinates, the Schwarzschild metric has the form

$$ds^2 = \left[\left(1+\frac{M}{\bar{r}}\right)^2\eta_{\alpha\beta} + \left(\frac{\bar{r}+M}{\bar{r}-M}\right)\frac{M^2}{\bar{r}^4}x_\alpha x_\beta\right]dx^\alpha dx^\beta - \left(\frac{\bar{r}-M}{\bar{r}+M}\right)c^2 dt^2, \quad (10.65)$$

where

$$\bar{r} = r - M. \quad (10.66)$$

The interior field of a hollow (non-rotating) sphere The Schwarzschild line element (10.19) follows from the requirement of spherical symmetry alone, and it therefore holds also in the matter-free interior of a hollow, non-rotating sphere. But in this case the metric must be finite at $r=0$, that is, M must vanish; the space inside a hollow sphere is field-free (flat) as in the Newtonian gravitation theory.

10.7 Experiments to verify the Schwarzschild metric

The gravitational fields of the Earth and the Sun constitute our natural environment and it is in these fields that the laws of gravity have been investigated and described by equations. Both fields are to good approximation spherically symmetric and, as a result, suitable objects to test the Einstein theory as represented in the Schwarzschild metric.

10 The Schwarzschild solution

The Einstein theory contains the Newtonian theory of gravitation as a first approximation and in this sense is of course also confirmed by Kepler's laws. What chiefly interest us here, however, are the – mostly very small – corrections to the predictions of the Newtonian theory. In very exact experiments one must distinguish carefully between the following three sources of deviation from the Newtonian spherically symmetric field.

(a) relativistic corrections to the spherically symmetric field,
(b) Newtonian corrections, due to deviations from spherical symmetry (flattening of the Earth or Sun, taking into account the gravitational fields of other planets),
(c) relativistic corrections due to deviations from spherical symmetry and staticity.

While one can almost always ignore the relativistic corrections of category (c), the Newtonian corrections (b) are often larger than the relativistic effects (a) which are really of interest to us, and can be separated from them only with difficulty.

In discussions of measurements and experiments in spherically symmetric gravitational fields one often uses a metric more general than the Schwarzschild one, for example, the latter, written in isotropic coordinates with higher order terms neglected:

$$ds^2 = \left(1 + \gamma \frac{2M}{\bar{r}} + \cdots\right)(d\bar{x}^2 + d\bar{y}^2 + d\bar{z}^2) - \left(1 - \frac{2M}{\bar{r}} + \beta \frac{2M^2}{\bar{r}^2} + \cdots\right)c^2 dt^2. \tag{10.67}$$

The free parameters β and γ (two of the so-called PPN-parameters, see section 29.2) are found as 'best-fit' parameters to the observational data. Since they both have the value 1 for the Schwarzschild metric, $\beta - 1$ and $\gamma - 1$ are a measure of the agreement between Einstein's theory and observation. Perihelion precession $\Delta\phi_P$, light deflection $\Delta\phi$ and light travel time Δt for the metric (10.67) differ from the Einstein values (shown with a suffix E) by

$$\Delta\phi_P = \tfrac{1}{3}(2 + 2\gamma - \beta)\Delta\phi_{PE}, \quad \Delta\phi = \tfrac{1}{2}(1 + \gamma)\Delta\phi_E, \quad \Delta t = \tfrac{1}{2}(1 + \gamma)\Delta t_E. \tag{10.68}$$

Perihelion precession and planetary orbits Einstein's theory predicts, (10.38), the following values for the perihelion precession per century:

$$\begin{array}{ll} \text{Mercury} & 42.98'', \\ \text{Venus} & 8.6'', \\ \text{Mars} & 1.35'', \end{array} \quad \begin{array}{ll} \text{Earth} & 3.8'', \\ \text{Satellite} & \leq 1000''. \end{array} \tag{10.69}$$

Fig. 10.5. How the Sun deflects light.

Because the deviation from spherical symmetry of the earth's gravitational field is so large that one can determine the density distribution of the earth from observed irregularities in satellite orbits, artificial satellites are far from ideal test objects.

In the first decade of relativity theory the most promising evidence came from data on Mercury's orbit. Astronomers before Einstein were already perturbed because although most of the observed perihelion precession, 5600" per century, could be ascribed to the influence of other planets, there remained an unexplained 41" now seen to be in good agreement with Einstein's theory.

The survey of the orbit of Mars on the Viking Mission (1976–82) and radar measurements of the distances to Venus and Mercury have furnished data of substantially increased accuracy. More measurements and a comprehensive computer analysis (e.g. the influence of the larger asteroids on Mars' orbit was taken into account) has produced the values:

$$\beta - 1 = 3(\pm 3.1) \times 10^{-3}, \qquad \gamma - 1 = 0.7(\pm 1.7) \times 10^{-3}. \qquad (10.70)$$

The unknown quadrupole moment of the sun is the main source of uncertainty in the reduction of the data. The independently obtained data for Mars and Mercury suggest, however, that it can be neglected.

In 1974 a pulsar (PSR1913+16) was discovered that forms a binary system with a smaller star whose nature (white dwarf, neutron star or black hole) is unknown. The elliptic orbit of the pulsar shows an unusually large rotation (periastron precession) of $4.22(\pm 0.04)°$ per year, that is, 271 times the total value for Mercury. It is highly probable that this is a purely relativistic effect.

Light deflection by the Sun Maximum deflection occurs when the light ray grazes the surface of the sun, see Fig. 10.5, giving

$$\Delta \phi = 1.75". \qquad (10.71)$$

To measure this effect one compares the appearance of the night sky in some region with the appearance of the same region during a solar eclipse. (Without the eclipse the Sun's luminosity would swamp that of the stars.) The effect of the Sun appears to move the closest stars away from it. Although experimental problems, caused, for example, by the distortion of the photographic emulsion during the developing process, have produced values between 1.43″ and 2.7″ this effect, and Eddington's experimental verification in 1919, brought Einstein public recognition.

Another possibility arises from the shielding of strong radio sources, quasars, by the Sun during the Earth's orbit. The use of sufficiently large arrays of interferometers and measurements at different frequencies to eliminate the effect of the solar plasma on the deflection makes possible substantially greater accuracy than in optical measurements. Current measurements (Fomalont and Svamak, 1976) confirm the Einsteinian value to within 1 percent:

$$\gamma - 1 = 0.007\,(\pm 0.009). \tag{10.72}$$

Gravitational lenses and caustics Although light deflection already shows the focussing effect of the gravitational field, gravitational lensing refers to the crossing of light rays, causing, for example, a double image, see Fig. 10.4. In the solar gravitational field this effect occurs first at a distance $d = R_\odot/\Delta\phi = R_\odot^2/4M \approx 8.2 \times 10^{10}$ km from the Sun, that is, practically outside the solar system. When we see from the earth a double image of a star the gravitational lens is not the Sun but a distant star or a galactic core.

A double quasar was found in 1979 whose two images are so close (6″) and have such similar spectra that it is probable that there is only *one* quasar, and a galaxy is acting as a gravitational lens.

The increase of intensity at a caustic of a gravitational lens has not yet been observed.

Redshifts The red shift produced by the Earth's gravitational field was measured first by Pound and Rebka (1960) using the Mössbauer effect. The ^{57}Fe source in the basement of a tower was set in motion so that the resultant Doppler shift corresponded exactly to the energy loss at the receiver, 22.5 m higher. The relation

$$\Delta\lambda/\lambda = gh \tag{10.73}$$

was confirmed with 1 percent accuracy.

A somewhat more accurate confirmation, 7×10^{-5}, has been made in an experiment in which a hydrogen maser was taken, with the aid of a rocket, to a height of 10000 km (Vessot, 1980).

118   *Foundations of Einstein's theory of gravitation*

These results can also be considered as evidence for the assertion that atomic and molecular clocks measure proper rather than coordinate time.

Measurements of the travel time of radar signals   The time taken by a radar signal that has been reflected by a planet (e.g. Mercury, Venus) or that has been emitted by a satellite can be compared with the relativistic formula

$$\Delta t = \frac{1}{c} \int \sqrt{\frac{g_{\alpha\beta} \, dx^\alpha dx^\beta}{-g_{44}}} \tag{10.74}$$

in order to verify Einstein's theory. This so-called fourth test of general relativity was first proposed by Shapiro in 1964. The main data come from the Viking mission to Mars, giving

$$\gamma - 1 = 0(\pm 0.002). \tag{10.75}$$

Precession of a top in a satellite   The spin vector S^a of a top which is transported along a geodesic inside a satellite satisfies, because of (9.31)–(9.36), the equations

$$\frac{DS^a}{D\tau} = 0, \quad S^a S_a = \text{constant}, \quad S^a u_a = 0. \tag{10.76}$$

Since the unit vectors $h^a{}_{(\nu)}$, which are used in the satellite and point towards the fixed stars, are not parallelly transported during the motion, the components $S_{(\nu)} = S_a h^a{}_{(\nu)}$ of the spin vector in the rest system of the satellite change as

$$\frac{dS_{(\nu)}}{d\tau} = \frac{DS_{(\nu)}}{D\tau} = S_a \frac{Dh^a{}_{(\nu)}}{D\tau}. \tag{10.77}$$

The components of the tetrad vectors in isotropic Schwarzschild coordinates are obtained from the three orthogonal unit vectors used in the rest system of the satellite by carrying out a Lorentz transformation with the speed $-v^\alpha = -u^\alpha d\tau/dt$ and a change of scale

$$\left. \begin{array}{l} dx^{\alpha'} = (1 + M/2\bar{r})^2 dx^\alpha \approx (1 - U/c^2) dx^\alpha, \\[4pt] dt' = \left(\dfrac{1 - M/2\bar{r}}{1 + M/2\bar{r}}\right) dt \approx (1 + U/c^2) dt. \end{array} \right\} \tag{10.78}$$

Including only terms of first order, we then have

$$\left. \begin{array}{l} h^\alpha{}_{(\nu)} = (1 + U/c^2) \delta^\alpha_\nu + v^\alpha v_\nu / 2c^2, \\[4pt] h^4{}_{(\nu)} = v_\nu, \quad S_{(\alpha)} = S_\alpha. \end{array} \right\} \tag{10.79}$$

Substituting this expression into (10.77), and, bearing in mind that

$$\frac{dv_\alpha}{d\tau} \approx -U_{,\alpha}, \tag{10.80}$$

we have finally

$$\left.\begin{array}{l}\dfrac{dS_{(\nu)}}{d\tau} \approx S_{(\alpha)} \dfrac{3}{2c^2}[U_{,(\alpha)}v_{(\nu)} - U_{,(\nu)}v_{(\alpha)}],\\[6pt] U_{,\nu} \approx U_{(\nu)}, \quad v_\alpha = v_{(\alpha)},\end{array}\right\} \tag{10.81}$$

that is, the spin vector rotates in the satellite's system with the angular velocity ω,

$$\frac{dS}{dt} = \omega \times S, \quad \omega = -\frac{3}{2c^2} v \times \operatorname{grad} U. \tag{10.82}$$

For an Earth satellite the angular velocity amounts to about $8''$ per year; no experiments have as yet been made.

Bibliography to section 10

Textbooks   Straumann (1984)

Monographs and collected works   Hawking and Israel (1987), International Conference on Relativistic Theories of Gravitation (1965)

Review and research articles   Darwin (1959), Duff (1974), Hellings (1984), Herlt and Stephani (1976), Nobili and Will (1986), O'Connell (1972), Plebanski and Mielnik (1962), Reasenberg (1983), Shapiro (1980), Thorne (1973), Will (1979, 1981)

11 The interior Schwarzschild solution

11.1 *The field equations*

If we want to determine the gravitational field inside a celestial body then we need a model for this body, that is, we must say something about its energy–momentum tensor. Ignoring thermodynamic effects, such as heat conduction and viscosity, the ideal fluid medium (8.84)

$$T_{mn} = (\mu + p/c^2)u_m u_n + p g_{mn} \tag{11.1}$$

is a useful approximation.

We seek a spherically symmetric, static solution (ignoring radial matter currents in the stars), and thus require that the general line element (10.2) is independent of time, that is,

$$ds^2 = e^{\lambda(r)}dr^2 + r^2(d\vartheta^2 + \sin^2\vartheta\, d\varphi^2) - e^{\nu(r)}c^2 dt^2 \tag{11.2}$$

holds; the matter is at rest in this coordinate system,

$$u^m = (0, 0, 0, ce^{-\nu/2}), \tag{11.3}$$

and μ and p are functions purely of the radius r.

In setting up the field equations we can use the components of the Ricci tensor already calculated in (10.10):

$$\left.\begin{aligned}
R_{11} &= -\frac{\nu''}{2} - \frac{\nu'^2}{4} + \frac{\nu'\lambda'}{4} + \frac{\lambda'}{r}, \\
R_{44} &= e^{\nu-\lambda}\left[\frac{\nu''}{2} + \frac{\nu'^2}{4} - \frac{\nu'\lambda'}{4} + \frac{\nu'}{r}\right], \\
R_{22} &= -e^{-\lambda}\left[1 + \frac{r}{2}(\nu'-\lambda')\right] + 1 = \frac{R_{33}}{\sin^2\vartheta}, \\
R &= R^n_n = -2e^{-\lambda}\left[\frac{\nu''}{2} + \frac{\nu'^2}{4} - \frac{\nu'\lambda'}{4} + \frac{\nu'-\lambda'}{r} + \frac{1}{r^2}\right] + \frac{2}{r^2}.
\end{aligned}\right\} \tag{11.4}$$

The Einstein equations,

$$R^n_m - \tfrac{1}{2} R g^n_m = \kappa T^n_m, \tag{11.5}$$

hence assume the form

$$\kappa p = R^1_1 - \frac{R}{2} = e^{-\lambda}\left[\frac{\nu'}{r} + \frac{1}{r^2}\right] - \frac{1}{r^2}, \tag{11.6a}$$

$$\kappa p = R^2_2 - \frac{R}{2} = e^{-\lambda}\left[\frac{\nu''}{2} + \frac{\nu'^2}{4} - \frac{\nu'\lambda'}{4} + \frac{\nu'-\lambda'}{2r}\right], \tag{11.6b}$$

and

$$-\kappa\mu c^2 = R^4_4 - \frac{R}{2} = -e^{-\lambda}\left[\frac{\lambda'}{r} - \frac{1}{r^2}\right] - \frac{1}{r^2}. \tag{11.6c}$$

The four functions λ, ν, p and μ are to be determined from these three equations and an equation of state $f(\mu, p) = 0$ yet to be formulated.

11.2 The general solution of the field equations

As we have discussed in detail in section 9.1, the field equations are only integrable if the balance equations of energy and momentum $T^{mn}{}_{;n} = 0$ are satisfied. These conservation laws often give – analogously to the first

11 The interior Schwarzschild solution

integrals of classical mechanics – an important indication of how to solve the field equations. Since for static distributions of matter and pressure we have

$$\mu_{,n}u^n = 0, \quad p_{,n}u^n = 0, \quad \text{and} \quad u^n{}_{;n} = 0, \tag{11.7}$$

the equations

$$T^{mn}{}_{;n} = [pg^{mn} + (\mu - p/c^2)u^m u^n]_{;n} = 0 \tag{11.8}$$

simplify to the one equation

$$p' + (\mu + p/c^2)u_{1;4}u^4 = p' - (\mu + p/c^2)\Gamma^4_{14}u_4 u^4 = 0,$$

which leads to

$$p' = -\frac{\nu'}{2}(p + \mu c^2). \tag{11.9}$$

This equation is a consequence of the field equations (11.6) and can be used in place of one of these three equations.

Since one can write the field equation (11.6c) in the form

$$\kappa \mu c^2 r^2 = -(e^{-\lambda}r)' + 1,$$

it can easily be integrated to give

$$re^{-\lambda} = r - 2m(r) + C. \tag{11.10}$$

The function $m(r)$, defined by

$$m(r) = \frac{\kappa c^2}{2} \int_0^r \mu(x) x^2 \, dx, \tag{11.11}$$

is called the *mass function*. Equation (11.11) can also be interpreted as showing that $m(r)$ is proportional to the total mass contained within the sphere of radius r, but then one must be careful to note that r is only the coordinate radius, the true radius $R(r)$ of the sphere being given by

$$R(r) = \int_0^r e^{\lambda(x)/2} \, dx. \tag{11.12}$$

The constant of integration C in (11.10) must be set equal to zero in order that $g^{11} = e^{-\lambda}$ remain finite at $r = 0$; hence we have

$$e^{-\lambda(r)} = 1 - 2m(r)/r. \tag{11.13}$$

While in (11.13) we have succeeded in specifying one of the metric functions independently of a special equation of state, for the further integration of the field equations we must fix the equation of state. The most simple possibility is to assume a constant rest-mass density,

$$\mu = \text{constant}. \tag{11.14}$$

This equation of state certainly does not give a particularly good stellar model; a constant mass density is a first approximation only for small stars in which the pressure is not too large. The spherically symmetric, static solution with the special equation of state (11.14) is called the *interior Schwarzschild solution*.

For constant mass density (11.13) becomes

$$e^{-\lambda} = 1 - Ar^2, \quad A = \tfrac{1}{3}\kappa\mu c^2, \tag{11.15}$$

and (11.9), namely,

$$(p + \mu c^2)' = -\frac{\nu'}{2}(p + \mu c^2),$$

can be integrated to give

$$p + \mu c^2 = B e^{-\nu/2}. \tag{11.16}$$

As the third field equation to be solved, we choose the combination

$$\kappa(\mu c^2 + p) = e^{-\lambda}\left(\frac{\lambda'}{r} + \frac{\nu'}{r}\right) = \kappa B e^{-\nu/2}$$

of (11.6a) and (11.6c). Upon substitution for $e^{-\lambda}$ it goes over to

$$e^{\nu/2}\left(\frac{\nu'}{r} - Ar\nu' + 2A\right) = \kappa B,$$

and, through the intermediate step

$$2(1 - Ar^2)^{3/2}[e^{\nu/2}(1 - Ar^2)^{-1/2}]' = \kappa Br,$$

can be easily solved to give

$$e^{\nu/2} = \frac{\kappa B}{2A} - D\sqrt{1 - Ar^2}. \tag{11.17}$$

The equations (11.14) to (11.17) give the general solution for the case of constant mass density. They contain two constants of integration, B and D, which are to be determined by the matching conditions.

11.3 Matching conditions and connection to the exterior Schwarzschild solution

From the Maxwell theory one knows that matching conditions for certain field components must be satisfied at the interface between two media; these matching conditions follow from the Maxwell equations. In a completely analogous way, we must here ensure certain continuity properties

11 The interior Schwarzschild solution

of the metric at the surface of the star if we want to construct the complete gravitational field from the solution in the interior of the star (the interior Schwarzschild solution) and the solution in the exterior space (the Schwarzschild solution). As we shall show in section 16, the appropriate matching conditions follow from the Einstein field equations. Since in our simple example physical plausibility considerations will answer the purpose, we shall limit ourselves here to some brief remarks on the matching conditions.

Continuity properties of the metric and its derivatives can obviously be destroyed by coordinate transformations and inappropriate choice of coordinates. We therefore formulate the matching conditions most simply in a special coordinate system in which the boundary is a coordinate surface $x^4 =$ constant (in our example $r = r_0$) and Gaussian coordinates are employed in the neighbourhood of the boundary, so that

$$ds^2 = \epsilon(dx^4)^2 + g_{\alpha\beta}dx^\alpha dx^\beta, \quad \epsilon = \pm 1 \tag{11.18}$$

($\epsilon = +1$, if x^4 is a spacelike coordinate). Since second derivatives of the metric appear in the field equations, their existence must be ensured, that is, we demand that:

$g_{\alpha\beta}$ and $g_{\alpha\beta,4}$ are continuous on the boundary surface $x^4 =$ constant.
$$\tag{11.19}$$

By these requirements we have excluded layer structure in the surface (δ-function singularities in the energy–momentum tensor).

In order to connect interior and exterior Schwarzschild solutions to one another on the surface of the star $r = r_0$, we ought first of all to introduce (separate interior and exterior) Gaussian coordinates by

$$dx^4 = dr e^{\lambda(r)/2}, \tag{11.20}$$

and try to satisfy the conditions (11.19) through choice of the still unspecified integration parameters B and D (the internal solution) and of M (the external solution). But here we want to deal with the problem in a more intuitive fashion; the reader can reflect on the equivalence of the two methods.

We require that the metric g_{mn} is continuous for $r = r_0$ and that the pressure p vanishes on the surface of the star. Because of (11.16) and (11.17), the pressure depends upon r according to

$$p = Be^{-\nu/2} - \mu c^2 = \frac{1}{\kappa} \frac{3AD\sqrt{1-Ar^2} - \frac{\kappa}{2}B}{\frac{\kappa B}{2A} - D\sqrt{1-Ar^2}}, \tag{11.21}$$

and hence this requirement corresponds to the three equations

$$
\begin{aligned}
e^\lambda \text{ continuous: } & 1 - Ar_0^2 = 1 - 2M/r_0, \\
e^\nu \text{ continuous: } & \left(\frac{\kappa B}{2A} - D\sqrt{1-Ar_0^2}\right)^2 = 1 - 2M/r_0, \\
\text{and } p = 0: \quad & 3AD\sqrt{1-Ar_0^2} = \frac{\kappa}{2} B.
\end{aligned}
\qquad (11.22)
$$

They have the solution

$$
\begin{aligned}
M &= \frac{A}{2} r_0^3 = \frac{\kappa \mu c^2}{6} r_0^3 = \frac{4\pi}{3} r_0^3 f \frac{\mu}{c^2}, \\
D &= \frac{1}{2}, \\
\text{and} \quad \kappa B &= 3A\sqrt{1-Ar_0^2} = \kappa \mu c^2 \sqrt{1 - \tfrac{1}{3}\kappa \mu c^2 r_0^2},
\end{aligned}
\qquad (11.23)
$$

by means of which all the constants of integration are related to the mass density μ and the stellar radius r_0.

The spherically symmetric gravitational field of a star with mass density $\mu = $ constant and radius r_0 is thus described (Schwarzschild, 1916) by the interior Schwarzschild solution

$$
\begin{aligned}
ds^2 &= \frac{dr^2}{1-Ar^2} + r^2(d\vartheta^2 + \sin^2\vartheta\, d\varphi^2) \\
&\quad - \left[\frac{3}{2}\sqrt{1-Ar_0^2} - \frac{1}{2}\sqrt{1-Ar^2}\right]^2 c^2 dt^2, \\
\mu &= \text{constant}, \quad A = \frac{1}{3}\kappa\mu c^2, \quad \kappa p = 3A \frac{\sqrt{1-Ar^2} - \sqrt{1-Ar_0^2}}{3\sqrt{1-Ar_0^2} - \sqrt{1-Ar^2}}
\end{aligned}
$$

(11.24)

inside the star and by the exterior (vacuum) Schwarzschild solution

$$
\begin{aligned}
ds^2 &= \frac{dr^2}{1-2M/r} + r^2(d\vartheta^2 + \sin^2\vartheta\, d\varphi^2) - \left(1 - \frac{2M}{r}\right)c^2 dt^2, \\
2M &= Ar_0^3
\end{aligned}
\qquad (11.25)
$$

outside. We should point out that in the coordinates used here $\partial g_{rr}/\partial r$ is discontinuous on the boundary surface $r = r_0$, but that this discontinuity can be removed by making a coordinate transformation.

11.4 A discussion of the interior Schwarzschild solution

In interpreting the constant M, one must note that it is a measure of the total effective mass of the star as seen from outside (field-producing mass). As (11.23) shows, this is in fact proportional to the *coordinate volume* $4\pi r_0^3/3$, but not, however, to the true three-dimensional volume of the star.

While the mass density μ is constant, the pressure p increases inwards; the solution is non-singular as long as p is finite. At $r=0$, where p takes its maximum value, this is only possible, because of (11.24), for

$$3\sqrt{1-Ar_0^2} > 1;$$

that is, for

$$r_0 > \tfrac{9}{8} 2M. \tag{11.26}$$

This inequality is to be interpreted as saying that for given total mass M the interior solution is regular (exists) only if the stellar radius r_0 is large enough, and in any case larger than the Schwarzschild radius $2M$. For normal stars like our Sun this is always the case, but for stars with very dense matter (nuclear matter) it may not be possible to realize (11.26). There is then no interior Schwarzschild solution and, as we shall describe in detail later (section 23.2), no stable interior solution at all.

The three-dimensional space

$$d^{(3)}s^2 = \frac{dr^2}{1-Ar^2} + r^2(d\vartheta^2 + \sin^2\vartheta\, d\varphi^2) \tag{11.27}$$

of the interior Schwarzschild solution has an especially simple geometry. One sees this best by introducing a new coordinate χ via $r = A^{-1/2}\sin\chi$ and transforming the line element (11.27) into the form

$$d^{(3)}s^2 = \frac{1}{A}[d\chi^2 + \sin^2\chi(d\vartheta^2 + \sin^2\vartheta\, dc^2)]. \tag{11.28}$$

The metric (11.28) is that of a three-dimensional hypersphere of radius $R = A^{-1/2}$, which can be represented in a flat, four-dimensional, Euclidean embedding-space according to

$$\left.\begin{array}{ll} z_1 = R\cos\chi, & z_3 = R\sin\chi\sin\vartheta\sin\varphi, \\ z_2 = R\sin\chi\cos\vartheta, & z_4 = R\sin\chi\sin\vartheta\cos\varphi, \\ R^2 = z_1^2 + z_2^2 + z_3^2 + z_4^2 = \text{constant}. \end{array}\right\} \tag{11.29}$$

We are dealing with a three-dimensional space of constant curvature, whose curvature tensor has the simple form

$$R_{\alpha\mu\beta\nu} = A(g_{\alpha\beta}g_{\mu\nu} - g_{\alpha\nu}g_{\mu\beta}) \tag{11.30}$$

(cf., for example, (1.38) and section 21.2). In this space all points are geometrically equivalent; of course the points in the star are physically 'distinguishable,' because g_{44} (that is, in essence the pressure p) is position dependent.

12 The Reissner–Nordström (Reissner–Weyl) solution

The Reissner–Nordström solution (Reissner, 1916) is the spherically symmetric, static, exterior field of a charged distribution of mass. We state without proof that the gravitational field is described by the metric

$$ds^2 = \frac{dr^2}{1 - \frac{2M}{r} + \frac{\kappa e^2}{2r^2}} + r^2(d\vartheta^2 + \sin^2\vartheta d\varphi^2) - \left(1 - \frac{2M}{r} + \frac{\kappa e^2}{2r^2}\right)c^2 dt^2, \tag{12.1}$$

and the electromagnetic field by the four-potential

$$A_\alpha = 0, \quad U = -A_4 = e/r. \tag{12.2}$$

The potential (12.2) is a solution of the source-free Maxwell equations

$$[\sqrt{-g}\, g^{ma} g^{nb}(A_{b,a} - A_{a,b})]_{,n} = 0 \tag{12.3}$$

in the Riemannian space of the metric (12.1), and the metric (12.1) satisfies the Einstein field equations

$$R_{in} - \tfrac{1}{2} R g_{in} = \kappa T_{in}, \tag{12.4}$$

with the energy–momentum tensor of the Maxwell field (12.2) on the right-hand side. The system (12.1) and (12.2) is thus an exact solution of the coupled Einstein–Maxwell equations.

Since for large r the term $\kappa e/2r^2$ in the metric can be ignored, an observer situated at great distance will interpret $m = 8\pi M/\kappa c^2$ as the total mass of the source (cf. (10.21)). From (12.2) one deduces that $Q = 4\pi e$ is the total charge of the source.

In practice celestial bodies are uncharged, and so the influence of the electromagnetic field on their metric can be ignored, the Reissner–Nordström solution being replaced by its special case, the Schwarzschild solution. Originally the hope was that in the Reissner–Nordström solution one had found a useful model of the electron. But even for the electron,

12 The Reissner-Nordström (Reissner-Weyl) solution

the particle with the largest charge per unit mass, $\kappa e^2/M$ has the value of only 2.8×10^{-13} cm. The influence of the term $\kappa e^2/2r^2$ therefore only becomes important at such dimensions that the effects of quantum mechanics and quantum field theory dominate, and the theories of general relativity and classical electrodynamics are no longer adequate to describe the properties of matter.

The Reissner-Nordström solution thus has only slight physical significance. It deserves attention, however, as a simple example of an exact solution of the Einstein-Maxwell equations.

Linearized theory of gravitation, far fields and gravitational waves

13 The linearized Einstein theory of gravitation

13.1 *Justification for a linearized theory and its realm of validity*

One speaks of a linearized theory when the metric deviates only slightly from that of flat space,

$$g_{mn} = \eta_{mn} + f_{mn}, \quad f_{mn} \ll 1, \tag{13.1}$$

and therefore all terms in the Einstein equations which are non-linear in f_{mn} or its derivatives can be discarded and the energy–momentum tensor T^{ik} can be replaced by its special-relativistic form.

This energy–momentum tensor then satisfies the special relativistic equation

$$T^{ik}{}_{,k} = 0. \tag{13.2}$$

Since no covariant derivatives occur in (13.2), in the linearized theory the gravitational field has no influence upon the motion of the matter producing the field. One can specify the energy–momentum tensor arbitrarily, provided only that (13.2) is satisfied, and calculate the gravitational field associated with it. This property of the linearized theory, which appears to be wholly advantageous, has however the consequence that the gravitational field corresponding to the exact solution can deviate appreciably from that of the linearized theory if the sources of the field (under the influence of their own gravitational field) move in a manner quite different from that supposed. It is therefore quite possible that there is no exact solution whose essential features agree with those of a particular solution of the linearized theory. Since, however, one would like to use approximation procedures at precisely those places where the exact solution is unknown, one must beware of conclusions drawn from the results of linearized theory.

Statements made in the linearized theory will be reliable then if we have a good knowledge of the motion of the sources and if these sources

13 The linearized Einstein theory of gravitation

are not too massive (if the field produced by them is weak). This is the case, for example, in the planetary system. The linearized theory can also be used to analyze the fields, due to sources regarded as known, at great distance from these sources, or to describe the metric and the gravitational field in the neighbourhood of a point at which we have introduced a locally geodesic coordinate system. As (13.1) already shows, the linearized theory is applicable only as long as one can introduce approximately Cartesian coordinates. From the standpoint of the (curved) Universe this means we shall always be dealing with local applications.

13.2 The fundamental equations of the linearized theory

As we have already shown in section 9.2 (9.14), the curvature tensor associated with the metric (13.1) has the form

$$R^a{}_{mbn} = \tfrac{1}{2}\eta^{as}(f_{sn,mb} + f_{mb,sn} - f_{mn,bs} - f_{bs,mn}), \tag{13.3}$$

obtained by ignoring all non-linear terms in f_{mn}.

For further calculations we should make clear that indices in the f_{mn} and its derivatives are always moved up or down with the flat-space metric η_{ab}, so that we have

$$f^{ab} = \eta^{am}\eta^{bn}f_{mn}, \quad f^a{}_a = \eta^{ab}f_{ab}, \ldots. \tag{13.4}$$

From (13.3) we thus obtain the linearized field equations

$$R_{mn} - \frac{R}{2}\eta_{mn} = -\frac{1}{2}[f_{mn}{}^{,a}{}_{,a} + f^a{}_{a,mn} - f^a{}_{n,ma} - f^a{}_{m,na} - \eta_{mn}f^{i,a}{}_{i,a}$$
$$+ \eta_{mn}f^{ab}{}_{,ab}] = \kappa T_{mn}. \tag{13.5}$$

The following considerations lead, by means of suitable definitions and subsidiary conditions (coordinate transformations), to a simpler and mathematically clearer formulation of the field equations. The analogy between this procedure and that usual in electrodynamics is summarized in Table 13.1.

We first of all introduce in the place of the quantities f_{mn} new field functions $\bar{f}_{mn}$, which occur in the expansion

$$\sqrt{-g}\,g^{mn} = \eta^{mn} - \bar{f}^{mn} \tag{13.6}$$

of the density of the metric tensor and which are linked to the old functions through the equations

$$\bar{f}_{mn} = f_{mn} - \tfrac{1}{2}\eta_{mn}f^a{}_a, \quad f_{mn} = \bar{f}_{mn} - \tfrac{1}{2}\eta_{mn}\bar{f}^a{}_a, \quad f^a{}_a = -\bar{f}^a{}_a. \tag{13.7}$$

The field equations (13.5) then read

$$\bar{f}_{mn}{}^{,a}{}_{,a} + \eta_{mn}\bar{f}^{ab}{}_{,ab} - \bar{f}^{a}{}_{n,ma} - \bar{f}^{a}{}_{m,na} = -2\kappa T_{mn}. \quad (13.8)$$

We shall now simplify them further by means of coordinate transformations of the form

$$\tilde{x}^n = x^n + b^n(x^i) \quad (13.9)$$

(these transformations are the analogue of the gauge transformations of electrodynamics). From (13.9) we obtain

$$\begin{aligned}\tilde{g}^{mn} &= g^{as}(\delta_a^n + b^n{}_{,a})(\delta_s^m + b^m{}_{,s}), \\ \tilde{g} &= |\tilde{g}^{mn}|^{-1} = g(1 + 2b^a{}_{,a})^{-1},\end{aligned} \quad (13.10)$$

and hence

$$\tilde{\bar{f}}^{mn} = \bar{f}^{mn} - b^{n,m} - b^{m,n} + \eta^{mn}b^a{}_{,a}. \quad (13.11)$$

The four functions $b^n(x^i)$ can be chosen arbitrarily; of course the transformation (13.9) must not take us outside the framework of the linearized theory, that is to say, $\tilde{\bar{f}}^{mn} \ll 1$ must hold. If we substitute (13.11) into the field equation (13.8), then we see, upon making the choice

$$\Box b^n = \eta^{rs}b^n{}_{,rs} = \bar{f}^{mn}{}_{,m}, \quad (13.12)$$

that the field equations become particularly simple. The field variables $\bar{f}_{mn}$ (we now drop the tilde) then satisfy the equation

$$\bar{f}^{mn}{}_{,n} = (\sqrt{-g}\,g^{mn})_{,n} = 0 \quad (13.13)$$

(thus we use the harmonic coordinates defined in section 10.6), and the Einstein field equations reduce to the inhomogeneous wave equation

$$\Box \bar{f}_{mn} = \bar{f}_{mn}{}^{,a}{}_{,a} = -2\kappa T_{mn}. \quad (13.14)$$

Of course, one may take only those solutions of the field equation (13.14) which satisfy the subsidiary conditions (13.13); the existence of such solutions is guaranteed by (13.2).

13.3 A discussion of the fundamental equations and a comparison with special-relativistic electrodynamics

The fundamental equations (13.14) and (13.8) have just the usual appearance of the equations of a classical field theory in Minkowski space. They are linear and, after introduction of the subsidiary condition (13.13), they

13 The linearized Einstein theory of gravitation

Table 13.1

	Maxwell theory	Linearized Einstein theory
Fundamental field variables	four-potential A_m	$\bar{f}_{mn}$
General field equations	$A_m{}^{,a}{}_{,a} - A^a{}_{,m,a} = -\frac{1}{c} j_m$	$\bar{f}_{mn}{}^{,a}{}_{,a} + \eta_{mn} \bar{f}^{ab}{}_{,ab} - +\bar{f}^a{}_{n,am}$ $- \bar{f}^a{}_{m,na} = -2\kappa T_{mn}$
Field equations are invariant under	gauge transformations $\tilde{A}_m = A_m + b_{,m}$ (field tensor invariant)	coordinate transformations $\tilde{\bar{f}}_{mn} = \bar{f}_{mn} - b_{n,m} - b_{m,n}$ $+ \eta_{mn} b^a{}_{,a}$ (Christoffel symbols changed, curvature tensor invariant)
Subsidiary conditions	$A^a{}_{,a} = 0$	$\bar{f}^{mn}{}_{,n} = 0$
Form of the field equations simplified by these conditions	$\Box A_m = -\frac{1}{c} j_m$	$\Box \bar{f}_{mn} = -2\kappa T_{mn}$
Further possible gauge transformations are restricted by	$\Box b = 0$	$\Box b^n = 0$

are even uncoupled.[‡] One can completely dispense with the idea of a Riemannian space and regard the $\bar{f}_{mn}$ as the components of a tensor field by means of which gravitation is described in a flat space. The action of this field upon a test particle is then given (according to the geodesic equation) by

$$\frac{d^2 x^a}{d\tau^2} = -\frac{1}{2} \eta^{ab} (f_{bm,n} + f_{bn,m} - f_{nm,b}) \frac{dx^n}{d\tau} \frac{dx^m}{d\tau}. \qquad (13.15)$$

Field equations and equations of motion are Lorentz invariant.

Although this kind of gravitational theory is very tempting (and has hence occasionally been interpreted as the correct theory of gravitation), it does however have a serious shortcoming; the gravitational force does not react back on the sources of the field. If one tries to correct this, one is led back to the Einstein theory.

The striking analogy between the linearized Einstein equations and electrodynamics is shown in Table 13.1.

[‡] *Editor's note:* The equations are only uncoupled provided Cartesian (and not, for example, spherical polar) coordinates are used.

132   Linearized gravitation, far fields and gravitational waves

In the linearized theory of gravitation, too, we can represent the solution to the field equations in terms of the sources, namely, in the form of a 'retarded potential'

$$\bar{f}_{mn}(r, t) = \frac{\kappa}{2\pi} \int \frac{T_{mn}\left(\bar{r}, t - \frac{|r-\bar{r}|}{c}\right)}{|r-\bar{r}|} \, d^3\bar{x}. \tag{13.16}$$

To this particular solution one can still always add solutions of the homogeneous equations

$$\Box \bar{f}_{mn} = 0, \quad \bar{f}^{mn}{}_{,n} = 0, \tag{13.17}$$

and thus, for example, go over from the retarded to the advanced solutions.

Sometimes it is more convenient to simplify the metric by a coordinate transformation

$$\tilde{f}_{mn} = f_{mn} - b_{m,n} - b_{n,m} \tag{13.18}$$

and not to use *any* harmonic coordinates. Such a transformation can remove purely coordinate effects, that is, terms which give no contribution to the curvature tensor.

13.4 The far field due to a time-dependent source

In electrodynamics one learns that in general the following components dominate in the far field of an arbitrary distribution of charge and current (the characteristic r-dependence is in brackets): electrostatic monopole (r^{-1}), electrostatic and magnetostatic dipole (r^{-2}), electrostatic quadrupole (r^{-3}), oscillating electric and magnetic dipole and electric quadrupole (all r^{-1}). For charges which are not moving too quickly, the spacelike contribution to the four-current density is smaller than the timelike contribution by a factor of c, because $j^n = (\rho v, \rho c)$, and therefore the electromagnetic radiation emerging from a system is essentially that due to an oscillating electric dipole.

In a similar manner we now want to investigate and characterize the gravitational far field of a matter distribution. The calculations are simple, but somewhat tedious. To keep a clear view we divide them into three steps.

Step 1: Power series expansion of the integrand of (13.16)   We assume that we are dealing with an isolated distribution of matter; that is, T_{mn} is non-zero only within a finite spatial region (Fig. 13.1). In the far field we can then replace $|r-\bar{r}|$ by the first term of a power series expansion:

13 The linearized Einstein theory of gravitation

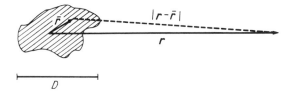

Fig. 13.1. Isolated matter distribution.

$$|r-\bar{r}| = \sqrt{r^2 - 2r\bar{r} + \bar{r}^2} = r - \frac{x^\alpha \bar{x}^\alpha}{r} - \frac{1}{2}\frac{x^\alpha x^\beta}{r^3}(\bar{x}^\alpha \bar{x}^\beta - \bar{r}^2 \delta^{\alpha\beta}) + \cdots,$$
$$\frac{1}{|r-\bar{r}|} = \frac{1}{r} + \frac{x^\alpha \bar{x}^\alpha}{r^3} + \frac{1}{2}\frac{x^\alpha x^\beta}{r^5}(3\bar{x}^\alpha \bar{x}^\beta - \bar{r}^2 \delta^{\alpha\beta}) + \cdots.$$
(13.19)

Imagining the series (13.19) to have been substituted into the argument $t - |r-\bar{r}|/c$ of the energy–momentum tensor and an expansion of the components T_{mn} carried out, we then have

$$T_{mn}(\bar{r}, t - |r-\bar{r}|/c) = T_{mn}(\bar{r}, t - r/c) + \frac{1}{c}\dot{T}_{mn}(\bar{r}, t - r/c)\{r - |r-\bar{r}|\}$$
$$+ \frac{1}{2c^2}\ddot{T}_{mn}(\bar{r}, t - r/c)\{r - |r-\bar{r}|\}^2 + \cdots. \quad (13.20)$$

For a motion of the matter periodic in time (frequency ω) it is justifiable to ignore higher time derivatives if the diameter of the matter distribution is small by comparison with c/ω, and thus small compared to the wavelength of the waves radiated out.

The integrand of (13.16) has, after substitutions of (13.19) and (13.20), the form

$$\frac{T_{mn}(\bar{r}, t - |r-\bar{r}|/c)}{|r-\bar{r}|} = T_{mn}\left[\frac{1}{r} + \frac{x^\alpha \bar{x}^\alpha}{r^3} + \frac{x^\alpha x^\beta}{2r^5}(3\bar{x}^\alpha \bar{x}^\beta - \bar{r}^2 \delta^{\alpha\beta})\right]$$
$$+ \frac{\dot{T}_{mn}}{c}\left[\frac{x^\alpha \bar{x}^\alpha}{r^2} + \frac{x^\alpha x^\beta}{2r^4}(3\bar{x}^\alpha \bar{x}^\beta - \bar{r}^2 \delta^{\alpha\beta})\right]$$
$$+ \frac{\ddot{T}_{mn}}{c^2}\left[\frac{1}{2}\frac{x^\alpha x^\beta}{r^3}\bar{x}^\alpha \bar{x}^\beta\right], \quad (13.21)$$

where on the right-hand side the argument $t - r/c$ in T_{mn} and its derivatives has been suppressed.

Step 2: Definition of the moments of the energy–momentum tensor and simplification by Lorentz transformations and conservation laws For matter not moving too quickly, the component T_{44} dominates the energy-momentum tensor, and we have

$$|T_{44}| \gg |T_{4\alpha}| \gg |T_{\alpha\beta}|. \tag{13.22}$$

Accordingly, in substituting the integrand (13.21) into the formula (13.16) it is only necessary to evaluate the integrals

$$\left. \begin{array}{lll} \int T_{44} \, d^3\bar{x} \equiv m, & \int T_{44}\bar{x}^\alpha \, d^3\bar{x} \equiv d^\alpha, & \int T_{44}\bar{x}^\alpha\bar{x}^\beta \, d^3\bar{x} \equiv d^{\alpha\beta}, \\ \int T_{4\nu} \, d^3\bar{x} \equiv -p_\nu, & \int T_{4\nu}\bar{x}^\alpha \, d^3\bar{x} \equiv b_\nu{}^\alpha, & \int T_{\alpha\beta} \, d^3\bar{x} \equiv a_{\alpha\beta}. \end{array} \right\} \tag{13.23}$$

All these quantities are in principle functions of the retarded time $t - r/c$. From the energy law

$$T^{n4}{}_{,4} = -T^{n\mu}{}_{,\mu} \tag{13.24}$$

and the law of angular momentum

$$(T^{4m}\bar{x}^s - T^{4s}\bar{x}^m)_{,4} = -(T^{\nu m}\bar{x}^s - T^{\nu s}\bar{x}^m)_{,\nu} \tag{13.25}$$

we obtain, upon integration over the matter distribution and application of the Gauss law (all operations are carried out in a flat space),

$$m = \text{constant } (>0), \quad p^\nu = \text{constant}, \quad b^{\nu\alpha} - b^{\alpha\nu} = B^{\nu\alpha} = \text{constant}, \quad \dot{d}^\alpha/c = p^\alpha. \tag{13.26}$$

We can therefore transform the three-momentum p^α to zero by making a Lorentz transformation and then (because $m > 0$) transform away the matter dipole moment d^α by shifting the origin of the spatial coordinate system.

Further, from the two conservation laws (13.24) and (13.25), we obtain the equations

$$\left. \begin{array}{l} T^{4\alpha}\bar{x}^\beta + T^{4\beta}\bar{x}^\alpha = (T^{44}\bar{x}^\alpha\bar{x}^\beta)_{,4} + (T^{4\nu}\bar{x}^\alpha\bar{x}^\beta)_{,\nu} \\ \text{and} \\ (T^{44}\bar{x}^\alpha\bar{x}^\beta)_{,44} = (T^{\mu\nu}\bar{x}^\alpha\bar{x}^\beta)_{,\mu\nu} - 2(T^{\mu\alpha}\bar{x}^\beta + T^{\mu\beta}\bar{x}^\alpha)_{,\mu} + 2T^{\alpha\beta}, \end{array} \right\} \tag{13.27}$$

which upon integration lead to the relation

$$b^{\alpha\beta} + b^{\beta\alpha} = -\dot{d}^{\alpha\beta}/c, \quad a_{\alpha\beta} = \ddot{d}_{\alpha\beta}/2c^2 \tag{13.28}$$

between the moments of the energy-momentum tensor.

Taken together, all the moments in which we are interested can thus be expressed in terms of the mass m, the angular momentum $B_{\nu\alpha}$ and the mass quadrupole moment $d_{\alpha\beta}$ according to

$$\left. \begin{array}{l} m = \text{constant}, \quad p^\nu = 0, \quad d^\nu = 0, \quad B_{\nu\alpha} = \text{constant}, \\ b_{\nu\alpha} = \left(B_{\nu\alpha} - \dfrac{1}{c}\dot{d}_{\nu\alpha} \right)\bigg/2, \quad a_{\alpha\beta} = \ddot{d}_{\alpha\beta}/2c^2. \end{array} \right\} \tag{13.29}$$

13 The linearized Einstein theory of gravitation

Step 3: Writing down the metric and simplifying it by coordinate transformations Substituting the integrand (13.21) into the formula (13.16), using the results (13.23), and remembering the relations (13.29), we have

$$\begin{aligned}
\frac{2\pi}{\kappa}\bar{f}_{44} &= \frac{m}{r} + \frac{x^\alpha x^\beta}{2r^5}(3d_{\alpha\beta} - \eta_{\alpha\beta}d^\sigma_\sigma) + \frac{x^\alpha x^\beta}{2r^4 c}(3\dot{d}_{\alpha\beta} - \eta_{\alpha\beta}\dot{d}^\sigma_\sigma) + \frac{x^\alpha x^\beta}{2c^2 r^3}\ddot{d}_{\alpha\beta}, \\
\frac{2\pi}{\kappa}\bar{f}_{4\nu} &= \frac{B_{\nu\alpha}x^\alpha}{2r^3} - \frac{\dot{d}_{\nu\alpha}x^\alpha}{2cr^3} - \frac{\ddot{d}_{\nu\alpha}x^\alpha}{2c^2 r^2}, \\
\frac{2\pi}{\kappa}\bar{f}_{\nu\mu} &= \frac{\ddot{d}_{\nu\mu}}{2c^2 r}.
\end{aligned} \quad (13.30)$$

The conversion to the f_{mn} yields relatively complicated expressions which we shall not give explicitly. The reader can verify by direct calculation that after a coordinate transformation (13.18) with the generating functions

$$\begin{aligned}
\frac{2\pi}{\kappa}b_4 &= \frac{x^\alpha x^\beta}{8r^4}(3d_{\alpha\beta} - \eta_{\alpha\beta}d^\sigma_\sigma) + \frac{x^\alpha x^\beta}{8cr^3}(\dot{d}_{\alpha\beta} + \eta_{\alpha\beta}\dot{d}^\sigma_\sigma), \\
\frac{2\pi}{\kappa}b_\nu &= -\frac{3d_{\nu\alpha}x^\alpha}{4r^3} - \frac{\dot{d}_{\nu\alpha}x^\alpha}{2cr^2} + \frac{3d_{\alpha\beta}x^\alpha x^\beta x^\nu}{4r^5} + \frac{(\dot{d}_{\alpha\beta} + \eta_{\alpha\beta}\dot{d}^\sigma_\sigma)x^\alpha x^\beta x^\nu}{8cr^4}
\end{aligned} \quad (13.31)$$

no time derivatives are contained in f_{44} and $f_{4\nu}$. The far field of an isolated matter distribution then has, in the linearized theory, the metric

$$\begin{aligned}
g_{44} &= -1 + f_{44} = -1 + \frac{2M}{r} + \frac{x^\alpha x^\beta}{r^5}(3D_{\alpha\beta} - \eta_{\alpha\beta}D^\sigma_\sigma) + O(r^{-4}), \\
g_{4\nu} &= f_{4\nu} = \frac{2x^\alpha}{r^3}\epsilon_{\mu\nu\alpha}P^\mu + O(r^{-3}),
\end{aligned}$$

and

$$\begin{aligned}
g_{\mu\nu} &= \eta_{\mu\nu}\left[1 + \frac{2M}{r} + \frac{x^\alpha x^\beta}{r^5}(3D_{\alpha\beta} - \eta_{\alpha\beta}D^\sigma_\sigma)\right] \\
&+ \frac{2}{3c^2 r}(3\ddot{D}_{\nu\mu} - \eta_{\nu\mu}\ddot{D}^\lambda_\lambda) + \frac{\eta_{\nu\mu}}{3c^2 r^3}(3\ddot{D}_{\alpha\beta} - \eta_{\alpha\beta}\ddot{D}^\lambda_\lambda)x^\alpha x^\beta \\
&- \frac{(3\ddot{D}_{\nu\alpha} - \eta_{\nu\alpha}\ddot{D}^\lambda_\lambda)}{6c^2 r^3}x^\alpha x^\mu - \frac{(3\ddot{D}_{\mu\alpha} - \eta_{\mu\alpha}\ddot{D}^\lambda_\lambda)}{6c^2 r^3}x^\alpha x^\nu \\
&+ \frac{(3\ddot{D}_{\alpha\beta} - \eta_{\alpha\beta}\ddot{D}^\lambda_\lambda)}{12c^2 r^5}x^\alpha x^\beta x^\nu x^\mu + O(r^{-2}),
\end{aligned} \quad (13.32)$$

with the abbreviations

$$\text{mass:} \quad M = \frac{\kappa}{8\pi} \int T_{44}\left(\bar{r}, t - \frac{r}{c}\right) d^3\bar{x} = \text{constant},$$

$$\text{angular momentum:} \quad P^\mu = \epsilon^{\mu\nu}{}_\alpha \frac{\kappa}{8\pi} \int T_{4\nu}\left(\bar{r}, t - \frac{r}{c}\right) \bar{x}^\alpha d^3\bar{x} = \text{constant},$$

and

$$\text{quadrupole moment:} \quad D^{\alpha\beta} = \frac{\kappa}{8\pi} \int T_{44}\left(\bar{r}, t - \frac{r}{c}\right) \bar{x}^\alpha \bar{x}^\beta d^3\bar{x}$$

$$= D^{\alpha\beta}\left(t - \frac{r}{c}\right).$$

(13.33)

13.5 A discussion of the properties of the far field (linearized theory)

As we have already shown in section 9.2, $f_{44} = g_{44} + 1$ is essentially the Newtonian potential of the matter distribution. It contains here one mass term and one quadrupole term, but no dipole contribution (we have transformed this away by the choice of coordinate system). If we compare the linear approximation

$$g_{44} = -1 + \frac{2M}{r} + \frac{x^\alpha x^\beta}{r^5}(3D_{\alpha\beta} - \eta_{\alpha\beta}D^\sigma_\sigma) + O(r^{-4}) \qquad (13.34)$$

with the expansion

$$g_{44} = -1 + 2M/r - 2M^2/r^2 + O(r^{-3}) \qquad (13.35)$$

of the Schwarzschild metric (in isotropic coordinates) in powers of r^{-1}, see (10.63), (10.67), then we see that retention of the quadrupole term in g_{44} is only justified in exceptional cases. The higher non-linear terms with M^2 will almost always dominate; (13.34) is a good approximation only up to terms in r^{-1}. The same restriction holds for the part of $g_{\mu\nu}$ proportional to $\eta_{\mu\nu}$.

The occurrence of the angular momentum in $g_{4\nu}$ is interesting. In the Newtonian theory there is no dependence of the gravitational field upon the rotation of a celestial body. To appreciate the physical meaning of this term in the metric, we compare (13.32) with the coordinate system

$$ds^2 = \eta_{\alpha\beta} dx^\alpha dx^\beta - 2\epsilon_{\beta\nu\mu} x^\nu \frac{\omega^\mu}{c} dx^4 dx^\beta - \left(1 + \frac{2a_\nu x^\nu}{c^2}\right)(dx^4)^2 \qquad (13.36)$$

13 The linearized Einstein theory of gravitation

of an arbitrarily moving observer, which we introduced in (8.17); this coordinate system rotates with angular velocity

$$\omega^\alpha = -\frac{c}{2}\epsilon^{\alpha\beta\nu}g_{4\beta,\nu} \tag{13.37}$$

with respect to a local inertial system. The coordinate system used here, upon which we have based the linearized gravitational theory, and which we have to identify with the system in which the fixed stars are at rest, is thus locally a rotating coordinate system, or, conversely, the local inertial systems rotate with angular velocity

$$\Omega^\sigma = \frac{c}{2}\epsilon^{\sigma\beta\nu}g_{4\beta,\nu} = -\left(\frac{cP^\sigma}{r^3} - \frac{3x^\sigma x_\nu P^\nu c}{r^5}\right) \tag{13.38}$$

with respect to the fixed stars. In principle this could be demonstrated by the precession of a top, the *Lense–Thirring (1918) effect*. However for the Earth's rotation this is a very small correction to the geodetic precession (10.82). The analogue in electrodynamics of the components $g_{4\nu}$ of the metric and their effect is the magnetic field, which is created by currents and exerts a couple upon dipoles (Ω^σ has precisely the spatial structure of the force field of a dipole).

The most important terms for the far field of a source are those strongest at infinity, namely, those proportional to r^{-1}, that is to say, the parts of the metric (13.32) containing the mass M or the second derivative of the quadrupole moment $D_{\alpha\beta}(t-r/c)$. In electrodynamics the corresponding potentials

$$U = \frac{Q}{4\pi r} + \frac{x^\alpha}{4\pi r^2}\left[\frac{\dot{p}_\alpha(t-r/c)}{c} + \frac{p_\alpha(t-r/c)}{r}\right] + O(r^{-2}), \quad A_\nu = \frac{\dot{p}_\nu(t-r/c)}{4\pi r} \tag{13.39}$$

represent the far field of a charge Q and an oscillating electric dipole p_α, and the terms proportional to $\dot{p}_\alpha$ lead to the radiation of electromagnetic waves. We may therefore suppose that the occurrence in the metric of $\ddot{D}_{\alpha\beta}/r$ signifies that the system is emitting gravitational waves and that, in contrast to the possibility of dipole radiation from a charge distribution, the gravitational radiation is quadrupolar in character. Both suppositions can be to a certain extent verified. In order to be able to make more exact statements one must of course abandon the linearized theory. (In electrodynamics, too, the Poynting vector, which characterizes the radiation, is quadratic in the field strengths.)

Bibliography to section 13

Review and research articles   Campbell (1973), Stephani (1966)

14 Far fields due to arbitrary matter distributions and balance equations for momentum and angular momentum

14.1 *What are far fields?*

The linearized theory of gravitation is based on the presumption that over *whole* regions of space, at any rate in the vicinity of the sources of the field, the gravitational field is weak, and the metric deviates only slightly from that of a Minkowski space. As has already been mentioned in the introduction to the previous section, we often meet in nature the situation in which a distribution of matter (a satellite near the Earth, the Earth, the planetary system, our Galaxy) is surrounded by vacuum, and the closest matter is so far away that the gravitational field is weak in an *intermediate region,* where the metric deviates from Minkowski space only by terms in r^{-1} or still higher inverse powers of r. In the neighbourhood of the sources, however, the field can be strong.

If such an intermediate region exists, then we speak of the far field of the configuration in question (Fig. 14.1). Notice that here by contrast to most problems, for example, in electrodynamics, we may not simply assume an isolated matter distribution which is surrounded only by a vacuum. The assumption of a void (the 'infinite empty space') into which waves pass and disappear contradicts the basic conception of general relativity; also the fact that we orient our local inertial system towards the fixed stars indicates that we must always in principle take into account the existence of the whole Universe whenever we examine the properties of a part of the Universe.

While in the linearized theory we investigated *approximate solutions* to the field equations, their dependence upon the structure of the sources and their behaviour at great distances from the sources, now we are interested in *exact solutions* in regions where the gravitational fields are weak. Our goal here is to obtain statements about the system from a knowledge of the far field.

The simplest examples are gravitational fields whose far fields are independent of time. We assume that to good approximation the metric can be written as

14 Far fields

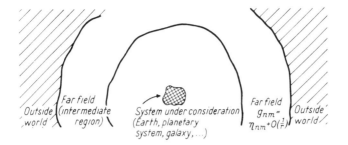

Fig. 14.1. How the far field is defined.

$$g_{mn} = \eta_{mn} + a_{mn}/r + b_{mn}/r^2 + O(r^{-3}). \tag{14.1}$$

As the region of the far field is an annular or shell-like region and there may be sources outside, we really ought to add onto (14.1) a series with increasing powers of r. We drop these terms, however; this corresponds to the model of an isotropic external environment.

The functions a_{mn} and b_{mn}, which are independent of r and t, are to be determined by substitution of (14.1) into the vacuum field equations

$$R_{mn} = 0. \tag{14.2}$$

We can simplify these calculations by first obtaining solutions $\bar{f}_{mn}$ to the linearized field equations

$$\Delta \bar{f}_{mn} = 0 \quad \text{and} \quad f^{nm}{}_{,n} = 0, \tag{14.3}$$

and then calculating the non-linear corrections.

Every solution of the potential equation can be represented in the form of a multipole expansion. Thus neglecting terms which are $O(r^{-3})$ we have

$$\left. \begin{array}{l} \bar{f}_{44} = A/r + A_\alpha x^\alpha/r^3, \quad \bar{f}_{4\nu} = B_\nu/r + B_{\nu\alpha} x^\alpha/r^3, \\ \bar{f}_{\mu\nu} = \bar{f}_{\nu\mu} = C_{\nu\mu}/r + C_{\nu\mu\alpha} x^\alpha/r^3, \end{array} \right\} \tag{14.4}$$

where, because of the subsidiary condition $\bar{f}^{mn}{}_{,n} = 0$, the constants are restricted by the algebraic conditions

$$\left. \begin{array}{ll} B_\nu = 0, & B_{\nu\alpha} = \eta_{\nu\alpha} B + \epsilon_{\nu\alpha\beta} F^\beta, \\ C_{\nu\mu} = 0, & C_{\nu\mu\alpha} = \delta_{\nu\mu} C_\alpha - \delta_{\mu\alpha} C_\nu - \delta_{\nu\alpha} C_\mu. \end{array} \right\} \tag{14.5}$$

B and $C_{\mu\nu\alpha}$ can be eliminated by a coordinate transformation (13.10) with

$$b^4 = B/r, \quad b^\alpha = -C^\alpha/r. \tag{14.6}$$

Experiment shows that A is always non-zero (mass is always positive), so that by a shift of the origin of coordinates A_α can be transformed away as well. The linear approximation thus gives the metric

$$ds^2 = (\eta_{mn} + \bar{f}_{mn} - \tfrac{1}{2}\eta_{mn}\bar{f}^a_a)dx^n dx^m$$

$$= \left(1 + \frac{A}{2r}\right)\eta_{\alpha\beta}dx^\alpha dx^\beta + \frac{2}{r^3}\epsilon_{\nu\alpha\beta}x^\alpha F^\beta dx^\nu dx^4 - \left(1 - \frac{A}{2r}\right)(dx^4)^2. \tag{14.7}$$

If we compare this expression with the metric (13.32), which we derived from the description of the fields in terms of the sources, then we see that the constants A and F^β can be identified with the mass M and the angular momentum P according to

$$A = 4M \quad \text{and} \quad F^\beta = 2P^\beta. \tag{14.8}$$

This identification is not merely a repetition of the linearized theory. In the linearized theory mass and angular momentum were defined through the integrals (13.33) over the source distribution. Now, in the investigation of the far field of an (unknown) source, we take as definitions of the mass and the angular momentum of the source just those coefficients in the expansion of the far field which act upon a test body or a top in exactly the same way as the mass or angular momentum, respectively, of a weak source.

We have now to put into the metric (14.7) the corrections arising from the non-linearity of the Einstein equations; (14.7) is not of course a solution of the field equations (14.2), even if we ignore terms in r^{-3}. Since we are taking terms only up to r^{-2} and corrections due to the non-linearity are always weaker by at least one power of r than the original terms, we need to take into account terms quadratic in the mass parameter. However, we can obtain these by series expansion from the exact Schwarzschild solution (10.63) without performing additional calculations.

We thus obtain the result that the far field of an arbitrary, time-independent source has, in suitable coordinates, the form

$$ds^2 = (1 + 2M/r + 3M^2/r^2)\eta_{\alpha\beta}dx^\alpha dx^\beta + \frac{4}{r^3}\epsilon_{\nu\alpha\beta}x^\alpha P^\beta dx^\nu dx^4$$

$$- (1 - 2M/r + 2M^2/r^2)(dx^4)^2 + O(r^{-3}), \tag{14.9}$$

in which M and P^α are regarded as the mass and the angular momentum because of the way they act upon test bodies and because of the analogy with the linearized theory.

14.2 The energy–momentum pseudo-tensor for the gravitational field

The problem Linearized gravitational theory and its strong analogy with electrodynamics leads one to suppose that time-varying gravitational systems emit gravitational waves. Is it possible by examining the far field to establish whether such waves exist, and, if so, under what conditions?

In a special-relativistic field theory one would probably try to answer this question in the following fashion. The decisive factor in the existence of waves or radiation is not merely that the fields are time dependent, but that energy, momentum and angular momentum are transported from one region of space to another. One therefore encloses the system under consideration in, for example, a spherical surface, writes down the balance equations for the above-mentioned quantities, and ascertains whether, for example, an energy current is flowing through the sphere. If this is the case, one can speak of radiation (provided that particles are not just flowing across the boundary surface and carrying with them energy, etc.). We therefore start out from the balance equations

$$T^{mn}{}_{,n} = 0 \quad \text{and} \quad (T^{mn}x^a - T^{ma}x^n)_{,n} = 0 \tag{14.10}$$

for momentum and angular momentum for the field under consideration.

If we want to perform a similar analysis in gravitation theory, we must construct an energy–momentum tensor of the gravitational field (not of the matter!) and derive balance equations from it. In general relativity there is, however, just one energy–momentum tensor, namely, that due to the matter. Although its divergence vanishes,

$$T^{mn}{}_{;n} = 0, \tag{14.11}$$

one cannot, however, deduce an integral conservation law from (14.11) in a Riemannian space, because integration is the reverse of partial, and not covariant, differentiation and one cannot apply the Gauss law to the divergence of a symmetric second-rank tensor (cf. section 7).

Since we are supposing the existence of gravitational waves, this negative statement should not really surprise us. If energy can be transported in the form of gravitational waves, the energy of the sources alone cannot remain conserved. Rather, one would expect that in place of (14.11) there is a differential balance equation, formulated in terms of partial derivatives, of the structure

$$[(-g)(T^{mn} + t^{mn})]_{,n} = 0, \tag{14.12}$$

142   Linearized gravitation, far fields and gravitational waves

which expresses the fact that a conservation law holds only for the sum of the matter (T^{mn}) and the gravitational field (t^{mn}).

The problem is, therefore, to construct a t^{mn} from the metric in such a way that (14.12) is satisfied as a consequence of the field equations

$$G_{nm} = R_{nm} - \tfrac{1}{2} R g_{nm} = \kappa T_{nm} \tag{14.13}$$

alone.

Before turning to this problem, we want to formulate clearly the alternatives confronting us. Either we wish to calculate only with tensors and allow only covariant statements, in which case we use (14.11) and can write down no balance equation for the energy transport by radiation. Or else we want such a balance equation (14.12), which can only be formulated in a non-covariant manner; as one can see from (14.12), t^{mn} is not a tensor, and we call it the *energy-momentum pseudo-tensor* of the gravitational field.

Since we pick out a Minkowski metric in the far field in a non-covariant fashion anyway, to begin with we shall accept the lack of covariance in (14.12), not going into its consequences until later. There are, however, good reasons for deciding to maintain covariance and to regard the question of energy transport by gravitational waves as inappropriate in the theory of gravitation, because the concept of energy has lost its meaning there (cf. section 14.4).

Construction of the energy-momentum pseudo-tensor t^{mn}   The programme just described perhaps sounds plausible, but already in the initial equation (14.12) something is lacking, namely, that t^{mn} is not uniquely determined. The addition of a term of the form $H^{ikl}{}_{,l}$ to $(-g)t^{ik}$ in no way affects the validity of (14.12), provided only that H^{ikl} is antisymmetric in k and l:

$$(-g)\tilde{t}^{ik} = (-g)t^{ik} + H^{i[kl]}{}_{,l}, \qquad [(-g)\tilde{t}^{ik}]_{,k} = [(-g)t^{ik}]_{,k}. \tag{14.14}$$

Thus, one finds in the extensive literature on this problem a whole series of different proposals, which finally, in the formulation of the conservation laws, give the same statements. We shall therefore not attempt to derive our preferred (Landau–Lifschitz) form of t^{mn}, but rather guess a trial substitution from seemingly plausible requirements and then verify its correctness.

In analogy to the properties of the energy–momentum tensors of all other fields, t^{mn} should be symmetric, it should be bilinear in the first derivatives of the metric, and it should contain no second derivatives.

14 Far fields

Furthermore, bearing in mind the field equations (14.13), (14.12) must be satisfied identically, that is,

$$[(-g)(G^{mn}+\kappa t^{mn})]_{,n} = 0 \tag{14.15}$$

must hold for every metric. Equation (14.15) can be satisfied most simply if we introduce a superpotential U^{mni} according to

$$\left.\begin{array}{l}(-g)(G^{mn}+\kappa t^{mn}) = U^{mni}{}_{,i},\\ U^{mni} = -U^{min}, \quad U^{mni}{}_{,i} = U^{nmi}{}_{,i}.\end{array}\right\} \tag{14.16}$$

Since second derivatives of the metric occur in G^{mn} on the left-hand side of (14.16), U^{mni} should contain at most first derivatives. We can ensure this by writing U^{mni} as the divergence of a quantity U^{mnik}:

$$U^{mni} = U^{mnik}{}_{,k}, \tag{14.17}$$

which depends only upon the metric, not on its derivatives. From the symmetry requirement, the form of U^{mnik} is uniquely determined up to a factor; we make the choice

$$U^{mnik} = \tfrac{1}{2}(-g)(g^{mn}g^{ik} - g^{mi}g^{nk}). \tag{14.18}$$

While the validity of (14.15) is ensured because of our construction procedure, we must examine explicitly whether t^{mn} does contain no second derivatives of the metric. From (14.16), that is, from

$$\kappa(-g)t^{mn} = \tfrac{1}{2}[(-g)(g^{mn}g^{ik} - g^{mi}g^{nk})]_{,ik} - (-g)G^{mn}, \tag{14.19}$$

one obtains, after a rather long calculation,

$$\begin{aligned}2\kappa(-g)t^{mn} = &\; \mathfrak{g}^{mn}{}_{,k}\mathfrak{g}^{ik}{}_{,i} - \mathfrak{g}^{mi}{}_{,i}\mathfrak{g}^{nk}{}_{,k} + \tfrac{1}{2}g^{mn}g_{li}\mathfrak{g}^{lk}{}_{,p}\mathfrak{g}^{pi}{}_{,k}\\ &- g_{ik}\mathfrak{g}^{ip}{}_{,l}(g^{ml}\mathfrak{g}^{nk}{}_{,p} + g^{nl}\mathfrak{g}^{mk}{}_{,p}) + g_{li}g^{kp}\mathfrak{g}^{nl}{}_{,k}\mathfrak{g}^{mi}{}_{,p}\\ &+ \tfrac{1}{8}(2g^{ml}g^{nk} - g^{mn}g^{lk})(2g_{ip}g_{qr} - g_{pq}g_{ir})\mathfrak{g}^{ir}{}_{,l}\mathfrak{g}^{pq}{}_{,k},\end{aligned} \tag{14.20}$$

where the abbreviation

$$\mathfrak{g}^{mn} \equiv \sqrt{-g}\, g^{mn} \tag{14.21}$$

has been used. The energy–momentum pseudo-tensor t^{mn} therefore really does have the desired properties. That we have succeeded so simply in expressing the second derivatives of the metric contained in G^{mn} by the derivatives of U^{mnik} is closely connected with the possibility mentioned in (9.45) of splitting up the Lagrangian of the gravitational field.

Properties of the energy–momentum pseudo-tensor The energy–momentum pseudo-tensor t^{mn} is not a tensor; one can see this property most clearly by noticing that at any point of the space-time the energy–momentum pseudo-tensor can be made to vanish by the introduction of locally geodesic coordinates $g_{mn} = \eta_{mn}$, $g_{mn,a} = 0$. Therefore, if our idea of associating energy and momentum with the pure gravitational field is at all meaningful, then the gravitational energy is on no account to be thought of as localizable; it is at best a quantity which one can associate with a whole spatial region, its value at any one point being arbitrarily alterable through choice of the coordinate system.

On the other hand, the energy–momentum pseudo-tensor does transform like a tensor under coordinate transformations which have the formal structure of a Lorentz transformation:

$$x^{n'} = L^{n'}_n x^n, \qquad L^{n'}_n L^a_{n'} = \delta^a_n,$$
$$t^{n'm'} = L^{n'}_n L^{m'}_m t^{nm}. \tag{14.22}$$

This property is important when the energy–momentum pseudo-tensor is used for the far field of an isolated matter distribution, where the space deviates only weakly from a Minkowski space and hence Lorentz transformations have a physical meaning.

14.3 The balance equations for momentum and angular momentum

We want now to use the energy-momentum pseudo-tensor and the superpotential to obtain global statements about the energy, momentum and angular momentum of the system under consideration, from the local balance equations for the four-momentum

$$[\kappa(-g)(T^{mn} + t^{mn})]_{,n} = U^{mni}{}_{,in} = U^{mnik}{}_{,kin} = 0, \tag{14.23}$$

and from the balance equation

$$\{[\kappa(-g)(T^{mn} + t^{mn})]x^a - [\kappa(-g)(T^{ma} + t^{ma})]x^n\}_{,m}$$
$$= (U^{mni}{}_{,i} x^a - U^{mai}{}_{,i} x^n)_{,m} = 0 \tag{14.24}$$

for angular momentum which follows from it.

To this end we integrate (14.23) and (14.24) over the region G_3 of the three-dimensional space $x^4 = $ constant indicated in Fig. 14.2, which contains the matter and which reaches into the far-field zone, and with the help of the Gauss law transform these integrals into surface-integrals over the surface Σ of G_3, giving

$$\frac{d}{dx^4} \int_{G_3} U^{m4i}{}_{,i} \, d^3x = -\int_{G_3} U^{m\nu i}{}_{,i\nu} \, d^3x = -\int_{\Sigma} U^{m\nu i}{}_{,i} \, df_\nu, \tag{14.25}$$

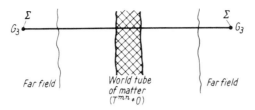

Fig. 14.2. Derivation of the balance equations.

and

$$\frac{d}{dx^4}\int_{G_3}(U^{4ni}{}_{,i}x^a - U^{4ai}{}_{,i}x^n)\,d^3x = -\int_{\Sigma}(U^{\nu ni}{}_{,i}x^a - U^{\nu ai}{}_{,i}x^n)\,df_\nu. \tag{14.26}$$

Because of the symmetry properties (14.16), $U^{m4i}{}_{,i}$ contains no time derivative, so that the left-hand sides of these equations can also be transformed into surface integrals. We designate these integrals as

$$p^m \equiv \int_{G_3}[(-g)(T^{4m} + t^{4m})]\,d^3x = \int_{\Sigma}U^{m4\nu}\,df_\nu, \tag{14.27}$$

and

$$I^{an} \equiv \int_{G_3}[(-g)(T^{4n}x^a - T^{4a}x^n + t^{4n}x^a - t^{4a}x^n)]\,d^3x$$

$$= \int_{\Sigma}(U^{4n\nu}x^a - U^{4a\nu}x^n + U^{4na\nu} - U^{4an\nu})\,df_\nu, \tag{14.28}$$

and notice that if T^{mn} vanishes on Σ then we obtain the relations

$$\frac{d}{dt}p^m = -c\int_{\Sigma}U^{m\nu i}{}_{,i}\,df_\nu = -c\int_{\Sigma}(-g)t^{m\nu}\,df_\nu, \tag{14.29}$$

and

$$\frac{d}{dt}I^{an} = -c\int_{\Sigma}(U^{\nu ni}{}_{,i}x^a - U^{\nu ai}{}_{,i}x^n)\,df_\nu = -c\int_{\Sigma}(-g)(t^{\nu n}x^a - t^{\nu a}x^n)\,df_\nu. \tag{14.30}$$

We can interpret these equations as balance equations for the momentum p^m and the angular momentum I^{an}. They state that the momentum and angular momentum of a source (of the region bounded by Σ) change when gravitational radiation is transported over the boundary surface Σ. Their particular advantage lies in the fact that all the quantities occurring need to be known only on Σ, that is, in the far-field region.

In practice we shall identify the surface Σ with a 'sphere' $r=$ constant; since its surface element is given by

$$df_\nu = x^\nu r \sin \vartheta \, d\vartheta \, d\varphi + O(r) \tag{14.31}$$

in the far-field region (r very large), we need take into account only those contributions to the integrands which tend to zero no faster than as r^{-2}.

If we are to test the physical content of the balance equations in the example of the stationary metric (14.9), then, according to (14.27), for the calculation of the momentum we need retain only the terms in the metric proportional to r^{-1} (which give terms in r^{-2} in $U^{m4\nu}$); that is, we can use the relation

$$U^{mni} = \tfrac{1}{2}(\mathfrak{g}^{mn}\mathfrak{g}^{ik} - \mathfrak{g}^{mi}\mathfrak{g}^{nk})_{,k} = \tfrac{1}{2}(\eta^{nk}\bar{f}^{mi}{}_{,k} - \eta^{ik}\bar{f}^{mn}{}_{,k}), \tag{14.32}$$

valid in the linearized theory. The result of this simple calculation is

$$p^\nu = 0, \qquad p^4 = 8M\pi/\kappa = mc^2 = \text{constant}; \tag{14.33}$$

that is, the spatial momentum is zero and the energy p^4 is connected with the mass m measured in the far field exactly as in the special-relativistic formula. On the other hand, only the term in $g_{4\nu}$ proportional to r^{-2} gives a contribution to the angular momentum I^{an}; one obtains

$$I^{4\nu} = 0, \qquad I^{\alpha\nu} = \frac{8\pi}{\kappa}\epsilon^{\alpha\nu\beta}P_\beta. \tag{14.34}$$

The results (14.33) and (14.34) thus confirm our ideas, in particular the interpretation of p^m as the momentum and I^{an} as the angular momentum of the system comprising matter plus gravitational field.

For time-dependent fields, momentum and angular momentum will not remain conserved. We examine this in the energy balance equation of the far field (13.32) of a time-dependent source. Since the energy–momentum pseudo-tensor is quadratic in the first derivatives of the metric (cf. (14.20)), we have to take into account in the energy law

$$\frac{dp^4}{dt} = -c\int(-g)t^{4\nu}x_\nu r \sin \vartheta \, d\vartheta \, d\varphi \tag{14.35}$$

only the terms of the metric whose first derivatives go like r^{-1}, namely,

$$\left.\begin{aligned}
\mathfrak{g}^{44} &\approx \eta^{44} - \bar{f}^{44} \approx -1 - \frac{\kappa}{4\pi}\frac{\ddot{D}_{\alpha\beta}x^\alpha x^\beta}{c^2 r^3}, \\
\mathfrak{g}^{4\nu} &\approx -\bar{f}^{4\nu} \approx \frac{\kappa}{4\pi}\frac{\ddot{D}_{\nu\alpha}x^\alpha}{c^2 r^2}, \qquad D_{\alpha\beta} = D_{\alpha\beta}\!\left(t - \frac{r}{c}\right), \\
\text{and} \quad \mathfrak{g}^{\mu\nu} &\approx \eta^{\mu\nu} - \bar{f}^{\mu\nu} \approx \eta^{\mu\nu} - \frac{\kappa}{4\pi}\frac{\ddot{D}_{\nu\mu}}{c^2 r}.
\end{aligned}\right\} \tag{14.36}$$

14 Far fields

After a simple, but rather lengthy, calculation one obtains

$$\frac{dp^4}{dt} = -\frac{1}{5c^5}(\dddot{D}_{\alpha\beta} - \tfrac{1}{3}\eta_{\alpha\beta}\dddot{D}^{\sigma}_{\sigma})(\dddot{D}^{\alpha\beta} - \tfrac{1}{3}\eta^{\alpha\beta}\dddot{D}^{\tau}_{\tau}); \qquad (14.37)$$

that is, the energy of the system always decreases. In most cases, for example, in the planetary system, this loss of energy through gravitational quadrupole radiation can certainly be ignored, since it is proportional to the sixth power of the frequency ω of the system.

The weakness of this application of the balance equations comes to light when one tries to calculate not the loss of energy but the total energy of the system emitting quadrupole waves: the corresponding integrals diverge for $r \to \infty$ if the system emits continuously (the whole space is filled with radiation). This diverging of the total energy is possible because in the linearized theory we have ignored the back reaction produced by the emission of radiation upon the motion of the sources, and consequently the system can give up energy continuously without exhausting the supply. It would be desirable of course to test the balance equations in the far field of an exact solution. Unfortunately, however, no exact solution is known which describes the emission of radiation by a physically reasonable system.

14.4 Is there an energy law for the gravitational field?

Because of the significance of the law of conservation of energy and (for systems which are not closed) the energy balance equation in many areas of physics, we shall examine their rôle in the theory of gravitation again, to some extent repeating the discussion of section 14.3.

In special relativity, electrodynamics, thermodynamics, quantum mechanics and quantum field theory it is always the case that a quantity 'energy' can be defined for a system, and is constant if the system is isolated; if the system interacts with its surroundings, a balance equation can be written down so that the energy of the whole system (system plus surroundings) is again constant.

By analogy one would expect that, for example, electrical energy and energy of the gravitational field could transform into one another, their sum remaining constant (if there are no other types of interaction). The Einstein gravitational theory gives a completely different answer, however. In a *general* gravitational field there is indeed a conservation law in the neighbourhood of a point,

$$T^{mn}{}_{,n} = 0,$$

for the energy and momentum of the field-producing matter obtained upon introduction of the inertial coordinate system there (locally geodesic system), but it holds only so long as (in a region of space so small that) the curvature of the space, that is, the real gravitational effects, can be ignored. In this sense, and with this restriction, the theory of gravitation corroborates the conservation laws of special-relativistic physics.

Over larger spatial regions when the gravitational field is properly included there is no energy balance equation. It is incorrect to regard this as a violation of energy conservation; there exists in general no local covariant quantity 'energy,' with which one can associate the property of conservation or non-conservation. None of the foundations of physics is thereby destroyed; energy is only a (very important) auxiliary quantity for describing interactions, but the interaction of all parts of the Universe is quite essential for the theory of gravitation.

The situation is rather more favourable if the gravitational field is not completely general, but possesses certain additional properties. Thus one can associate energy and momentum with a system that is separated from the rest of the Universe by a far-field zone, in the sense of section 14.1, and for which the integrals (14.27) exist. Unlike the general system, in which, for example, the superpotential also exists, these integrals assume an invariant significance through the use of Minkowski coordinates. Then balance equations can be formulated, which for real systems, whose far fields of course do not reach to infinity, are only approximations. A localization of the energy in the interior of the system is in principle impossible.

Another important possibility for applying the concepts of energy, momentum or angular momentum occurs when the gravitational field possesses symmetries. While the local inertial system is invariant under the full Lorentz group and possible translations (rotations) just correspond to the usual energy–momentum (angular-momentum) conservation law of physics, the whole space-time has symmetry properties only in exceptional cases. If, however, symmetries are present, they always correspond to conservation laws. We shall return to this problem in chapter 19.

Bibliography to section 14

Textbooks   Landau and Lifschitz (1975), Misner, Thorne and Wheeler (1973), Schmutzer (1968)

Review and research articles   Bondi *et al.* (1962), Bonnor (1966), Geroch (1979), Newman and Tod (1980), Sachs (1962, 1963), Trautman (1962)

15 Gravitational waves

15.1 *Do gravitational waves really exist?*

The existence of gravitational waves was for a long time disputed, but in recent years, however, their existence has been generally accepted. As often in the history of a science, the cause of the variance of opinions is to be sought in a mixture of ignorance and inexact definitions. Probably in the theory of gravitation, too, the dispute will only be completely settled when an exact solution, for example, of the two-body problem, has been found, from which one can see in what sense such a double-star system emits waves and in what sense it does not, and when the existence of such waves has been demonstrated.

Waves in the most general sense are time-dependent solutions of the Einstein equations; of course such solutions exist. But this definition of waves is, as we can see from our experience with the Maxwell theory, rather too broad, for a field which changes only as a result of the relative motion of the source and the observer (motion past a static field) would not be called a wave. Almost all other additional demands which a gravitational wave should satisfy lead, however, to the characterization 'radiation or transport of energy,' and this is where the difficulties begin, as explained in the previous chapter, starting with the definition of energy.

In order to make the situation relatively simple, in spite of the non-linearity of the field equations, one can restrict attention to those solutions which possess a far-field zone in the sense of section 14.1. Thus imagine the planetary system as seen from a great distance. Does this system emit gravitational waves as a consequence of the motion of the Sun and the planets? The linearized theory answers this question in the affirmative, but ignores the back reaction of the radiation upon the motion of the bodies. The general opinion of physicists is, however, that such a system tries to adjust its state (the Sun captures planets which have lost their kinetic energy by radiation) and thereby emits waves. There is little to be said against this supposition if one imagines the planetary system in an otherwise empty space. If, however, one is thinking of an exact solution which includes the planetary system and the rest of the Universe (the rest is the real Cosmos!) then the process must be regarded in the following way (cf. Fig. 15.1). From an initially non–spherically-symmetric field inside the far-field zone and the external Universe which (as a consequence) is also not spherically symmetric there develops a Schwarzschild solution in the interior and a Friedman universe in the exterior (cf. section 26). Both parts of the Universe strive to adjust their state, but whether,

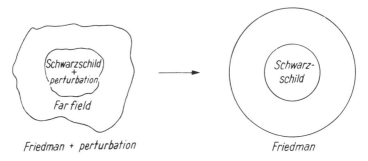

Fig. 15.1. Settling down of a perturbed gravitational system.

and in which direction, energy transport occurs through the far-field zone is unclear – neither of the two partners in the interaction is preferred in principle. It is therefore not at all certain whether a freely gravitating system (a system with exclusively gravitational interaction) emits gravitational waves.

The situation is clearer when the properties (the matter distribution) of a system are changed discontinuously by intervention from outside, that is, by non-gravitational interaction, for example, by the explosion of a bomb. The change thus produced in the gravitational field propagates out in the form of gravitational shock waves with the velocity of light as measured by every observer in his local inertial system. We shall go into this again in the discussion of the initial value problem in section 16.5.

This theoretical discussion of definitions will not interest the experimental physicist as much as the questions as to whether and how one can produce gravitational waves and demonstrate their existence. Because the gravitational constant is so small there seems no prospect at the present time of producing gravitational waves of measurable intensity by forced motion of masses. The question therefore reduces to whether stars, stellar systems or other objects in our neighbourhood are emitting gravitational waves and with what experimental arrangements one could detect these waves. We shall go into this problem briefly in section 15.4.

Exact solutions describing the interactions between the motion of the sources and the emission of radiation are not known. The considerations of the next few pages almost always deal with an analysis of the local properties of possible solutions. One introduces a local inertial system in the far field or in the neighbourhood of a point and considers (small) deviations from the Minkowski metric caused by the space curvature which have wavelike character. Even when, in section 15.3, we discuss exact solutions, we are really dealing with an inadmissible idealization and

15 Gravitational waves

generalization of local properties of the gravitational fields, just as for exactly plane electromagnetic waves in the Maxwell theory, which also of course can only be realized in an approximate fashion (locally).

15.2 Plane gravitational waves in the linearized theory

The waves and their degrees of freedom The simplest solutions of the linearized field equations in matter-free space,

$$\left.\begin{array}{ll} \Box \bar{f}_{mn} = \eta^{ab}\bar{f}_{mn,ab} = 0, & \bar{f}^{mn}{}_{,n} = 0, \\ g_{mn} = \eta_{mn} + f_{mn}, & \bar{f}_{mn} = f_{mn} - \tfrac{1}{2}\eta_{mn} f^a{}_a, \end{array}\right\} \quad (15.1)$$

are the plane, monochromatic gravitational waves

$$\bar{f}_{mn} = \hat{a}_{mn} e^{ik_r x^r}, \quad \hat{a}_{mn} = \text{constant}, \quad k_r k^r = 0, \quad \hat{a}_{nm} k^m = 0, \quad (15.2)$$

from which (in the sense of a Fourier synthesis) all solutions of (15.1) can be obtained by superposition. (In this section indices are again shifted with the flat-space metric η_{mn}.)

The independent components $\hat{a}_{mn}$, ten in number because of the symmetry, are restricted by the four subsidiary conditions $\hat{a}_{mn} k^n = 0$. One might therefore suppose that a plane, monochromatic gravitational wave has six degrees of freedom (generalization of degrees of freedom due to polarization). But the waves (15.2) contain some pure coordinate waves; these are waves whose curvature tensor vanishes identically, so that they can be eliminated by coordinate transformations. It is therefore convenient for many calculations to eliminate these physically meaningless degrees of freedom. To this end we have at our disposal the coordinate transformations (13.9), whose generating functions,

$$b^n(x^m) = -i\hat{b}^n e^{ik_m x^m}, \quad (15.3)$$

satisfy the wave equation

$$\Box b^n = 0, \quad (15.4)$$

and which effect a change of gauge

$$a_{nm} = \hat{a}_{nm} - \hat{b}_n k_m - \hat{b}_m k_n + \eta_{nm} \hat{b}_r k^r. \quad (15.5)$$

The four constants $\hat{b}_n$ can now be chosen so that, in addition to (15.2),

$$a_{4m} = 0 = a_{n4} \quad (15.6)$$

(because $a_{mn} k^n = 0$ these are three additional conditions) and

$$a_\mu{}^\mu = 0 = a_m{}^m \quad (15.7)$$

are satisfied. The remaining two independent components of a_{mn} cannot be transformed away, and therefore are of true physical significance.

The conditions (15.2), (15.6) and (15.7) have an interpretation which can be easily visualized. Let us choose the spatial coordinate system in such a manner that the wave propagates along the z direction, that is, k^r has only the components

$$k^r = (0, 0, \omega/c, \omega/c). \tag{15.8}$$

Then because of (15.2), (15.6) and (15.7) only the amplitudes a_{xx}, a_{xy} and a_{yy} of the matrix a_{mn} are non-zero, and in addition we have $a_{xx} = -a_{yy}$. The gravitational wave is therefore transverse and, corresponding to the two degrees of freedom of the wave, there are two linearly independent polarization states, which when (15.8) holds can be realized, for example, by

$$a_{xx} = -a_{yy}, \quad a_{nm} = 0 \quad \text{otherwise}, \tag{15.9}$$

and

$$a_{xy} = a_{yx}, \quad a_{nm} = 0 \quad \text{otherwise} \tag{15.10}$$

('linear polarization').

The result of this analysis is thus the following. Gravitational waves propagate with the speed of light (k^r is a null vector). They are transverse and possess two degrees of freedom of polarization. In the preferred coordinates (15.8) they have the metric

$$\left. \begin{array}{l} \mathrm{d}s^2 = (1+f_{xx})\mathrm{d}x^2 + 2f_{xy}\mathrm{d}x\,\mathrm{d}y + (1-f_{xx})\mathrm{d}y^2 + \mathrm{d}z^2 - c^2\mathrm{d}t^2, \\ f_{xx} = a_{xx}\cos\left(\dfrac{\omega}{c}z - \omega t + \varphi\right), \quad f_{xy} = a_{xy}\cos\left(\dfrac{\omega}{c}z - \omega t + \psi\right) \end{array} \right\} \tag{15.11}$$

(from now on we always take the real part of the complex quantities f_{mn}).

The curvature tensor of plane gravitational waves Independent of any special gauge the curvature tensor

$$R_{ambn} = \tfrac{1}{2}(f_{an,mb} + f_{mb,an} - f_{mn,ba} - f_{ab,mn}) \tag{15.12}$$

of a plane gravitational wave always has the property

$$R_{ambn}k^n = 0, \tag{15.13}$$

as a result of the relations

$$f_{mn}k^n = \tfrac{1}{2}k_m f_a^a, \quad f_{mn,ab} = -k_a k_b f_{mn} \tag{15.14}$$

which follow from (15.1) and (15.2). The null vector k^n characterizing the wave is an eigenvector of the curvature tensor.

In the special gauge of the metric (15.11) all non-vanishing components of the curvature tensor can be expressed by

15 Gravitational waves

$$R_{\alpha 4\beta 4} = -\frac{1}{2}\frac{d^2 f_{\alpha\beta}}{c^2 dt^2}. \quad (15.15)$$

The motion of test particles in a plane, monochromatic gravitational wave
If one writes down the equation of motion

$$\frac{d^2 x^a}{d\tau^2} + \Gamma^a_{nm}\frac{dx^n}{d\tau}\frac{dx^m}{d\tau} = 0 \quad (15.16)$$

of a test particle in the coordinate system (15.11), then one finds that

$$x^\alpha = \text{constant}, \qquad x^4 = c\tau \quad (15.17)$$

is also a solution of the geodesic equation, because

$$\Gamma^a_{44} = \tfrac{1}{2}\eta^{ab}(2f_{b4,4} - f_{44,b}) = 0. \quad (15.18)$$

Particles initially at rest always remain at the same place; they appear to be completely uninfluenced by the gravitational wave. This initially surprising result becomes comprehensible when we remember that the curvature of the space enters the relative acceleration of two test particles, and the action of the gravitational waves should therefore be detectable in this relative acceleration (and not in the relative positions).

Now which acceleration is measured by an observer at rest at the origin (O, ct) of the spatial coordinate system, who observes a particle which is at rest at the point (x^α, ct)? For the interpretation of his measurement the observer will use not the coordinate system (15.11), but rather a local inertial system which he carries with him,

$$ds^2 = \eta_{\alpha\beta}d\hat{x}^\alpha d\hat{x}^\beta - c^2 dt^2 + \text{higher terms}, \quad (15.19)$$

which arises from (15.11) by the transformation

$$\hat{x}^\alpha = x^\alpha + \tfrac{1}{2}f^\alpha{}_\beta(O, ct)x^\beta. \quad (15.20)$$

In this inertial system the test particle has the time-varying position $\bar{x}^\alpha$, and its acceleration is

$$\frac{d^2\bar{x}^\alpha}{dt^2} = \frac{1}{2}\frac{d^2 f^\alpha{}_\beta(O, ct)}{dt^2}x^\beta = -c^2 R^\alpha{}_{4\beta 4}x^\beta \quad (15.21)$$

(cf. (15.15)).

Since $f_{\alpha\beta}$ has components only in the xy-plane, test particles also are only accelerated relative to one another in this plane, perpendicular to the direction of propagation of the wave. In this physical sense too, the gravitational wave is transverse. Figure 15.2 shows the periodic motion of a ring of test particles under the influence of the linearly polarized wave (15.9).

154   *Linearized gravitation, far fields and gravitational waves*

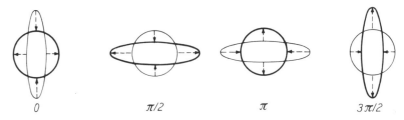

Fig. 15.2. Motion of a ring of test particles in a gravitational wave.

The energy–momentum pseudo-tensor of the plane wave In the case of the plane wave the energy–momentum pseudo-tensor (14.20) has the simple form

$$t^{mn} = \frac{\omega^2}{4} a_{ir}\bar{a}^{ir} k^m k^n \sin^2 k_s x^s. \tag{15.22}$$

Its proportionality to $k^m k^n$ is typical of a plane wave. It is found also for the electromagnetic wave (8.45) and expresses the fact that the total energy flows with the velocity of light, and that there is no static component.

15.3 Plane waves as exact solutions of Einstein's equations

Can one obtain exact solutions to the Einstein equations which have properties similar to those of the plane waves in the linearized theory? Before this question can be answered, it is necessary to characterize the required solutions in a covariant manner and thereby define in what sense one wants to make a generalization.

If one scrutinizes the results of the previous section, then one sees that only (15.13) is a covariant statement. And in fact one could try to start from there and characterize plane waves by demanding that they are solutions of the system

$$R_{mn} = 0, \quad R_{ambn} k^n = 0, \quad k_n k^n = 0. \tag{15.23}$$

However, only the stronger conditions,

$$R_{mn} = 0, \quad k_{a;n} = 0, \quad k_n k^n = 0, \tag{15.24}$$

actually give the restriction to plane waves. (One can convince oneself that (15.23) follows from (15.24)!) The requirement that $k_{a;n} = 0$ generalizes the property of plane waves in flat space, that they possess parallel rays with which are associated a null vector k^n. These waves are therefore called *plane-fronted waves with parallel rays*.

15 Gravitational waves

We shall encounter the more general class of solutions (15.23) again in section 18.

Choice of a suitable coordinate system Since $k_{[a;b]}=0$ holds, the null vector k_a can be written as the gradient of a function u. If we identify u with the coordinate x^4, then we have

$$k_a = u_{,a}, \quad u = x^4, \quad k_a = (0,0,0,1). \tag{15.25}$$

For a plane wave in flat space, which is propagating in the z direction, u is proportional to $ct-z$. Since k_a is a null vector, g^{44} vanishes, and by coordinate transformations $x^{\alpha'} = x^{\alpha'}(x^a)$, $u' = u$ one can arrive at

$$g^{14} = g^{24} = g^{44} = 0, \quad g^{34} = 1, \tag{15.26}$$

and, because $g^{4a}g_{am} = \delta^4{}_m$,

$$g_{31} = g_{32} = g_{33} = 0, \quad g_{34} = 1. \tag{15.27}$$

The reader may verify for himself that this and the following transformations really do exist (existence theorems for partial differential equations) and do not destroy the form of the metric already obtained previously.

The null vector field k_a is covariantly constant, and from this and (15.23)–(15.27) it follows that

$$k_{a;n} = \Gamma^m_{an} k_m = -\tfrac{1}{2} g_{an,3} = 0. \tag{15.28}$$

If we label the coordinates as $x^1 = x$, $x^2 = y$, $x^3 = v$, $x^4 = u$ and introduce conformally Euclidean coordinates into the two-dimensional xy-subspace (which is always possible on a surface) then we arrive at the metric

$$\mathrm{d}s^2 = p^2(x,y,u)(\mathrm{d}x^2 + \mathrm{d}y^2) + 2m_1(x,y,u)\mathrm{d}x\,\mathrm{d}u + 2m_2(x,y,u)\mathrm{d}y\,\mathrm{d}u$$
$$+ 2m_4(x,y,u)\mathrm{d}u^2 + 2\mathrm{d}u\,\mathrm{d}v,$$

$$g_{ab} = \begin{pmatrix} p^2 & 0 & 0 & m_1 \\ 0 & p^2 & 0 & m_2 \\ 0 & 0 & 0 & 1 \\ m_1 & m_2 & 1 & 2m_4 \end{pmatrix},$$

$$g^{ab} = \begin{pmatrix} p^{-2} & 0 & -m_1 p^{-2} & 0 \\ 0 & p^{-2} & -m_2 p^{-2} & 0 \\ -m_1 p^{-2} & -m_2 p^{-2} & -2m_4 + \dfrac{m_1^2 + m_2^2}{p^2} & 1 \\ 0 & 0 & 1 & 0 \end{pmatrix}.$$

$$\tag{15.29}$$

Solution of the field equations   Because we have

$$k_{m;b;n} - k_{m;n;b} = 0 = R^a{}_{mbn} k_a, \tag{15.30}$$

the components $R^4{}_{mbn}$ and R_{3mbn} of the curvature tensor vanish identically, so that the field equations reduce to the five equations

$$R_{mn} = R^1{}_{m1n} + R^2{}_{m2n} = 0. \tag{15.31}$$

Upon substitution of

$$\left.\begin{array}{l}\Gamma^1_{12} = \Gamma^2_{22} = -\Gamma^2_{11} = (\ln p)_{,2}, \quad \Gamma^2_{12} = \Gamma^1_{11} = -\Gamma^1_{22} = (\ln p)_{,1},\\ \Gamma^A_{3B} = 0, \quad \Gamma^4_{AB} = 0, \quad A, B = 1, 2,\end{array}\right\} \tag{15.32}$$

into the defining equation for the curvature tensor,

$$R^a{}_{mhn} = \Gamma^a_{mn,b} - \Gamma^a_{mb,n} + \Gamma^a_{rn}\Gamma^r_{mb} - \Gamma^a_{rb}\Gamma^r_{mn}, \tag{15.33}$$

it follows that $R_{11} = 0$ and $R_{22} = 0$ are equivalent to

$$\Delta(\ln p) = \left(\frac{\partial^2}{\partial x^2} + \frac{\partial^2}{\partial y^2}\right) \ln p = 0. \tag{15.34}$$

$\ln p$ is therefore the real part of an analytic function of $x + iy$, so that we can achieve

$$p = 1, \quad \Gamma^A_{BC} = 0, \quad A, B = 1, 2, \tag{15.35}$$

by a coordinate transformation in the xy-plane.

If we now calculate the components $R^1{}_{412}$ and $R^2{}_{421}$ using (15.35) and

$$\Gamma^a_{3b} = 0, \quad \Gamma^2_{14} = \tfrac{1}{2}(m_{2,1} - m_{1,2}) = -\Gamma^1_{24}, \tag{15.36}$$

then we see that the relation

$$m_{1,2} - m_{2,1} = F'(u) \tag{15.37}$$

follows from the field equations $R_{14} = R_{24} = 0$. With the aid of the coordinate transformations

$$\left.\begin{array}{l}v = \bar v - \int m_1 \,dx + \tfrac{1}{2} F'(u) xy,\\ \bar u = F(u), \quad \bar v = \bar v F'^{-1}(u),\\ \bar x = x \cos u + y \sin u, \quad \bar y = -x \sin u + y \cos u,\end{array}\right\} \tag{15.38}$$

this enables us to introduce the simplified form

$$ds^2 = dx^2 + dy^2 + 2\,du\,dv + H(x, y, u)\,du^2 \tag{15.39}$$

of the metric (the dash on the coordinates will from now on be dropped). The remaining field equation yet to be satisfied, $R_{44} = 0$, then reads

15 Gravitational waves

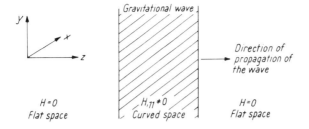

Fig. 15.3. A special wave-packet.

$$\Delta H = \left(\frac{\partial^2}{\partial x^2} + \frac{\partial^2}{\partial y^2}\right) H(x, y, u) = 0. \qquad (15.40)$$

Metrics (15.39) which satisfy this relation are the most general plane-fronted waves with parallel rays.

Properties of plane-fronted waves with parallel rays In the coordinates

$$u = \frac{1}{\sqrt{2}}(z - ct), \qquad v = \frac{1}{\sqrt{2}}(z + ct) \qquad (15.41)$$

flat space has the line element

$$ds^2 = dx^2 + dy^2 + 2 du\, dv. \qquad (15.42)$$

If one compares this expression with the gravitational wave (15.39), then one can see that the wave is plane also in the intuitive sense, that the characteristic function H depends upon the time only in the combination $z - ct$.

The general manifold of solutions also contains special wave-packets which are so constructed that the space before and after passage of the wave is flat (cf. Fig. 15.3).

Since there is no potential function which is regular over the whole xy-plane, H always possesses singularities (the only exception $H = H(u)$ leads to a flat four-dimensional space). To avoid such singularities it is in many cases convenient to use another coordinate system. For the simplest form of a wave

$$H = (x^2 - y^2) h(u), \qquad (15.43)$$

for example, the coordinate transformation

$$\left.\begin{array}{l} x = \bar{x} a(u), \quad y = \bar{y} b(u), \quad v = \bar{v} - \tfrac{1}{2} a' a \bar{x}^2 - \tfrac{1}{2} bb' \bar{y}^2, \\ h(u) = a''/a = -b''/b, \end{array}\right\} \qquad (15.44)$$

leads to the line element (the dash on the coordinates has been suppressed after the transformation)

$$\left.\begin{array}{l} ds^2 = a^2(u)dx^2 + b^2(u)dy^2 + 2du\,dv, \\ a''b + ab'' = 0, \end{array}\right\} \qquad (15.45)$$

with a metric regular over the whole xy-plane.

To end this discussion we shall again compare the exact solution with the plane waves of the linearized theory. If we once more go through the calculations for the derivation of the metric (15.39) of the exact solution, or if we substitute this metric immediately into the field equations, then surprisingly we can establish that no quadratic expressions of any kind in H or its derivatives occur. The exact solution (15.39) is therefore also a solution of the linearized field equations, and it even satisfies the gauge conditions $(\sqrt{-g}\,g^{mn})_{,n} = 0$, but not always of course the requirement $|H| \ll 1$. If we want to compare exact solutions and approximate solutions in detail, then we must linearize the exact solution; in the case of the solution (15.45) this can be done by carrying out the substitution

$$a = 1 + \alpha/2, \quad b = 1 - \beta/2, \quad \alpha \ll 1, \quad \beta \ll 1, \qquad (15.46)$$

and ignoring higher terms in α and β. Because of (15.45) we have $\alpha'' = -\beta''$, and hence $\alpha = -\beta + c_1 u + c_2$, but then c_1 must be zero (the coordinate u can become arbitrarily large!) and c_2 can be eliminated by a coordinate transformation. Therefore the linearized form of (15.45) with (15.43) is

$$ds^2 = [1 + \alpha(u)]dx^2 + [1 + \alpha(u)]dy^2 + 2du\,dv. \qquad (15.47)$$

As a comparison with (15.9)–(15.11) shows, we are dealing with a linearly polarized packet of plane waves of differing frequencies.

15.4 The experimental evidence for gravitational waves

The curvature of space caused by gravitational waves can in principle be detected experimentally by the change in the trajectories of particles or in the states of oscillation of mechanical or electromagnetic systems. Up until now, however, attempts at experimental detection have not been successful.

Only when large masses are rapidly accelerated does one expect that the resulting gravitational waves will be of detectable strengths. Such processes could occur, for example, in gravitational collapse (cf. section 23) or a supernova-like explosion of a star in our galaxy, or a rapidly rotating

15 Gravitational waves

binary stellar system. At present one can only give order-of-magnitude estimates of the intensity of the resultant gravitational waves. For two particles on the earth, they might cause relative displacements h of the order 10^{-17} in the frequency range 1–10^3 Hz, although 10^{-21} might be more likely.

The first experimental search for gravitational waves was initiated by J. Weber (1961). His 'aerial' consisted of an aluminium cylinder 1.53 m long and radius 0.33 m; waves arriving at the cylinder transversally would cause perturbations in its length. Initially a number of 'events' were observed. However, these could not be reproduced, in spite of greatly improved experimental procedures such as sapphire crystal strain gauges and a cryogenic environment. Current technology of this type can measure $h \approx 10^{-18}$, corresponding to length perturbations of 10^{-16} cm, that is, a thousandth of the radius of the nucleus of an atom.

An alternative type of gravitational wave detector relies on laser interferometry, using two or more mirrors to measure length changes. Already (in 1985) similar sensitivities have been achieved.

In principle it should be possible to determine the effect of gravitational waves on the orbits of satellites or planets, and to detect, or establish the non-existence of, ultra low frequency waves.

The very precise data arising from observations of the pulsar $PSR\,1913+16$ seem to have produced an indirect proof of the existence of gravitational waves. This rapidly rotating binary system might be expected to emit appreciable amounts of gravitational quadrupole radiation, and thereby lose energy and hence rotate faster. The observed change in period of $2(\pm 0.22) \times 10^{-12}$ is in remarkable agreement with the theoretical value of 2.4×10^{-12}.

Bibliography to section 15

Monographs and collected works   Breuer (1975), de Witt-Morette (1974), Hawking and Israel (1987), International Conference on Relativistic Theories of Gravitation (1965), Smarr (1979), Weber (1961), Zakharov (1973)

Review and research articles   Bondi, Pirani and Robinson (1959), Braginski and Thorne (1983), Branski and Manukin (1974), Douglas and Braginski (1980), Drever (1984), Jordan, Ehlers, Kundt, Sachs and Trümper (1962), Jordan, Ehlers and Sachs (1961), Kundt (1961), Reasenberg (1983), Schimming (1973), Weber (1980)

16* The Cauchy problem for the Einstein field equations

16.1 *The problem*

The basic physical laws mostly have a structure such that from a knowledge of the present state of a system its future evolution can be determined. In mechanics, for example, the trajectory of a point mass is fixed uniquely by specifying its initial position and initial velocity; in quantum mechanics, the Schrödinger equation determines the future state uniquely from the present value of the ψ function.

As we shall see in the following sections, the equations of the gravitational field also have such a causal structure. In order to appreciate this we must first clarify what we mean by 'present' and 'present state.' As a preliminary to this we examine the properties of a three-dimensional surface in a four-dimensional space. In the later sections we shall concern ourselves with the initial value problem mainly in order to improve our understanding of the structure of the field equations.

16.2 *Three-dimensional hypersurfaces and reduction formulae for the curvature tensor*

Metric and projection tensor Suppose we are given a three-dimensional hypersurface in a four-dimensional Riemannian space which can be imagined as the element of a family of surfaces; the normal vectors n^a to this family of surfaces must not be null:

$$n_a n^a = \epsilon = \pm 1. \tag{16.1}$$

Let us take these surfaces as the coordinate surfaces $x^4 = $ constant of a coordinate system that is not necessarily orthogonal and denote the components of the normal vectors by

$$\left. \begin{array}{l} n_a = (0, 0, 0, \epsilon N), \quad n^a = (-N^\alpha/N, 1/N), \\ a, b, \ldots = 1, \ldots, 4, \quad \alpha, \beta, \ldots = 1, \ldots, 3. \end{array} \right\} \tag{16.2}$$

Then the metric tensor $g_{\alpha\beta}$ of the hypersurface,

$$\overset{(3)}{\mathrm{d}s^2} = g_{\alpha\beta} \mathrm{d}x^\alpha \mathrm{d}x^\beta, \tag{16.3}$$

and the metric tensor g_{ab} of the four-dimensional space are related by

16 The Cauchy problem for the Einstein field equation

$$\overset{(4)}{\mathrm{d}s^2} = g_{ab}\mathrm{d}x^a\mathrm{d}x^b = g_{\alpha\beta}(\mathrm{d}x^\alpha + N^\alpha\mathrm{d}x^4)(\mathrm{d}x^\beta + N^\beta\mathrm{d}x^4) + \epsilon(N\mathrm{d}x^4)^2,$$
(16.4)

from which we obtain for the inverse tensors

$$g^{ab} = \begin{pmatrix} g^{\alpha\beta} + \epsilon N^\alpha N^\beta/N^2 & -\epsilon N^\alpha/N^2 \\ -\epsilon N^\beta/N^2 & \epsilon/N^2 \end{pmatrix}, \quad \left. \begin{matrix} g_{\alpha\beta}g^{\beta\nu} = \delta_\alpha^\nu, \\ N_\alpha = g_{\alpha\beta}N^\beta. \end{matrix} \right\}$$
(16.5)

With the help of the projection tensor $h_{ab} = g_{ab} - \epsilon n_a h_b$, which has the properties

$$h_{ab}h^b{}_c = h_{ac}, \quad h_{ab}n^a = 0, \quad h_{\alpha\beta} = g_{\alpha\beta}, \quad h^{\alpha\beta} = g^{\alpha\beta}, \quad h_b^4 = 0, \quad (16.6)$$

we can decompose every tensor into its components parallel or perpendicular to the vector normal to the hypersurface.

The extrinsic curvature tensor K_{ab} In making the splitting

$$n_{a;b} = n_{a;i}(\epsilon n^i n_b + h_b^i) \qquad (16.7)$$

of the covariant derivative of the normal vector we encounter the tensor K_{ab} defined by

$$K_{ab} = -n_{a;i}h^i{}_b = -n_{a;b} + \epsilon n_a n_b. \qquad (16.8)$$

Since n_a is a unit vector and is proportional to the gradient of a family of surfaces, K_{ab} is symmetric; it has of course no components in the direction of the normal to the surface:

$$K_{ab} = K_{ba}, \quad K_{ab}n^a = 0. \qquad (16.9)$$

Its components are linear combinations of the Christoffel symbols $\overset{4}{\Gamma}{}^4_{\alpha\beta}$ of the four-dimensional space

$$\left. \begin{matrix} K_{ab} = -h_a^i h_b^k n_{i;k} = \epsilon N h_a^\alpha h_b^\beta \overset{4}{\Gamma}{}^4_{\alpha\beta}, \\ K_{\alpha\beta} = \dfrac{1}{2N}(N_{\alpha,\beta} + N_{\beta,\alpha} - 2N_\nu \overset{3}{\Gamma}{}^\nu_{\alpha\beta} - g_{\alpha\beta,4}). \end{matrix} \right\}$$
(16.10)

As one can see from Fig. 16.1, the tensor K_{ab} has a simple geometrical meaning; under a shift of the normal vector along the hypersurface we have

$$Dn_a = n_{a;b}\mathrm{d}x^b = -K_{ab}\mathrm{d}x^b. \qquad (16.11)$$

K_{ab} is therefore a measure of the *extrinsic* curvature of the surface, that is, of the curvature in relation to the surrounding space (in contrast to

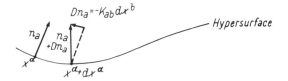

Fig. 16.1. Extrinsic curvature of a hypersurface.

the intrinsic curvature, which is characterized by the three-dimensional curvature tensor $\overset{3}{R}{}_{\alpha\beta\gamma\delta}$ of the surface alone). In the theory of surfaces the tensor K_{ab} is associated with the *second fundamental form*.

Decomposition of the derivative of a vector perpendicular to n_a For the covariant derivative of an arbitrary vector T_a orthogonal to the normal vector n_a which obeys

$$T_a n^a = 0, \qquad T^a = (T^\alpha, 0), \qquad T_a = (T_\alpha, T_\beta N^\beta), \qquad (16.12)$$

one obtains, after a short calculation using (16.9) and $n_{a;b} T^a = -n^a T_{a;b}$, the decomposition

$$T_{a;b} = h_a^i h_b^k T_{i;k} + \epsilon n_b \dot{T}_i h_a^i + \epsilon n_a T^i K_{ib} - n_a n_b T^i \dot{n}_i. \qquad (16.13)$$

Because $h_a^4 = 0$ and

$$h_a^i h_b^k T_{i;k} = h_a^\alpha h_b^\beta (T_{\alpha,\beta} - T_r \overset{4}{\Gamma}{}^r_{\alpha\beta})$$
$$= h_a^\alpha h_b^\beta [T_{\alpha,\beta} - \tfrac{1}{2} T^\rho (g_{\rho\beta,\alpha} + g_{\rho\alpha,\beta} - g_{\alpha\beta,\rho})],$$

the first term of this decomposition, which is wholly orthogonal to the normal vector, can be expressed in terms of the covariant derivative of the three-vector T_α with respect to the three-dimensional metric $g_{\alpha\beta}$:

$$\left. \begin{array}{c} h_a^i h_b^k T_{i;k} = h_a^\alpha h_b^\beta T_{\alpha\|\beta}, \\[4pt] T_{\alpha\|\beta} \equiv T_{\alpha,\beta} - \overset{3}{\Gamma}{}^\rho_{\alpha\beta} T_\rho. \end{array} \right\} \qquad (16.14)$$

In the derivation of this relation only the definition of the covariant derivative and the orthogonality of T_i to n_i have been used, and therefore analogous equations hold for the projections of the derivative of arbitrary tensors of higher rank perpendicular to the normal vector.

Reduction formulae for the curvature tensor The aim of the following calculations is to set up relations between the curvature tensor $\overset{4}{R}{}_{abmn}$ of the four-dimensional space on the one hand, and the properties of the

16 The Cauchy problem for the Einstein field equation

hypersurface, that is, of the three-dimensional curvature tensor $\overset{3}{R}_{\alpha\beta\mu\nu}$ and the quantities n_a and K_{ab}, on the other.

Because of (16.14), (16.13) and the equation

$$h_{ab;i} = -(\dot{n}_a n_b + n_a \dot{n}_b) n_i + \epsilon(K_{ai} n_b + K_{bi} n_a), \tag{16.15}$$

which follows from (16.8), we have

$$(T_{\beta|\mu|\nu} - T_{\beta|\nu|\mu}) h^\beta_b h^\mu_m h^\nu_n$$
$$= (T_{r;s} h^r_i h^s_k)_{;q} h^j_b (h^k_m h^q_n - h^k_n h^q_m)$$
$$= (T_{r;s;q} - T_{r;q;s}) h^r_b h^s_m h^q_n + T_{r;s} (h^r_i h^s_k)_{;q} h^j_b (h^k_m h^q_n - h^k_n h^q_m).$$

or

$$\overset{3}{R}_{\alpha\beta\mu\nu} T^\alpha h^\beta_b h^\mu_m h^\nu_n = \overset{4}{R}_{\alpha rsq} T^\alpha h^r_b h^s_m h^q_n + \epsilon(K_{nb} K_{m\alpha} - K_{bm} K_{n\alpha}) T^\alpha.$$

Since this equation holds for every vector T^α, the relation

$$\overset{4}{R}_{\alpha\beta\mu\nu} = \overset{3}{R}_{\alpha\beta\mu\nu} + \epsilon(K_{\beta\mu} K_{\alpha\nu} - K_{\beta\nu} K_{\mu\alpha}) \tag{16.16}$$

holds between the curvature tensors (remember that $h^\beta_\sigma = \delta^\beta_\sigma$, $h^4_a = 0$). In the theory of surfaces one refers to the analogous relation between the intrinsic and extrinsic curvatures of a surface as Gauss's equation.

We obtain expressions for the remaining components of the four-dimensional curvature tensor by making similar transformation of the second derivatives of the normal vector. From (16.8), (16.14) and (16.15) we have first of all

$$(n_{q;r;s} - n_{q;s;r}) h^q_b h^r_m h^s_n = [(n_{q;i} h^i_r)_{;s} - (n_{q;i} h^i_s)_{;r}] h^q_b h^r_m h^s_n$$
$$= (K_{qs;r} - K_{qr;s}) h^q_b h^r_m h^s_n = (K_{\beta\nu|\mu} - K_{\beta\mu|\nu}) h^\beta_b h^\mu_m h^\nu_n.$$

and from this follows the Codazzi equation:

$$\overset{4}{R}^a{}_{\beta\mu\nu} n_a = K_{\beta\nu|\mu} - K_{\beta\mu|\nu}. \tag{16.17}$$

Analogously, from

$$(n_{q;r;s} - n_{q;s;r}) h^q_b n^r h^s_n$$
$$= [(\epsilon \dot{n}_q n_r - K_{qr})_{;s} - (\epsilon \dot{n}_q n_s - K_{qs})_{;r}] n^r h^q_b h^s_n$$
$$= \dot{n}_{q;s} h^q_b h^s_n + K_{qr} n^r{}_{;s} h^q_b h^s_n - \epsilon \dot{n}_b \dot{n}_n + K_{bn;r} n^r - K_{qs} (h^q_b h^s_n)_{;r} n^r,$$

we obtain finally

$$\overset{4}{R}^a{}_{\beta m\nu} n_a n^m = \dot{n}_{(\beta;\nu)} + K_{\beta\mu} K^\mu{}_\nu - \epsilon \dot{n}_\beta \dot{n}_\nu + \underset{n}{\pounds} K_{\beta\nu}. \tag{16.18}$$

164   Linearized gravitation, far fields and gravitational waves

The reduction formulae (16.16) and (16.18) are frequently used for expressing the curvature tensor of a metric

$$ds^2 = g_{\alpha\beta}dx^\alpha dx^\beta + \epsilon N^2(dx^4)^2, \quad \epsilon = \pm 1, \qquad (16.19)$$

in terms of the three-dimensional subspace (the metric $g_{\alpha\beta}$) and the function N. In the special case ($N^\alpha = 0$) the equations simplify to

$$\left. \begin{aligned} & K_{\alpha\beta} = -g_{\alpha\beta,4}/2N, \\ & \overset{4}{R}{}_{\alpha\beta\mu\nu} = \overset{3}{R}{}_{\alpha\beta\mu\nu} + \epsilon(K_{\beta\mu}K_{\alpha\nu} - K_{\beta\nu}K_{\alpha\mu}), \\ & \overset{4}{R}{}^4{}_{\beta\mu\nu} = \epsilon(K_{\beta\nu\|\mu} - K_{\beta\mu\|\nu})/N, \\ & \overset{4}{R}{}^4{}_{\beta 4\nu} = \epsilon K_{\beta\nu,4}/N - N_{,\beta\|\nu}/N + \epsilon K_{\alpha\beta}K^\alpha{}_\nu. \end{aligned} \right\} \qquad (16.20)$$

16.3 The Cauchy problem for the Einstein vacuum field equations

We are now in a position to be able to answer more precisely the following question. Let a spacelike surface be given; that is, a surface with a timelike normal vector n_a: which initial values of the metric *can* one specify on this surface and which *must* one prescribe in order to able to calculate the subsequent evolution of the system with the aid of the vacuum field equations?

It is clear *ab initio* that, independent of the choice of the initial values, the *metric* of the space-time cannot be determined uniquely; we can carry out arbitrary coordinate transformations on the initial surface as in the whole four-dimensional space. Only certain characteristic geometrical properties will be specifiable which then evolve with time in a way which can be determined. For example, one can show that the quantities N^α of the metric (16.4) must be given not just on the initial surface $x^4 = $ constant, but in the whole space, in order to fix uniquely the metric. To simplify the calculations, in the following we shall start from $N^\alpha = 0$; that is, we shall restrict ourselves to the time-orthogonal coordinates (16.19) with $\epsilon = -1$. The essential results of the analysis of the initial value problem are unaffected by this specialization.

Vacuum field equations   Let us write down the Einstein field equations $R_{bm} = 0$ for the metric (16.19), that is, for

$$ds^2 = g_{\alpha\beta}dx^\alpha dx^\beta - N^2(dx^4)^2,$$

16 The Cauchy problem for the Einstein field equation

using

$$\overset{4}{R}_{bm} = (h^{ra} - n^r n^a) \overset{4}{R}_{abrm} = \overset{4}{R}{}^4{}_{b4m} + g^{av} \overset{4}{R}_{abvm},$$

and the reduction formulae (16.20). Then after making a useful rearrangement, we obtain

$$\left. \begin{array}{l} \overset{4}{R}{}^4_4 - g_{\beta\mu} \overset{4}{R}{}^{\beta\mu} = -\overset{3}{R} - K^\beta_\beta K^\mu_\mu + K_{\beta\mu} K^{\beta\mu} = 0, \\ N \overset{4}{R}{}^4_\mu = K^\beta{}_{\mu\|\beta} - K^\beta{}_{\beta\|\mu} = 0, \end{array} \right\} \quad (16.21)$$

and

$$\overset{4}{R}_{\beta\mu} = \overset{3}{R}_{\beta\mu} - 2K_{\alpha\beta} K^\alpha{}_\mu + K^\alpha_\alpha K_{\beta\mu} - K_{\beta\mu,4}/N - N_{,\beta\|\mu}/N = 0. \quad (16.22)$$

Initial values and dynamical structure of the field equations The Einstein field equations are second-order differential equations; accordingly one would expect to be able to specify the metric and its first derivatives with respect to time (x^4) on an initial surface $x^4 = 0$ and hence calculate the subsequent evolution of the metric with time. As examination of the field equations (16.21) and (16.22) shows, this surmise must be made precise in the following way:

(*a*) In order to be able to calculate the highest time derivatives occurring in equation (16.22), namely, $K_{\beta\mu,4}$, one must know the metric ($g_{\alpha\beta}, N$) and its first time derivatives ($K_{\alpha\beta}$); that is, one must specify these quantities on the hypersurface $x^4 = 0$.

(*b*) The field equations (16.21) contain only spiral derivatives of $g_{\alpha\beta}$ and $K_{\alpha\beta}$, and consequently the initial values $g_{\alpha\beta}(x^\nu, 0)$ and $K_{\alpha\beta}(x^\nu, 0)$ cannot be freely chosen. The equation (16.21) thus play the rôle of subsidiary conditions, limiting the degrees of freedom contained in the initial value data, namely, the intrinsic and the extrinsic curvature of the three-dimensional space.

(*c*) It is not possible to determine the time derivative of N with the aid of the field equations from the initial values $g_{\alpha\beta}(x^\nu, 0)$, $K_{\alpha\beta}(x^\nu, 0)$ and $N(x^\nu, 0)$. Rather, the function $N = \sqrt{-g_{44}}$ must be specified for all times (had we used the N_α of the general form (16.4) of the metric, we would have found the same for them). Since one can always achieve $N = 1$ by coordinate transformations (introduction of Gaussian coordinates), this special rôle played by N becomes understandable: N does not correspond to a true dynamical degree of freedom.

(*d*) If one has specified N and the initial values of $g_{\alpha\beta}$ and $K_{\alpha\beta}$, bearing in mind the four subsidiary conditions (16.21), then from the six field

166   Linearized gravitation, far fields and gravitational waves

equations (16.22) one can calculate the subsequent time evolution of the metric. The equations (16.22) are therefore also called the real dynamical field equations.

The Bianchi identities,

$$\overset{4}{R}{}^a{}_{b;a} = [(h^a_i - n^a n_i)\overset{4}{R}{}^i{}_b]_{;a} = 0, \qquad (16.23)$$

ensure, because of the equation

$$-(n_i \overset{4}{R}{}^i{}_b)_{;a} n^a = \frac{1}{N}(N\overset{4}{R}{}^4{}_b)_{;4} = NK^\nu_\nu \overset{4}{R}{}^4{}_b + (\overset{4}{R}{}^\alpha{}_b)_{;\alpha} \qquad (16.24)$$

that follows from them, that the subsidiary conditions (16.21) are satisfied not just for $x^4 = 0$ but for all times. That is to say, if the dynamical equations $\overset{4}{R}_{\beta\mu} = 0$ are satisfied for all times and if $R^4{}_b = 0$ for $x^4 = 0$, then because of (16.24) the time derivative of $R^4{}_b$ also vanishes (and with it all higher time derivatives), and so $R^4{}_b = 0$ holds always.

The splitting of the field equations into subsidiary conditions and dynamical equations, and the question of which variables of the gravitational field are independent of one another, play an important part in all attempts at quantizing the gravitational field.

16.4 The characteristic initial value problem

From the initial values of the metric and its first derivatives in the direction normal to the surface we were able in principle to calculate the metric in the whole space-time, because the vacuum field equations gave us the *second* derivatives in the direction normal to the surface as functions of the initial value data. In this context it was of only secondary importance that the surface normal was timelike, that is, that ϵ had the value -1.

The situation is completely different, however, if the initial surface $u = x^4 = $ constant is a null surface, that is, a surface whose vector normal $k_a = u_{,a}$ is a null vector. Because

$$k_a = u_{,a} = (0, 0, 0, 1), \qquad k_a k^a = g^{ab} k_a k_b = 0 \qquad (16.25)$$

we have $g^{44} = 0$ and furthermore, by the coordinate transformation $x^{\nu'} = x^\nu(x^\alpha, u)$, $u' = u$, we can also achieve $g^{41} = 0 = g^{42}$. In these preferred coordinates we therefore have (remembering that $k^a = (0, 0, k^3, 0)$ and $g_{ab} g^{bi} = \delta^i_a$)

$$g^{44} = g^{41} = g^{42} = 0, \qquad g_{31} = g_{32} = g_{33} = 0, \qquad g_{34} g^{34} = 1. \qquad (16.26)$$

16 The Cauchy problem for the Einstein field equation

Since second derivatives enter the curvature tensor only in the combination

$$\overset{4}{R}_{mabn} = \tfrac{1}{2}(g_{ab,mn} + g_{mn,ab} - g_{an,mb} - g_{mb,an}) + \cdots, \tag{16.27}$$

second derivatives with respect to $u = x^4$ occur only in the field equation

$$\overset{4}{R}_{44} = -\tfrac{1}{2}g^{\mu\nu}g_{\mu\nu,44} + \cdots = 0, \tag{16.28}$$

while the remaining nine field equations,

$$\overset{4}{R}_{4\alpha} = 0, \qquad \overset{4}{R}_{\alpha\beta} = 0, \tag{16.29}$$

contain at most first derivatives with respect to x^4.

Although through the choice (16.26) of coordinates we have more or less eliminated the unphysical degrees of freedom tied up with possible coordinate transformations, the field equations are in no way sufficient to calculate the second derivatives of the metric from the metric and its first derivatives. The *characteristic* initial value problem, that is, the initial value problem for a null hypersurface, differs fundamentally from the usual Cauchy problem, that is, from the initial value problem for a spacelike surface.

We shall not go into details here, but instead just clarify the physical reasons for this difference by reference to the example of plane waves

$$\left.\begin{array}{l}ds^2 = dx^2 + dy^2 + 2\,du\,dv + H\,du^2, \\[4pt] \left(\dfrac{\partial^2}{\partial x^2} + \dfrac{\partial^2}{\partial y^2}\right)H = 0, \quad H_{,n}k^n = \dfrac{\partial H}{\partial v} = 0\end{array}\right\} \tag{16.30}$$

discussed in section 15.3. If for $u = 0$ we were to specify the initial values of $h = g_{44}$ as a function $H(x, y, v, 0)$ which is initially arbitrary, then we would be unable to determine the subsequent behaviour of the function H from these values; nor would the additional specification of derivatives with respect to u change anything. The field equations (16.30) only give conditions for the initial values, the dependence of the metric upon u remaining undetermined. Physically this indeterminacy is connected with the possible occurrence of gravitational shock waves, and thus of waves whose amplitude H is zero outside a finite region of u. For an observer in an inertial system a null surface $u = z - ct = 0$ is of course a two-dimensional surface which moves at the speed of light. A wave-front of a gravitational shock wave which is parallel to this surface will not be noticed on the surface (there is no point of intersection; cf. Fig. 16.2). An observer who knows only the metric on the surface $u = 0$ cannot predict the arrival of the shock wave. A spacelike surface, on the other hand,

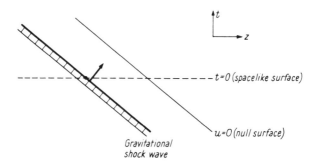

Fig. 16.2. A gravitational shock wave and the characteristic and usual initial value problems.

would intersect the shock wave somewhere; that is to say, from the initial data on such a surface the subsequent course of the wave can be determined (if one knows H on the surface $t = t_0$ for *all* values of z, then H is known as a function of $u = z - ct$).

16.5 Matching conditions at the boundary surface of two metrics

In solving the field equations one is often faced in practice with the problem of joining together two metrics obtained in different regions of space-time; for example, of joining a solution of the field equations $R_{ab} - Rg_{ab}/2 = \kappa T_{ab}$, valid inside a star, with that of the vacuum equations $R_{ab} = 0$, appropriate to the region outside.

Clearly it is not necessary for all components of the energy–momentum tensor to be continuous on the boundary surface. But what continuity properties must the energy–momentum tensor and the metric and its derivatives have in order that one can meaningfully speak of a solution to the Einstein equations?

We shall now deal with this problem under two restrictions: the boundary surface should not be a null surface (where even in the vacuum, pure discontinuities of the metric, that is, gravitational shock waves, can occur), and the energy–monetum tensor can indeed be discontinuous but should contain no δ-function singularities (surface layer structure should not occur). Further, we want to simplify the formulae by using a coordinate system of the form (cf. section 16.2)

$$ds^2 = g_{\alpha\beta} dx^\alpha dx^\beta + \epsilon N^2 (dx^4)^2 \tag{16.31}$$

on both sides of the boundary surface $x^4 = $ constant.

16 The Cauchy problem for the Einstein field equation

We can obtain a qualitative statement about the results to be expected by the following consideration. If certain components of the energy-momentum tensor are discontinuous, then, because of the field equations, the components of the curvature tensor are at worst discontinuous. But if the second derivatives of the metric are at most discontinuous, then the metric and its first derivatives must be continuous.

When making this statement quantitative, one must note that by a clumsy choice of coordinates artificial discontinuities can be produced in the metric. The boundary surface between the two spatial regions I and II should of course be a reasonable surface, that is, whether it be approached from I or II it must always show the same metrical properties. To avoid unnecessary singularities we shall introduce the same coordinate system on both sides of the boundary surface, that is, on this surface we demand that

$$[g_{\alpha\beta}] \equiv \underset{\mathrm{I}}{g_{\alpha\beta}} - \underset{\mathrm{II}}{g_{\alpha\beta}} = 0. \tag{16.32}$$

Clearly all derivatives $g_{\alpha\beta,\nu\mu...}$ of this metric should also be continuous in the surface, particularly the curvature tensor $\overset{3}{R}_{\alpha\beta\mu\nu}$. We can make the function $N = \sqrt{\epsilon g_{44}}$ continuous as well by suitable coordinate transformations, or even transform it to unity: here, however, we shall allow discontinuities, but no singularities.

Further statements about the continuity behaviour of the metric can be obtained from the field equations. As the reduction formulae (16.20) show, second derivatives of the metric in the direction of the surface normal are contained only in the components $\overset{4}{R}{}^4{}_{\beta 4\nu}$ of the curvature tensor; they consequently enter the spatial part of the field equations in the combination

$$G^\alpha_\beta = \frac{\epsilon}{N}(K^\alpha_\beta - \delta^\alpha_\beta K^\nu_\nu)_{,4} + \hat{G}^\alpha_\beta(K_{\mu\nu}, g_{\mu\nu}, g_{\mu\nu,\lambda}, K_{\mu\nu,\lambda}, N, N_{,\lambda}, ...) = \kappa T^\alpha_\beta. \tag{16.33}$$

Since in (16.33) neither T^α_β nor $\hat{G}^\alpha_\beta$ will be singular on the boundary surface, $K^\alpha_\beta - \delta^\alpha_\beta K^\nu_\nu$, and hence $K_{\alpha\beta}$ itself, must be continuous:

$$[K_{\alpha\beta}] = \underset{\mathrm{I}}{K_{\alpha\beta}} - \underset{\mathrm{II}}{K_{\alpha\beta}} = 0. \tag{16.34}$$

While (16.32) ensures the equality of the intrinsic curvature on both sides of the boundary surface, (16.34) demands equality of the extrinsic curvature too.

When the two matching conditions (16.32) and (16.34) are satisfied, then, because of (16.20), $\overset{4}{R}_{\alpha\beta\mu\nu}$ and $N\overset{4}{R}{}^{4}{}_{\beta\mu\nu}$ are also continuous; because of the field equations we must then also have

$$[T^4_4] = 0, \quad [NT^4_\alpha] = 0. \tag{16.35}$$

To summarize: if on the boundary surface $x^4 =$ constant of two metrics of the form (16.31) the energy–momentum tensor is non-singular, then the metric $g_{\alpha\beta}$ and the extrinsic curvature $K_{\alpha\beta} = -g_{\alpha\beta,4}/2N$ of the surface, as well as the components $T^4_4 = \epsilon n_a n^b T^a_b$ and $NT^4_\alpha = \epsilon n_a T^a_\alpha$ of the energy–momentum tensor, must all be continuous there. While a possible discontinuity of $N = \sqrt{\epsilon g_{44}}$ can be eliminated by a coordinate transformation, $T_{\alpha\beta}$ can be completely discontinuous; although we must of course have

$$[G^\alpha_\beta] = \kappa[T^\alpha_\beta]. \tag{16.36}$$

When in section 11.3 we joined together the interior and exterior Schwarzchild solutions, we satisfied (16.32) by requiring continuity of the metric and (16.35) by the condition $p = 0$; the matching conditions (16.34) are then automatically satisfied. N also turns out to be continuous in this case, whilst $N_{,4}$ (note that $r = x^4$!) is discontinuous.

Bibliography to section 16

Textbooks   Eisenhart (1949)

Monographs and collected works   Hawking and Ellis (1973), Petrov (1964), Treder (1962), Zakharov (1973)

Review and research articles   Bruhat (1962), Choquet-Bruhat and York (1980), Dautcourt (1963), Fischer and Marsden (1979), Sachs (1963)

Invariant characterization of exact solutions

Suppose that a solution of the Einstein equations is offered to a theoretical physicist with the request that he test it and establish whether it is already known, what physical situation it describes, what symmetries are present, and so on. Because of the freedom in the choice of coordinate system, such questions cannot usually be answered by merely looking at the solution. Thus one only establishes with certainty that the solution

$$ds^2 = dx^2 - x \sin y \, dx \, dy + x^2(\tfrac{5}{4} + \cos y) dy^2$$
$$+ x^2(\tfrac{5}{4} + \cos y - \tfrac{1}{4} \sin^2 y) \sin^2 y \, dt^2 - dz^2$$

describes flat Minkowski space (in inappropriate coordinates) by determining the curvature tensor. There exists, however, a series of methods for characterizing solutions invariantly (independently of the choice of coordinates), by means of which it has been possible to provide insight into the structure of solutions and hence often find ways of obtaining new solutions.

These methods, the most important of which we shall discuss in the following chapters, are at first sight of a purely mathematical nature. But, as often in theoretical physics, understanding of the mathematical structure simultaneously makes possible a deeper insight into the physical properties.

17 Preferred vector fields and their properties

17.1 *Special simple vector fields*

With many problems and solutions in general relativity, preferred vector fields occur. Their origin may be of a more physical nature (velocity field of a matter distribution, direction of light rays) or of a more mathematical nature (eigenvectors of the Weyl tensor, Killing vectors). One can use a knowledge of the properties of such vector fields for the purpose of classifying solutions or to simplify calculations by the introduction of coordinate systems which are adapted to the preferred vector field. We shall now discuss some special vector fields and coordinates appropriate to them.

Congruences of world lines The vector fields investigated in the following should have the property that at every point precisely one vector is defined. A family of world lines (congruence of world lines) is equivalent to such a vector field $a^n(x^i)$, its tangent vectors having the direction of a^n and covering the region of space under consideration smoothly and completely. This association is not unique, since not only a^n but also λa^n points in the direction of the tangent.

One obtains an especially simple form of the vectors $a^n(x^i)$ by taking these world lines as coordinate lines (for example, $x^\alpha = $ constant, x^4 variable); the vector field then has the normal form

$$a^n(x^i) = (0, 0, 0, a^4(x^i)). \tag{17.1}$$

By means of a coordinate transformation $x^{4'} = x^{4'}(x^i)$ one can set $a^4 = 1$. If a^n is the four-velocity of the matter, then in (17.1) we are dealing with a comoving coordinate system.

Hypersurface-orthogonal fields A vector field is called hypersurface orthogonal (or rotation-free) if it is possible to construct a family of surfaces $f(x^i) = $ constant across the congruence of world lines in such a way that the world lines, and with them the vectors of the field, are perpendicular to the surfaces (Fig. 17.1).

The vector field a^n must therefore point in the direction of the gradient to the family of surfaces,

$$f_{,n} = \lambda a_n,$$

and hence must also satisfy the equations

$$a_{n;m} - a_{m;n} = \frac{\lambda_{,n}}{\lambda} a_m - \frac{\lambda_{,m}}{\lambda} a_n \tag{17.2}$$

and hence

$$\omega^i \equiv \tfrac{1}{2} \epsilon^{imnr} a_{[m;n]} a_r = 0. \tag{17.3}$$

A vector field a^n can be hypersurface orthogonal only if its rotation ω^i as defined in (17.3) vanishes. One can show that this condition is also sufficient: a vector field is hypersurface orthogonal if (17.3) holds.

17 Preferred vector fields and their properties

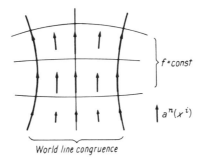

Fig. 17.1. Hypersurface-orthogonal vector field.

While the contravariant components a^n of a vector can always be transformed to the normal form (17.1), a corresponding transformation of the covariant components to the form

$$a_n(x^i) = (0, 0, 0, a_4(x^i)) \tag{17.4}$$

is only possible in a region of space if the vector field is hypersurface orthogonal. One can see this immediately from the fact that (17.4) is equivalent to $a_n = a_4 x^4{}_{,n}$. If one takes the surfaces $f = $ constant as coordinate surface, then provided a^n is not a null vector one can simultaneously with (17.1) and (17.4) bring the metric to the form

$$ds^2 = g_{\alpha\beta} dx^\alpha dx^\beta + g_{44}(dx^4)^2. \tag{17.5}$$

We shall examine the case of the null vector in section 17.3.

Geodesic vector fields A vector field is called geodesic when the world lines $x^i(s)$ of the associated congruence satisfy the geodesic equation:

$$t_{i;n}t^n = 0, \qquad t^i \equiv dx^i/ds. \tag{17.6}$$

Since $t^i = \lambda(x^m) a^i$ should certainly hold, this implies for the vector field a^n that

$$a_{[m}a_{i];n}a^n = 0. \tag{17.7}$$

This condition is also sufficient; that is to say, if it is satisfied, then one can always determine a function λ which, when multiplied by a^i, gives a t^i which satisfies (17.6).

If the vector field is hypersurface orthogonal and geodesic, then because of (17.1) and (17.4) we have in the metric (17.5)

174   *Invariant characterization of exact solutions*

$$a_{\alpha;n}a^n = 0 = \Gamma^4_{\alpha 4} = \tfrac{1}{2}g_{44,\alpha}g^{44}, \quad \alpha = 1, 2, 3;$$

that is, g_{44} depends only upon x^4 and can be brought to the value ± 1 by a coordinate transformation $x^{4\prime} = x^{4\prime}(x^4)$:

$$ds^2 = g_{\alpha\beta}(x^i)dx^\alpha dx^\beta \pm (dx^4)^2. \tag{17.8}$$

Killing vector fields   Killing vector fields are vector fields which satisfy the condition

$$a_{i;n} + a_{n;i} = 0. \tag{17.9}$$

Because of their great importance we shall discuss them in more detail in section 19.

Covariantly constant vector fields   A vector field is covariantly constant if its covariant derivative vanishes:

$$a_{i;n} = 0. \tag{17.10}$$

From the definition (6.4) of the curvature tensor we have immediately

$$a^k R_{kinm} = 0. \tag{17.11}$$

The curvature tensor and with it the metric are restricted if such a vector field exists.

If a^i is not a null vector, then in the metric (17.8) we have

$$a_{\alpha;\beta} = 0 = \Gamma^4_{\alpha\beta};$$

that is, $g_{\alpha\beta}$ is independent of x^4. Because of (16.19), the tensor $K_{\alpha\beta}$ of the extrinsic curvature of the surface $x^4 = $ constant vanishes, and the reduction formulae (16.15) and equation (17.11) lead to

$$\overset{4}{R}_{4\beta\nu\mu} = 0, \quad \overset{3}{R}_{\alpha\beta\nu\mu} = \overset{4}{R}_{\alpha\beta\nu\mu}.$$

For vacuum solutions of the Einstein field equations we have accordingly

$$\overset{4}{R}_{\alpha\beta} = \overset{3}{R}_{\alpha\beta} = 0,$$

and, since the curvature tensor of the three-dimensional subspace can be constructed from its Ricci tensor alone, according to (6.26), then the curvature tensor of the four-dimensional space completely vanishes.

We thus have the law that, if a vacuum solution of the Einstein field equations possesses a covariantly constant vector field, then either we are dealing with a null vector field or else the space is flat.

17 Preferred vector fields and their properties

The vacuum solutions with covariantly constant null vector field are just the plane gravitational waves investigated in section 15.3.

17.2 Timelike vector fields

The invariant decomposition of $u_{m;i}$ and its physical interpretation One of the most important examples of a timelike vector field is the velocity field $u^i(x^n)$ of a matter distribution; for example, that of the matter inside a star or that of the stars or galaxies (imagined distributed continuously) in the Universe. The properties of this velocity field are best recognized by examining the covariant derivative $u_{i;n}$. The idea consists essentially in decomposing that portion of the covariant derivative which is perpendicular to the four-velocity u^i, namely, the quantity $u_{i;n} + u_{i;m}u^m u_n/c^2$ (notice that the relation $u^i u_{i;n} = 0$ follows from $u^i u_i = -c^2$), into its antisymmetric part, its symmetric trace-free part, and the trace itself:

$$\begin{aligned} u_{i;n} &= -\dot{u}_i u_n/c^2 + \omega_{in} + \sigma_{in} + \Theta h_{in}/3, \\ \dot{u}_i &= u_{i;n}u^n = Du_i/D\tau, & \dot{u}_i u^i &= 0, \\ \omega_{in} &= u_{[i;n]} + \dot{u}_{[i} u_{n]}/c^2, & \omega_{in} u^n &= 0, \\ \sigma_{in} &= u_{(i;n)} + \dot{u}_{(i} u_{n)}/c^2 - \Theta h_{in}/3, & \sigma_{in} u^n &= 0, \\ \Theta &= u^i{}_{;i}, \\ h_{in} &= g_{in} + u_i u_n/c^2, & h_{in} u^n &= 0. \end{aligned} \quad (17.12)$$

Since this splitting is covariant, the individual components characterize the flow field invariantly; they have the names

$$\begin{aligned} &\dot{u}_i && \text{acceleration,} \\ &\omega_{in} && \text{rotation (rotational velocity),} \\ &\sigma_{in} && \text{shear (shear velocity),} \\ \text{and} \quad &\Theta && \text{expansion (expansion velocity).} \end{aligned} \quad (17.13)$$

We shall now clarify the physical meaning of these quantities (and thereby also justify the names (17.13)).

The congruence of world lines

$$x^a = x^a(y^\alpha, \tau), \quad (17.14)$$

which is associated with the velocity field

$$u^a(x^i) = \partial x^a/\partial \tau, \quad (17.15)$$

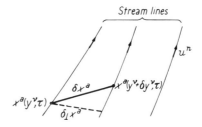

Fig. 17.2. Stream kinematics.

obviously has the physical significance of being a family of streamlines (cf. Fig. 17.2). Along the world line of every particle (every volume element) the y^α are constant and τ varies; y^α labels the different world lines (particles). Keeping the parameter τ fixed one passes from the world line (y^α) to the neighbouring world line $(y^\alpha + \delta y^\alpha)$ on advancing by

$$\delta x^a = \frac{\partial x^a}{\partial y^\alpha} \delta y^\alpha. \tag{17.16}$$

Since we have

$$\frac{D}{D\tau} \delta x^a = \frac{d}{d\tau} \delta x^a + \Gamma^a_{bc} \frac{dx^b}{d\tau} \delta x^c = \frac{\partial^2 x^a}{\partial \tau \partial y^\alpha} \delta y^\alpha + \Gamma^a_{bc} u^b \delta x^c$$

$$= \frac{\partial u^a}{\partial y^\alpha} \delta y^\alpha + \Gamma^a_{bc} u^b \delta x^c = \frac{\partial u^a}{\partial x^n} \delta x^n + \Gamma^a_{bc} u^b \delta x^c,$$

this difference vector changes with advance along the world line by

$$(\delta x^a)^\cdot = u^a_{;n} \delta x^n. \tag{17.17}$$

An observer comoving with the flow, however, will define as displacement to the neighbouring fluid elements not δx^a, but rather the projection of this quantity into his three-dimensional space, that is,

$$\delta_\perp x^a = (g^a_b + u^a u_b/c^2) \delta x^b = h^a_b \delta x^b. \tag{17.18}$$

Since this observer will use as his 'natural,' comoving local coordinate system one whose axes are Fermi–Walker transported (cf. sections 5.4 and 8.2), he will define as the velocity of the neighbouring matter elements the Fermi derivative of $\delta_\perp x^a$. Using (17.17) and (17.18), and remembering that $(\delta_\perp x^a) u_a = 0$, we obtain for this velocity

$$\frac{D}{D\tau}(\delta_\perp x^a) - \frac{1}{c^2}(\delta_\perp x^n)(u^a \dot{u}_n - \dot{u}^a u_n) = (\delta_\perp x^n)^\cdot h^a_n,$$

17 Preferred vector fields and their properties

and from this, with (17.12), finally

$$(\delta_\perp x^n)\dot{}\, h_n^a = (u^a{}_{;n} + \dot{u}^a u_n/c^2)(\delta_\perp x^n)$$
$$= (\omega^a{}_n + \sigma^a{}_n + \Theta \delta_n^a/3)(\delta_\perp x^n). \quad (17.19)$$

Equation (17.19) gives the connection between the velocity $(\delta_\perp x^n)\dot{}\, h_n^a$ of the neighbouring particle to the observer (velocity relative to the observer) and the (infinitesimal) position vector $\delta_\perp x^n$ pointing from the observer to the particle. From it we can deduce the following:

(a) The expansion Θ leads to a radially directed velocity field whose magnitude is independent of direction; a volume element is thereby magnified ($\Theta > 0$) or diminished ($\Theta < 0$) in size with its form preserved.

(b) Since the antisymmetric tensor ω_{mi} of the rotation can be mapped onto the vorticity vector ω^a according to (17.3) through

$$\omega^a = \tfrac{1}{2}\epsilon^{abmi} u_b \omega_{mi}, \qquad \omega_{mi} = \epsilon_{miab}\omega^a u^b \quad (17.20)$$

the velocity field described by it has the form

$$(\delta_\perp x^n)\dot{}\, h_n^a = \epsilon^a{}_{nmi}\omega^m u^i \delta_\perp x^n.$$

The velocity is perpendicular to the position vector $\delta_\perp x^n$ and to the vorticity vector ω^m, and thus we are dealing with a rotation about the axis defined by ω^m.

(c) The symmetric tensor σ_{an} of the shear leads to a direction-dependent velocity field which produces an ellipsoid out of a sphere of particles. Since the trace $\sigma^n{}_n$ vanishes, this ellipsoid has the same volume as the original sphere, and thus we here have a change in shape at constant volume.

Special cases and statements about possible coordinate systems To make calculations one often uses the comoving coordinate system

$$u^i = (0, 0, 0, u^4). \quad (17.21)$$

If the rotation ω_{mi} vanishes, then the flow given by u^i is hypersurface orthogonal and the metric can be brought to the form

$$ds^2 = g_{\alpha\beta}(x^i)dx^\alpha dx^\beta - u_4^2 dt^2, \qquad u_i = (0,0,0,u_4). \quad (17.22)$$

If one writes down the covariant derivative $u_{a;b}$ explicitly in this metric and compares the result with (17.12), then one can show that:

(a) For $\omega_{mi} = 0$ and $\sigma_{mi} = 0$ the metric $g_{\alpha\beta}$ of the three-space contains the time only in a factor common to all elements:

$$g_{\alpha\beta}(x^i) = V^2(x^\nu, t)\bar{g}_{\alpha\beta}(x^\mu). \quad (17.23)$$

(b) For $\omega_{mi}=0$ and $\Theta=0$ the determinant of the three-metric $g_{\alpha\beta}$ does not depend upon the time.

(c) For $\omega_{mi}=0$ and $\dot{u}_i=0$ one can transform u_4 to c.

If the expansion and the shear vanish ($\Theta=0$ and $\sigma_{mi}=0$), but not the rotation ($\omega_{mi}\neq 0$), then for the comoving observer the distances to neighbouring matter elements do not change, and we have a *rigid rotation*. In the comoving coordinate system (17.21) one can see this from the fact that, because

$$u^4 h_{ab,4} = \underset{u}{\pounds}(g_{ab}+u_a u_b/c^2) = u_{a;b}+u_{b;a}+(\dot{u}_a u_b+u_a \dot{u}_b)/c^2 = 0, \quad (17.24)$$

the purely spatial metric h_{ab} does not change with time.

17.3* Null vector fields

Null vector fields $k^n(x^i)$ can be characterized in a similar fashion to timelike vector fields by the components of their covariant derivative $k_{i;n}$. In this case some peculiarities arise from the fact that because $k_n k^n = 0$ one cannot simply decompose a vector, for example, into its components parallel and perpendicular to k^n: if we put $a_n = \lambda k_n + \hat{a}_n$, then λ and $\hat{a}_n$ cannot be uniquely determined. It is therefore preferable to use projections onto a two-dimensional subspace associated with the vector k^n (which is spanned by the vectors t_a and $\bar{t}_a$).

Null tetrads (Sachs tetrads)‡ The system of two real and two complex null vectors introduced according to (4.67) to (4.71), which satisfies the equations

$$\left. \begin{array}{l} k_a l^a = -1, \quad t_a \bar{t}^a = 1, \quad k_a k^a = k_a t^a = l_a l^a = l_a t^a = t_a t^a = 0, \\ g_{ab} = t_a \bar{t}_b + \bar{t}_a t_b - k_a l_b - k_b l_a, \end{array} \right\} \quad (17.25)$$

is called a null tetrad or Sachs tetrad.

The vectors of a null tetrad are of course not uniquely determined by specifying k^n: for a fixed direction k^a the relations (17.25) are preserved under the linear transformations

$$\left. \begin{array}{l} k'^a = A k^a, \\ l'^a = A^{-1} l^a + B\bar{B} k^a + \bar{B} t^a + B \bar{t}^a, \\ t'^a = e^{iC}(t^a + B k^a), \\ A>0, \quad C \text{ real}, \quad B \text{ complex}. \end{array} \right\} \quad (17.26)$$

‡ *Editor's note:* Although the author is historically correct, null tetrads are now often called NP tetrads after Newman and Penrose (1962). The notation has changed with $(k^a, l^a, t^a, \bar{t}^a) \to (l^a, n^a, m^a, \bar{m}^a)$, and there is also a change of signature.

17 Preferred vector fields and their properties

Later we shall also use transformations which link to one another the null tetrads with different direction k^a. One special transformation which converts a null tetrad with a prespecified direction k^a into a null tetrad with direction k'^a is

$$\left.\begin{aligned} k'^a &= k^a - E\bar{E}l^a + \bar{E}t^a + E\bar{t}^a, \\ l'^a &= l^a, \qquad E \text{ complex}, \\ t'^a &= t^a + El^a. \end{aligned}\right\} \tag{17.27}$$

This transformation works except when k'^a lies precisely in the old direction of l^a. In that case we set

$$k'^a = l^a, \qquad l'^a = k^a, \tag{17.28}$$

(exchange of k^a and l^a).

The general transformation between two null tetrads comprises (17.27) or (17.28) and (17.26). Accordingly, it contains six real parameters, and thus precisely the same number as a Lorentz transformation. And indeed we are here dealing with a particularly simple description of the Lorentz transformations; the parameter A on its own produces a special Lorentz transformation (a pseudo-rotation in the xt-plane which is spanned by m^a and k^a), the parameter C a rotation in the $t^a \bar{t}^a - (yz)$-plane, and the parameter B a so-called *null rotation*.

Geodesic null congruences and decomposition of $k_{a;b}$ Light rays and, as we shall see later, also the null vector fields induced by gravitational fields, lead to *geodesic* null vector fields. In the following we shall consider therefore only such fields.

We can describe a family of null geodesics by

$$x^a = x^a(y^\nu, v). \tag{17.29}$$

Here y distinguishes the different geodesics and v is an affine parameter along a fixed geodesic, that is, a parameter under the use of which the tangent vector

$$k^a = \frac{\partial x^a}{\partial v}, \qquad k^a k_a = 0, \tag{17.30}$$

satisfies the equation

$$\dot{k}_a \equiv k_{a;b} k^b = 0. \tag{17.31}$$

The affine parameter v is not determined uniquely by this requirement; a linear transformation

$$v' = A^{-1}(y^\alpha) v + D(y^\alpha) \tag{17.32}$$

is still possible along every geodesic, corresponding to a transformation

$$k'^a = A k^a. \tag{17.33}$$

We now decompose the covariant derivative $k_{i;n}$ of a geodesic null-vector field with the help of the projection tensor

$$p_{ab} = t_a \bar{t}_b + \bar{t}_a t_b = g_{ab} + l_a k_b + l_b k_a \tag{17.34}$$

into the components *in* the plane spanned (locally) by t_a and $\bar{t}_a$ and the perpendicular components. Taking into account (17.30) and (17.31), we obtain

$$k_{i,n} = A_{in} + a_i k_n + k_i b_n \tag{17.35}$$

with

$$\left.\begin{array}{l} A_{in} = k_{a;b} p_i^a p_n^b = 2\,\mathrm{Re}[(\theta + i\omega)\bar{t}_i t_n - \sigma \bar{t}_i \bar{t}_n], \\ a_i k^i = 0, \quad b_i k^i = 0, \quad \theta, \omega \text{ real}, \sigma \text{ complex}. \end{array}\right\} \tag{17.36}$$

In spite of the choice available in the introduction of the vectors t_a and $\bar{t}_a$ (for fixed k^a), the invariants

$$\left.\begin{array}{l} \omega = \sqrt{\tfrac{1}{2} A_{[nr]} A^{nr}} = \sqrt{\tfrac{1}{2} k_{[n;r]} k^{n;r}}, \\ |\sigma| = \sqrt{\tfrac{1}{2}[A_{(nm)} A^{nm} - \tfrac{1}{2}(A_n^n)^2]} = \sqrt{\tfrac{1}{2}[k_{(n;r)} k^{n;r} - \tfrac{1}{2}(k^i{}_{;i})^2]}, \\ \theta = \tfrac{1}{2} A^n{}_n = \tfrac{1}{2} k^i{}_{;i} \end{array}\right\} \tag{17.37}$$

formed from the antisymmetric part, the symmetric trace-free part, and the trace of A_{in}, respectively, characterize the vector field in a unique fashion, since they can be expressed solely in terms of $k_{n;i}$. For a fixed congruence of world lines, under a gauge transformation of the associated null vector field according to (18.33), the invariants (17.37) contain the factor A too.

The physical interpretation of the decomposition of $k_{i;n}$ – the optical scalars θ, ω and σ We fix attention on one element of the family of null geodesics (17.29), which we shall now call light rays, and consider the connecting vector

$$\delta x^a = \frac{\partial x^a(y^\nu, v)}{\partial y^\alpha} \delta y^\alpha \tag{17.38}$$

to neighbouring light rays. The neighbouring light rays clearly form a three-parameter family. From this family we single out a two-parameter family by the condition

$$k_a \delta x^a = 0. \tag{17.39}$$

17 Preferred vector fields and their properties 181

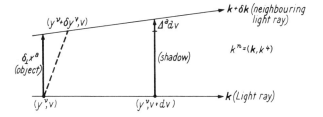

Fig. 17.3. Three-dimensional sketch for interpreting the optical scalars.

Equation (17.39) expresses the fact that δx^a is a spacelike vector. In the rest system of this vector we have, because $\delta x^a = (\delta r, 0)$ and $k^a = (k, k^4)$,

$$k \delta r = 0; \tag{17.40}$$

that is, the three-dimensional light rays of this family are perpendicular to the connecting vector δr. In an arbitrary coordinate system we can define the portion of the connecting vector restricted by (17.39) which is also perpendicular to the light rays by

$$\delta_\perp x^a = p^a_b \delta x^b. \tag{17.41}$$

We shall now calculate how $\delta_\perp x^a$ changes along the light rays, which we can visualize as how the shadow which the light rays throw onto a screen at right-angles to them differs from the 'object' $\delta_\perp x^a$ (cf. Fig. 17.3).

The required quantity

$$\Delta^a = p^a_b (p^b_i \delta x^i)_{;n} k^n \tag{17.42}$$

can be easily calculated by using the equation

$$(\delta x^i)_{;n} k^n = k_{i;n} \delta x^n, \tag{17.43}$$

which follows from

$$(\delta x^i)_{,n} k^n = \frac{\partial \delta x^i}{\partial v} = \frac{\partial^2 x^i}{\partial y^\alpha \partial v} \delta y^\alpha = \frac{\partial k^i}{\partial y^\alpha} \delta y^\alpha = \frac{\partial k^i}{\partial x^n} \delta x^n,$$

remembering the relations (17.25), (17.31), (17.35), (17.36) and (17.39). We obtain

$$\Delta^a = A^a{}_i \delta_\perp x^i = 2\operatorname{Re}[(\theta + i\omega)(\bar{t}_i t^a - \sigma \bar{t}_i \bar{t}^a)] \delta_\perp x^i$$
$$= \theta \delta_\perp x^a + i\omega(\bar{t}_i t^a - t_i \bar{t}^a) \delta_\perp x^i - 2\operatorname{Re}[\sigma \bar{t}_i \bar{t}^a] \delta_\perp x^i. \tag{17.44}$$

The three optical scalars θ, ω and σ can thus be visualized in the following way:

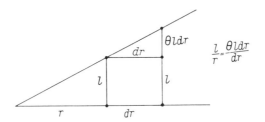

Fig. 17.4. How distance is defined with the aid of θ.

(a) The antisymmetric part of A_{ai}, associated with ω, produces a difference vector Δ^a which is perpendicular to $\delta_\perp x^a$. Since the shadow is then rotated with respect to the object, ω is called the *torsion* or the *rotation* of the light rays. Within the realm of validity of geometrical optics k_a is always hypersurface orthogonal, so that ω vanishes; systems with $\omega \neq 0$ can therefore be realized in a simple manner only by a twisted bundle of light rays.

(b) The symmetric trace-free part of A_{ai}, associated with σ, produces an ellipse of equal area as the shadow of a circle. The shrinking (stretching) of the axes is determined by $|\sigma|$, whilst the direction of the axes of the ellipse follows from the phase of σ. σ is called the *shear* of the null congruence.

(c) The trace of A_{ai}, which is associated with θ, produces a shadow which is diminished or magnified in size with respect to the object independently of direction. θ is therefore called the *expansion* of the light rays. The light rays emitted from a pointlike source of light constitute the standard example of a family with $\theta \neq 0$. Since in flat space we have $\theta = 1/r$ for these rays (cf. Fig. 17.4), one uses θ in curved space to define a parallax distance by

$$\theta = 1/r_P. \tag{17.45}$$

Special cases and appropriate coordinate systems For making calculations with null vector fields one often uses coordinate systems in which

$$k^i = (0, 0, k^3, 0). \tag{17.46}$$

If k_i is hypersurface orthogonal (ω vanishes), then as well as (17.46) one can set

$$k_i = \lambda u_{,i} = (0, 0, 0, k_4)$$

and

$$ds^2 = g_{AB} dx^A dx^B + 2 m_i dx^i du, \quad A, B = 1, 2. \tag{17.47}$$

A comparison of (17.47) with the form $k_i = (0, 0, \lambda, -\lambda)$ of a null vector in Minkowski space shows that $-u$ signifies a retarded time, for example, $u = z - ct$.

For plane waves with $k_{a;b} = 0$ all three optical scalars vanish in agreement with the intuitive interpretation of these quantities.

Bibliography to section 17

Textbooks   Eisenhart (1949)

Review and research articles   Ehlers (1961), Greenberg (1970), Kramer, Stephani, MacCallum and Herlt (1980), Sachs (1963)

18* The Petrov classification

18.1 *What is the Petrov classification?*

The Petrov classification is the classification of Riemannian spaces according to the algebraic properties of the Weyl tensor (conformal curvature tensor) defined by

$$C^{ai}{}_{sq} = R^{ai}{}_{sq} - \tfrac{1}{2}(g^a_s R^i_q + g^i_q R^a_s - g^i_s R^a_q - g^a_q R^i_s) + \tfrac{1}{6} R(g^a_s g^i_q - g^i_s g^a_q). \quad (18.1)$$

From other areas of physics, one knows that algebraic properties of tensors are linked with important physical properties. Thus, for example, in crystal optics the classification of media according to the number of distinct eigenvalues of the ϵ-tensor leads to the division into optically biaxial, uniaxial or isotropic crystals. Therefore we may also hope to find physically interesting relations by investigating the algebraic structure of the curvature tensor.

The examination of the conformal tensor does not suffice of course if one wants to determine all algebraic properties of the curvature tensor. The information lacking is hidden in the Ricci tensor or (because of the field equations) in the energy–momentum tensor. Here, however, we shall restrict our discussion to the Weyl tensor, which anyway coincides with the curvature tensor for vacuum fields: the Petrov classification is the classification of vacuum gravitational fields according to the algebraic properties of the curvature tensor.

We shall begin with the algebraic classification of electromagnetic fields (antisymmetric second-rank tensors), which has many analogies with the Petrov classification, both formal and as regards content, but which is

18.2 The algebraic classification of electromagnetic fields

If, by analogy with the classification of symmetric tensor fields above, one attempts also to classify antisymmetric tensors F_{in} by their eigenvalues λ and eigenvectors s_i,

$$F_{in}s^n = \lambda s_i, \tag{18.2}$$

then one soon sees (by contracting with s^i) that either the eigenvalues vanish or the eigenvectors must be null vectors; see section 4.5. This at once points to the fact that null vectors play a special rôle and suggests their use.

Null tetrads and self-dual bivectors All antisymmetric second-rank tensors (bivectors) can be constructed from the two real and two complex null vectors k_i, l_i, and t_i, $\bar{t}_i$ of a null tetrad (17.25). A special rôle, however, is played by the combinations

$$\left.\begin{aligned} U_{ab} &= \bar{t}_a l_b - \bar{t}_b l_a, \\ V_{ab} &= k_a t_b - k_b t_a, \\ W_{ab} &= t_a \bar{t}_b - t_b \bar{t}_a - k_a l_b + k_b l_a, \end{aligned}\right\} \tag{18.3}$$

of the null tetrad vectors. These bivectors are self-dual in the sense that under dualization they reproduce themselves up to a factor:

$$\left.\begin{aligned} \tilde{U}_{ab} &= \tfrac{1}{2}\epsilon_{abpq}U^{pq} = -\mathrm{i}U_{ab}, \\ \tilde{V}_{ab} &= -\mathrm{i}V_{ab}, \qquad \tilde{W}_{ab} = -\mathrm{i}W_{ab}. \end{aligned}\right\} \tag{18.4}$$

One can verify these relations by using, for example, the fact that the antisymmetric tensor $\epsilon_{abpq}l^p t^q$ yields zero upon contraction with l_a or t_a, and therefore because of (17.25) must be constructed from the vectors l_a and t_a, and that for the vectors of the null tetrad (verifiable in the local Minkowski system $k^a\sqrt{2} = (0,0,1,1)$, $l^a\sqrt{2} = (0,0,-1,1)$, and $t^a\sqrt{2} = (1,\mathrm{i},0,0)$), we have

$$\epsilon_{abpq}k^a l^b t^p \bar{t}^q = -\mathrm{i}. \tag{18.5}$$

Because of their definition (18.3) and the properties (17.25) of null vectors, the self-dual bivectors have the 'scalar products'

$$\left.\begin{aligned} W_{ab}V^{ab} &= W_{ab}U^{ab} = V_{ab}V^{ab} = U_{ab}U^{ab} = 0, \\ W_{ab}W^{ab} &= -4, \qquad U_{ab}V^{ab} = 2. \end{aligned}\right\} \tag{18.6}$$

18 The Petrov classification

Expansion of the field tensor with respect to the self-dual bivectors The non-self-dual field tensor F_{ab} of a Maxwell field cannot of course be expanded in terms of the self-dual bivectors (18.3). One can, however, associate a complex field tensor Φ_{ab} with this real field tensor F_{ab} according to

$$\Phi_{ab} = F_{ab} + i\tilde{F}_{ab} = F_{ab} + \frac{i}{2}\epsilon_{abpq}F^{pq}. \tag{18.7}$$

Because of the relation $\tilde{\tilde{F}}_{ab} = -F_{ab}$ which is valid for every antisymmetric tensor, Φ_{ab} is self-dual in the above sense,

$$\tilde{\Phi}_{ab} = \tilde{F}_{ab} - iF_{ab} = -i\Phi_{ab}, \tag{18.8}$$

and consequently can be expanded with respect to the bivectors U, V and W:

$$\Phi_{ab} = \varphi_0 U_{ab} + \varphi_1 W_{ab} + \varphi_2 V_{ab}. \tag{18.9}$$

The expansion coefficients φ_i can be calculated from the field tensor, because of (18.6), according to

$$\left.\begin{array}{l}\varphi_0 = \tfrac{1}{2}\Phi_{ab}V^{ab} = F_{ab}V^{ab} = B_y - E_x + i(E_y + B_x),\\ \varphi_1 = -\tfrac{1}{4}\Phi_{ab}W^{ab} = -\tfrac{1}{2}F_{ab}W^{ab} = E_z - iB_z,\\ \varphi_2 = \tfrac{1}{2}\Phi_{ab}U^{ab} = F_{ab}U^{ab} = E_x + B_y + i(E_y - B_x),\end{array}\right\} \tag{18.10}$$

(the last equation in each row holds in the local Minkowski system already used many times). Corresponding to the six independent components of a second-rank antisymmetric tensor there occur three complex coefficients.

The classification of electromagnetic fields – first formulation Symmetric tensors become particularly simple when one carries out a transformation to principal axes. Setting up the analogous problem for antisymmetric tensors consists in simplifying the expansion (18.9) by choice of the direction k^a of the null tetrad; that is, by adapting the null tetrad to the antisymmetric tensor under consideration.

Since the self-dual bivectors transform under a rotation (17.27) of the direction k^a, that is, under

$$k'^a = k^a + E\bar{E}l^a + \bar{E}t^a + E\bar{t}^a,$$
$$l'^a = l^a, \qquad t'^a = t^a + El^a,$$

according to

$$\left.\begin{array}{l}W'_{ab} = W_{ab} - 2EU_{ab},\\ V'_{ab} = V_{ab} - EW_{ab} + E^2 U_{ab},\\ U'_{ab} = U_{ab},\end{array}\right\} \tag{18.11}$$

we have for the expansion coefficients

$$\varphi_0 = \varphi'_0 - 2E\varphi'_1 + E^2\varphi'_2, \quad \varphi_1 = \varphi'_1 - E\varphi'_2, \quad \varphi_2 = \varphi'_2. \tag{18.12}$$

We can therefore make one of the coefficients φ_0 or φ_1 vanish by suitable choice of E, that is, by suitable choice of the new direction k^a, and thereby simplify the expansion (18.9). Since only φ_0 remains invariant under the transformations (17.26), which leave k^a fixed but alter l^a and t^a, and since we seek the most invariant choice possible, we demand that φ_0 vanish:

$$\varphi'_0 - 2\varphi'_1 E + \varphi'_2 E^2 = 0. \tag{18.13}$$

According to the number of roots E of this equation (and taking into account a few special cases) one can divide the electromagnetic fields into two classes.

Fields for whose field tensor the inequality

$$\varphi'^2_1 - \varphi'_2 \varphi'_0 \neq 0 \tag{18.14}$$

holds are called non-degenerate. They possess two different directions k^a for which φ_0 vanishes, since either (18.13) has two distinct roots E or (for $\varphi'_2 = 0$) in fact only one root E exists, but l^a is one of the null directions singled out and one likewise obtains $\varphi_0 = 0$ through the exchange of labels $l^a \leftrightarrow -k^a$, $U_{ab} \leftrightarrow \bar{V}_{ab}$.

Fields whose field tensor satisfies

$$\varphi'^2_1 - \varphi'_2 \varphi'_0 = 0 \tag{18.15}$$

are called degenerate fields or null fields. They possess only one null direction k^a with $\varphi_0 = 0$, since either (18.13) has a double root E or (for $\varphi'_2 = \varphi'_1 = 0$) k'^a already leads to $\varphi'_0 = 0$. If one has achieved $\varphi_0 = 0$, then because of (18.15), φ_1 also vanishes.

In the relation

$$\varphi^2_1 - \varphi_2 \varphi_0 = -4\Phi_{ab}\Phi^{ab} = -2(F_{ab}F^{ab} + \mathrm{i}F_{ab}\tilde{F}^{ab}), \tag{18.16}$$

which follows from (18.6) and (18.9), one sees particularly clearly that the classification of electromagnetic fields introduced above is independent of the choice of null tetrad and of the interpretation through the bivector expansion.

Equation (18.16) implies a simple prescription for establishing the type of an electromagnetic field: a Maxwell field is degenerate or a null field if and only if its two invariants vanish, that is if

$$F_{ab}F^{ab} = 0 = F_{ab}\tilde{F}^{ab}. \tag{18.17}$$

The classification of electromagnetic fields – second formulation   One can also translate the classification just set up into the more usual language of eigenvalue equations and eigenvectors. As one can deduce from (18.9) and (18.3), $\varphi_0 = 0$ is equivalent to

$$\Phi_{ab}k^b = (F_{ab} + i\tilde{F}_{ab})k^b = \varphi_1 k_a. \tag{18.18}$$

Non-degenerate fields thus possess two distinct, null eigenvectors k^a for which (18.18) holds or

$$k_{[c}F_{a]b}k^b = 0 = k_{[c}\tilde{F}_{a]b}k^b. \tag{18.19}$$

Degenerate fields (null fields), for which φ_0 and φ_1 vanish, possess only one null eigenvector k^a with

$$F_{ab}k^b = 0 = \tilde{F}_{ab}k^b. \tag{18.20}$$

Its field tensor has the simple structure $\Phi_{ab} = \varphi_2 V_{ab}$, that is,

$$F_{ab} = k_a p_b - k_b p_a, \qquad p_a k^a = 0 = p_a l^a. \tag{18.21}$$

18.3 The physical interpretation of null electromagnetic fields

The simplest example of a null electromagnetic field is a plane wave in Minkowski space

$$\left. \begin{array}{l} A_n = \text{Re}\{\hat{p}_n e^{ik_r x^r}\}, \qquad \hat{p}_n k^n = 0, \qquad k_n k^n = 0, \\ F_{nm} = \text{Re}\{(\hat{p}_m k_n - \hat{p}_n k_m) i e^{ik_r x^r}\}. \end{array} \right\} \tag{18.22}$$

One easily verifies that the necessary and sufficient condition (18.17) that the invariants vanish is satisfied.

Plane waves (null fields) also occur as far fields of isolated charge and current distributions. If one starts from the representation of the four-potential in terms of sources,

$$A_n(x^\alpha, t) = \frac{1}{4\pi c} \int \frac{j_n(y^\alpha, t - r'/c)}{r'} d^3y, \qquad r'^2 = (x^\alpha - y^\alpha)(x_\alpha - y_\alpha), \tag{18.23}$$

and expands the corresponding field tensor in powers of $1/r$,

$$F_{nm} = \overset{1}{F}_{nm}/r + \overset{2}{F}_{nm}/r^2 + \cdots, \qquad r^2 = x^\alpha x_\alpha, \tag{18.24}$$

then one sees that $\overset{1}{F}_{mn}$ has the structure

$$\overset{1}{F}_{nm} = p_m k_n - p_n k_m, \tag{18.25}$$

with

$$p_m = -\frac{1}{4\pi c}\frac{\partial}{\partial t}\int j_m(y^\alpha, t-r'/c)\,d^3y,$$

$$k_n = -(t-r'/c)_{,n} = \left(\frac{x_\alpha - y_\alpha}{r'c}, -\frac{1}{c}\right) \approx (x_\alpha/rc, -1/c),$$

$$k_n k^n = 0, \qquad A^n{}_{,n} \approx p^n k_n = 0.$$

(18.26)

The energy-momentum tensor (8.34) of a general null field has the simple form

$$T_{mn} = F^a{}_m F_{an} = p_i p^i k_m k_n.$$

(18.27)

In a local Minkowski system the relation $|S^\nu| = wc$ thus holds between the energy flux density (Poynting vector) $S^\nu = cT^{4\nu}$ and the energy density $w = T^{44}$. The energy flux density is as large as if the whole field energy moves with the velocity of light: null electromagnetic fields are 'pure' radiation fields.

The null eigenvector field $k_a(x^i)$ of a degenerate electromagnetic field has some special properties, which can be derived from the Maxwell equations. For degenerate fields we have of course $\Phi_{ab} = \varphi_2 V_{ab}$ and the field equations read

$$\Phi^{ab}{}_{;b} = [\varphi_2(k^a t^b - k^b t^a)]_{;b}$$
$$= (\varphi_2 k^a)_{;b} t^b + \varphi_2 k^a t^b{}_{;b} - (\varphi_2 k^b)_{;b} t^a - \varphi_2 k^b t^a{}_{;b} = 0.$$

(18.28)

If one contracts this equation with k_a and remembers that $k_a k^a = k^a k_{a;b} = 0$ and $k_a t^a{}_{;b} = -k_{a;b} t^a$, then one obtains

$$k_{a;b} k^b t^a = 0;$$

(18.29)

that is,

$$k_{a;b} k^b = \lambda k_a.$$

(18.30)

As a comparison with (17.7) shows, we are dealing with a geodesic vector field, because of (18.30). Contraction of (18.28) with $\bar{t}_a$ gives

$$k_{a;b} t^a t^b = 0,$$

(18.31)

the shear σ of the world line congruence vanishes.

In summary we can therefore make the following statements. Null fields (degenerate electromagnetic fields) are generalizations of plane waves; they occur as far fields of isolated sources of radiation and can be interpreted locally as pure radiation fields. Their eigenvectors are geodesic and shear-free.

18.4 The algebraic classification of gravitational fields

The expansion of the Weyl tensor in terms of self-dual bivectors With the Weyl tensor C_{arsq} of a gravitational field can be associated the tensor

$$\overset{*}{C}_{arsq} = C_{arsq} + \mathrm{i}\tilde{C}_{arsq} = C_{arsq} + \frac{\mathrm{i}}{2}\epsilon_{sqmn}C_{ar}{}^{mn} \qquad (18.32)$$

which is analogous to the complex field-strength tensor Φ_{ab}. This tensor is clearly self-dual with respect to the last two indices:

$$\overset{*}{\tilde{C}}_{arsq} = \tfrac{1}{2}\epsilon_{sqmn}\overset{*}{C}_{ar}{}^{mn} = -\mathrm{i}\overset{*}{C}_{arsq}. \qquad (18.33)$$

Because of the defining equation (18.1) for the Weyl tensor its contraction vanishes:

$$C^{ar}{}_{aq} = 0. \qquad (18.34)$$

Hence the Weyl tensor (in contrast to the curvature tensor of an arbitrary space) has the property that the dual tensors formed with respect to the first and to the last pairs of indices coincide:

$$\tilde{C}\tilde{}_{arsq} = \tfrac{1}{4}\epsilon_{arik}\epsilon_{sqmn}C^{ikmn}$$
$$= -C_{arsq} \qquad (18.35)$$

(cf. (4.28) and (4.53)). The tensor $\overset{*}{C}_{arsq}$ is therefore automatically also self-dual with respect to the front index pair and can be expanded entirely in terms of the self-dual bivectors (18.3):

$$\overset{*}{C}_{arsq} = \Psi'_0 U_{ar}U_{sq} + \Psi'_1(U_{ar}W_{sq} + U_{sq}W_{ar}) + \Psi'_2(U_{ar}V_{sq} + U_{sq}V_{ar} + W_{ar}W_{sq})$$
$$+ \Psi'_3(V_{ar}W_{sq} + V_{sq}W_{ar}) + \Psi'_4 V_{ar}V_{sq}. \qquad (18.36)$$

In this expansion account has already been taken of the symmetry properties of the Weyl tensor; the ten algebraically independent components are described by the five complex coefficients Ψ'_A.

The original Petrov classification (1969) consists essentially in classifying the types of the self-dual tensor $\overset{*}{C}_{arsq}$ according to the number of eigenbivectors W_{sq} defined by

$$\tfrac{1}{2}\overset{*}{C}_{arsq}Q^{sq} = \lambda Q_{ar}, \qquad \overset{*}{Q}_{ar} = -\mathrm{i}Q_{ar}. \qquad (18.37)$$

Here we shall take a rather different approach and therefore state the Petrov result without proof. Type *I* (special cases *D*, *O*) occurs when there are three eigenbivectors, type *II* (special case *N*) possesses two eigenbivectors and type *III* only one eigenbivector.

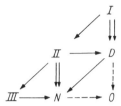

Fig. 18.1. The Penrose diagram.

The classification of gravitational fields – first formulation   By adapting the null tetrad, and along with it also for the Weyl tensor under investigation the self-dual bivectors, one can simplify the expansion (18.36) and set $\Psi_0 = 0$. To this end, because of (17.27) and (18.11), one has to line up the direction k^a (determine E) so that

$$\Psi_0 = \Psi'_0 - 4\Psi'_1 E + 6\Psi'_2 E^2 - 4\Psi'_3 E^3 + \Psi'_4 E^4 = 0. \qquad (18.38)$$

Equation (18.38) has, as an equation of fourth degree, precisely four roots E (this is true with corresponding interpretation also in the special cases: for $\Psi'_4 = 0$ and $\Psi'_0 \neq 0$ one obtains by the exchange of labels $l_a \leftrightarrow -k_a$ an equation of fourth degree, for $\Psi'_4 = \Psi'_0 = 0$ then $E = 0$ is a double root, and so on). To these four roots correspond four directions k^a (eigenvectors k^a) with $\Psi_0 = 0$. According to the multiplicity of these roots one can divide Riemannian spaces into the following types:

Non-degenerate fields:	Type *I*:	four distinct roots,
Degenerate fields:	Type *II*:	one double root and two simple roots,
	Type *D*:	two double roots,
	Type *III*:	one triple root and one simple root,
	Type *N*:	one four-fold root,
	Type *O*:	the Weyl tensor vanishes identically and all the Ψ'_A are zero.

The Penrose diagram (Fig. 18.1) provides a summary of the successive growth in degeneracy; every arrow signifies *one* additional degeneracy.

The classification of gravitational fields – second formulation   It is possible to avoid the detour through the tensor $\overset{*}{C}_{arsq}$ and pick out the types of gravitational fields directly from the Weyl tensor and its null eigenvectors k^a.

18 The Petrov classification

First of all one expresses the coefficients Ψ_A directly in terms of products of the Weyl tensor with the null-tetrad vectors, using (18.32) and (18.4):

$$\left.\begin{aligned}\Psi_0 &= \tfrac{1}{8}C_{arsq}V^{ar}V^{sq} = \tfrac{1}{4}C_{arsq}V^{ar}V^{sq} = C_{arsq}k^a t^r k^s t^q,\\ \Psi_1 &= -\tfrac{1}{16}C_{arsq}V^{ar}W^{sq} = C_{arsq}k^a t^r k^s l^q,\\ \Psi_2 &= \tfrac{1}{8}C_{arsq}U^{ar}W^{sq} = -C_{arsq}k^a t^r l^s \bar{t}^q,\\ \Psi_3 &= -\tfrac{1}{16}C_{arsq}U^{ar}W^{sq} = C_{arsq}l^a \bar{t}^r l^s k^q,\\ \Psi_4 &= \tfrac{1}{8}C_{arsq}U^{ar}U^{sq} = C_{arsq}l^a \bar{t}^r l^s \bar{t}^q.\end{aligned}\right\} \quad (18.39)$$

Ψ_0 vanishes for null eigenvectors k^a; the real symmetric tensor

$$S_{aq} = C_{arsq}k^r k^s \quad (18.40)$$

therefore contains no terms proportional to $t^r t^q$ and $\bar{t}^r \bar{t}^q$. Further, as a consequence of the symmetry properties of the Weyl tensor, the relations

$$S_{aq}k^q = 0, \qquad S^a_a = C^a{}_{rsa}k^r k^s = 0 \quad (18.41)$$

hold, and therefore it has the structure

$$S_{aq} = \alpha k_a k_q + \operatorname{Re}[\beta(k_a \bar{t}_q + k_q \bar{t}_a)]. \quad (18.42)$$

Eigenvectors of the Weyl tensor therefore have the property

$$k_{[b}C_{a]rs[q}k_{n]}k^r k^s = 0 \leftrightarrow \Psi_0 = 0. \quad (18.43)$$

If two eigenvectors coincide ($E=0$ is a double root of (18.38)), then Ψ_1 must also vanish, besides Ψ_0. Because of (18.39), it then follows that $\beta = 0$, and hence that

$$k_{[b}C_{a]rsq}k^r k^s = 0 \leftrightarrow \Psi_0 = \Psi_1 = 0. \quad (18.44)$$

By pursuing these considerations further one finally arrives at the results presented in Table 18.1 (in each case the last two columns hold for the null eigenvectors of higher degeneracy).

By rotations of the null tetrad (for fixed k^a) one can additionally set $\Psi_2 = 0$ for type I, $\Psi_3 = 0$ for type II, $\Psi_3 = \Psi_4 = 0$ for type D (for which $3\Psi_2\Psi_4 = 2\Psi_3^2$ always holds), and $\Psi_4 = 0$ for type III.

In order to determine the Petrov type of a given space one must first compute the Weyl tensor and hence, using (18.39) with an arbitrary null tetrad, the Ψ'_As. Equation (18.38) then gives the possible transformations E and thus the multiplicity of the null eigenvectors.

Table 18.1: *The Petrov types of Riemannian spaces and their properties*

Type	Multiplicity of the null eigenvectors	Vanishing coefficients	Criterion satisfied by C_{abcd}
I	(1,1,1,1)	Ψ_0	$k_{[b}C_{a]rs[q}k_{n]}k^r k^s = 0$
II	(2,1,1)	Ψ_0, Ψ_1	$k_{[b}C_{a]rsq}k^r k^s = 0$
D	(2,2)		
III	(3,1)	Ψ_0, Ψ_1, Ψ_2	$k_{[b}C_{a]rsq}k^r = 0$
N	(4)	$\Psi_0, \Psi_1, \Psi_2, \Psi_3$	$C_{arsq}k^a = 0$
O	no preferred null eigenvectors	all Ψ_A	$C_{arsq} = 0$

18.5 The physical interpretation of degenerate vacuum gravitational fields

For vacuum solutions of the Einstein field equations the Weyl tensor and the curvature tensor are identical, so that all statements in the previous section also hold for the curvature tensor.

The simplest example of a degenerate vacuum solution is provided by the plane gravitational waves dealt with in section 15.3. Because

$$R_{abmn}k^n = 0, \quad k^n k_n = 0, \tag{18.45}$$

they are of type N. One might therefore suppose that degenerate vacuum solutions are connected with gravitational radiation and gravitational waves. In fact one can show that the curvature tensor of an isolated matter distribution allows, under certain assumptions about the sources, at large distance an expansion

$$R_{abmn} = \frac{N_{abmn}}{r} + \frac{III_{abmn}}{r^2} + \frac{D_{abmn}}{r^3} + \cdots, \tag{18.46}$$

where the symbols N, III, and D refer to tensors of the respective algebraic types. The far field of every source of gravitational radiation (if such a field exists) is therefore a plane wave (type N) locally; if one approaches closer to the source, then the four initially coincident directions of the null eigenvectors separate (*peeling theorem*). Unfortunately this result is not so fruitful as the corresponding one in electrodynamics, because the

18 The Petrov classification

relation between N_{abmn}, III_{abmn}, and D_{abmn} and the properties of the sources of the field are not known.

Two simple properties of the null eigenvector fields of degenerate vacuum fields can be deduced from the Bianchi identities, which it is best to use here in the forms

$$\tilde{R}_{ar}{}^{sq}{}_{;q} = 0 \tag{18.47}$$

and since $R_{ab} = 0$,

$$\overset{*}{R}{}^{arsq}{}_{;q} = 0. \tag{18.48}$$

We shall show this explicitly for type II and type D ($\Psi_0 = \Psi_1 = 0$, $\Psi_2 \neq 0$); the calculations for the other types proceed analogously.

Because

$$V_{ab;q} W^{ab} = 4 k_{a;q} t^a, \tag{18.49}$$

we have from (18.48) and (18.36)

$$\begin{aligned} 0 = \overset{*}{R}{}^{arsq}{}_{;q} V_{ar} &= (\overset{*}{R}{}^{arsq} V_{ar})_{;q} - \overset{*}{R}{}^{arsq} V_{ar;q} \\ &= 4(\Psi_2 V^{sq})_{;q} - 8\Psi_3 V^{sq} k_{a;q} t^a - 2\Psi_2 U^{ar} V^{sq} V_{ar;q} - 8\Psi_2 W^{sq} k_{a;q} t^a. \end{aligned} \tag{18.50}$$

Contraction with k_s yields

$$V^{sq}{}_{;q} k_s + 2 k_{a;q} t^a k^q = k_{a;q} t^a k^q = 0; \tag{18.51}$$

that is, the vector field k^a is geodesic (cf. section 18.3). Contraction of (18.50) with t^s leads to

$$V^{sq}{}_{;q} t^s + 2 k_{s;q} t^s t^q = 3 k_{s;q} t^s t^q = 0; \tag{18.52}$$

that is, the vector field is also shear-free ($\sigma = 0$).

The (multiple) eigenvectors of degenerate vacuum solutions thus form a geodesic, shear-free congruence of world lines (as also do the eigenvectors of degenerate electromagnetic fields). The converse of this statement holds as well (Goldberg–Sachs theorem): if in a vacuum solution a shear-free, geodesic null congruence exists, then this solution is degenerate and the congruence is a (multiple) eigencongruence. This law is often used for determining the Petrov type, since then a statement can be made from knowledge of first derivatives ($k_{a;n}$) alone, while the type is generally only determinable from the curvature tensor (second derivatives).

As an example of this let us consider the Schwarzschild metric

$$\left.\begin{aligned} \mathrm{d}s^2 &= e^\lambda \mathrm{d}r^2 + r^2(\mathrm{d}\vartheta^2 + \sin^2\vartheta \, \mathrm{d}\varphi^2) - e^{-\lambda} c^2 \mathrm{d}t^2, \\ e^{-\lambda} &= 1 - 2M/r. \end{aligned}\right\} \tag{18.53}$$

A null tetrad suitable for the coordinate system and for the solution is formed by the vectors

$$\sqrt{2}k_a = (e^{\lambda/2}, 0, 0, e^{-\lambda/2}),$$
$$-\sqrt{2}l_a = -e^{\lambda/2}, 0, 0, -e^{-\lambda/2}),$$
and $$\sqrt{2}t_a = (0, r, -ir\sin\vartheta, 0),$$
(18.54)

for which one can easily verify the relation (17.25). Since of the partial derivatives of k_a only $k_{1,1}$ and $k_{4,1}$ are non-zero, we consequently have (using the Christoffel symbols (10.7) with $\lambda = -\nu$ and $\dot{\lambda} = 0$)

$$k_{a;q}t^a k^q = -\Gamma^c_{aq}t^a k^q k_c = 0,$$
$$k_{a;q}t^a t^q = -\Gamma^c_{aq}t^a t^q k_c = \Gamma^1_{22}t^2 t^2 k_1 + \Gamma^1_{33}t^3 t^3 k_1 = 0,$$
(18.55)

and we find the same relations for l_a as well: k_a and l_a always form a geodesic, shear-free null congruence and thus both are double null eigendirections of a degenerate solution. But the existence of two distinct, degenerate null eigenvectors implies that the Schwarzschild solution is of type D.

In this example one sees the advantages of using null vectors, but also the limitations in interpreting degenerate solutions as gravitational waves: the static Schwarzschild solution can only with difficulty be interpreted as a radiation field (standing, frozen-in wave).

Bibliography to section 18

Textbooks Synge (1965)

Monographs and collected works Petrov (1969), Zakharov (1973)

Review and research articles Ehlers (1971), Jordan, Ehlers and Kundt (1960), Jordan, Ehlers and Sachs (1961), Sachs (1963), Synge (1964)

19 Killing vectors and groups of motion

19.1 The problem

When we are handling physical problems, symmetric systems have not only the advantage of a certain simplicity, or even beauty, but also special physical effects frequently occur then. One can therefore expect in general relativity, too, that when a high degree of symmetry is present the field

19 Killing vectors and groups of motion

equations are easier to solve and that the resulting solutions possess special properties.

Our first problem is to define what we mean by a symmetry of a Riemannian space. The mere impression of simplicity which a metric might give is not of course on its own sufficient; thus, for example, the relatively complicated metric on page 171 in fact has more symmetries than the 'simple' plane wave (15.39). Rather, we must define a symmetry in a manner independent of the coordinate system. Here we shall restrict ourselves to continuous symmetries, ignoring discrete symmetry operations (for example, space reflections).

19.2 Killing vectors

The symmetry of a system in Minkowski space or in three-dimensional (Euclidean) space is expressed through the fact that under translation along certain lines or over certain surfaces (spherical surfaces, for example, in the case of spherical symmetry) the physical variables do not change. One can carry over this intuitive idea to Riemannian spaces and ascribe a symmetry to the space if there exists an s-dimensional ($1 \leq s \leq 4$) manifold of points in it which are physically equivalent: under a symmetry operation, that is, a motion which takes these points into one another, the metric does not change.

These ideas are made more precise by imagining a vector $\xi^i(x^a)$ at every point x^a of the space and asking for the conditions under which the metric does not change under a translation in the direction ξ^i. Since every finite motion can be constructed from infinitesimal motions, it is sufficient to ensure the invariance of the metric under the infinitesimal motion

$$\bar{x}^a = x^a + \xi^a(x^n)d\lambda = x^a + \delta x^a. \tag{19.1}$$

For such a transformation we have

$$\left. \begin{array}{l} \delta g_{ab} = g_{ab,n}\xi^n d\lambda, \\ \delta(dx^a) = d(\delta x^a) = \xi^a{}_{,n} dx^n d\lambda, \end{array} \right\} \tag{19.2}$$

so that the line elements at the point x^a and at the neighbouring point $\bar{x}^a$ are identical only if

$$\delta(ds^2) = \delta(g_{ab}dx^a dx^b)$$
$$= (g_{ab,n}\xi^n + g_{nb}\xi^n{}_{,a} + g_{an}\xi^n{}_{,b})dx^a dx^b d\lambda = 0. \tag{19.3}$$

A symmetry is present if and only if (19.3) is satisfied independently of the orientation of dx^a, that is, for

$$g_{ab,n}\xi^n + g_{nb}\xi^n{}_{,a} + g_{an}\xi^n{}_{,b} = 0. \tag{19.4}$$

For a given metric, (19.4) is a system of differential equations determining the vector field $\xi^i(x^n)$; if it has no solution, then the space has no symmetry. In spite of the fact that it contains partial derivatives, (19.4) is a covariant characterization of the symmetries present. One can see this by substituting the partial by covariant derivatives or formulating (19.4) with the help of the Lie derivative. (19.4) is equivalent to the equation

$$\xi_{b;a} + \xi_{b;a} = \underset{\xi}{\pounds} g_{ab} = 0, \tag{19.5}$$

which is clearly covariant.

Vectors ξ^i which are solutions of the equations (19.4) or (19.5) are called *Killing vectors*. They characterize the symmetry properties of Riemannian spaces in an invariant fashion (Killing, 1892).

If one chooses the coordinate system so that ξ^n has the normal form

$$\xi^n = (0, 0, 0, 1), \tag{19.6}$$

then (19.4) reduces to

$$\partial g_{ab}/\partial x^4 = 0: \tag{19.7}$$

the metric does not depend upon x^4. It is clear that in (19.4) the alternative definition of symmetry as 'independence of a coordinate' has been covariantly generalized.

The world line congruence associated with the Killing vector field, that is, the family of those curves which link points which can be carried into one another by symmetry operations, is obtained by integration of the equations

$$\frac{dx^n}{d\lambda} = \xi^n(x^i). \tag{19.8}$$

19.3 Killing vectors of some simple spaces

The Killing equations (19.5) constitute a system of first-order, linear differential equations for the Killing vectors $\xi^i(x^n)$; the number and type of solutions of these ten equations are dependent upon the metric and hence vary from space to space. Here we shall first of all determine the Killing vectors explicitly for two simple metrics, and only in the next section deduce some general statements about the diversity of solutions to the Killing equations.

19 Killing vectors and groups of motion

The Killing vectors of Minkowski space can without doubt be obtained most simply in Cartesian coordinates. Since all Christoffel symbols vanish, in these coordinates the Killing equations read

$$\xi_{a,b} + \xi_{b,a} = 0. \tag{19.9}$$

If one combines the equations

$$\xi_{a,bc} + \xi_{b,ac} = 0, \quad \xi_{b,ca} + \xi_{c,ba} = 0, \quad \xi_{c,ab} + \xi_{a,cb} = 0, \tag{19.10}$$

which result from (19.9) by differentiation, then one obtains

$$\xi_{a,bc} = 0, \tag{19.11}$$

with the general solution

$$\xi_a = c_a + \epsilon_{ab} x^b. \tag{19.12}$$

The Killing equations (19.9) are satisfied by (19.12), however, only if

$$\epsilon_{ab} = -\epsilon_{ba}. \tag{19.13}$$

Flat space thus possesses ten linearly independent Killing vectors; the four constants c_a correspond to four translations and the six constants ϵ_{ab} to six generalized rotations (three spatial rotations and three special Lorentz transformations).

One can also obtain relatively quickly the Killing vectors associated with the spherical surface

$$ds^2 = d\vartheta^2 + \sin^2 \vartheta \, d\varphi^2 = (dx^1)^2 + \sin^2 x^1 (dx^2)^2. \tag{19.14}$$

When written out separately the components of the Killing equation (19.4) are then

$$\left.\begin{array}{l} \xi^1_{,1} = 0, \quad \xi^1_{,2} + \sin^2 \vartheta \, \xi^2_{,1} = 0, \\ \xi^1 \cos \vartheta + \sin \vartheta \, \xi^2_{,2} = 0. \end{array}\right\} \tag{19.15}$$

The general solution

$$\xi^1 = A \sin(\varphi + a), \quad \xi^2 = A \cos(\varphi + a) \cot \vartheta + b \tag{19.16}$$

shows that there are three linearly independent Killing vectors, for example, the vectors

$$\left.\begin{array}{l} \underset{1}{\xi^a} = (\sin \varphi, \cos \varphi \cot \vartheta), \\[4pt] \underset{2}{\xi^a} = (\cos \varphi, -\sin \varphi \cot \vartheta), \\[4pt] \underset{3}{\xi^a} = (0, 1). \end{array}\right\} \tag{19.17}$$

19.4 Relations between the curvature tensor and Killing vectors

From the Killing equation (19.5)

$$\xi_{a;b} + \xi_{b;a} = 0$$

and the relation

$$\xi_{a;b;n} - \xi_{a;n;b} = R^m{}_{abn}\xi_m \tag{19.18}$$

valid for every vector, a series of relations can be derived which enable one to make statements about the possible number of Killing vectors in a given space.

Because of the symmetry properties of the curvature tensor, the identity

$$(\xi_{a;b} - \xi_{b;a})_{;n} + (\xi_{n;a} - \xi_{a;n})_{;b} + (\xi_{b;n} - \xi_{n;b})_{;a} = 0 \tag{19.19}$$

follows from (19.18) for every vector, which for Killing vectors yields

$$\xi_{a;b;n} + \xi_{n;a;b} + \xi_{b;n;a} = 0 \tag{19.20}$$

and together with (19.8) and (19.5) leads to

$$\xi_{n;b;a} = R^m{}_{abn}\xi_m. \tag{19.21}$$

This equation shows that from the Killing vector ξ_n and its first derivatives $\xi_{n;a}$ all higher derivatives can be calculated in a given Riemannian space. To determine a Killing vector field uniquely it therefore suffices to specify the values of ξ_n and $\xi_{n;a}$ at one point. Since one must of course at the same time remember that $\xi_{a;n} = -\xi_{n;a}$, then in an N-dimensional Riemannian space there are precisely $N + \binom{N}{2} = N(N+1)/2$ such initial values and, accordingly, a maximum of $N(N+1)/2$ linearly independent Killing vector fields. The physical space ($N=4$) thus has at most ten Killing vectors and, as we shall show, it has ten only if the space is one of constant curvature.

This maximum number cannot always be realized in a given space, since the Killing equations are not necessarily integrable for every combination of these initial values, and there even exist spaces without any symmetry. Thus, for example, from the combination of the equation

$$\xi_{n;b;a;i} - \xi_{n;b;i;a} = R^m{}_{nai}\xi_{m;b} + R^m{}_{bai}\xi_{m;n}, \tag{19.22}$$

19 Killing vectors and groups of motion

which holds for every tensor $\xi_{a;b}$, with (19.21) and the Killing equation, we obtain the relation

$$(R^m{}_{abn;i} - R^m{}_{ibn;a})\xi_m + (R^m{}_{abn}g^k_i - R^m{}_{ibn}g^k_a + R^m{}_{bai}g^k_n - R^m{}_{nai}g^k_b)\xi_{m;k} = 0, \quad (19.23)$$

which further restricts the freedom in specifying ξ_m and $\xi_{m;k}$. From the equations mentioned one can derive an algorithm for determining the number of possible Killing vector fields in a given space. We shall not describe the details here, however, but rather refer the reader to the specialist literature.

It is relatively easy to answer the question of which spaces possess precisely the maximum number $N(N+1)/2$ of Killing vectors. Clearly for such spaces (19.23) must imply no restrictions on the values of ξ_m and $\xi_{m;k}$, and therefore in this case we must have (remember that $\xi_{m;k} + \xi_{k;m} = 0$!)

$$R^m{}_{abp;i} = R^m{}_{ibp;a} \quad (19.24)$$

and

$$R^m{}_{abp}g^k_i - R^k{}_{abp}g^m_i - R^m{}_{ibp}g^k_a + R^k{}_{ibp}g^m_a$$
$$+ R^m{}_{bai}g^k_p - R^k{}_{bai}g^m_p - R^m{}_{pai}g^k_b + R^k{}_{pai}g^m_b = 0. \quad (19.25)$$

By contraction first just over i and k, and then both over i and k and over a and b, one obtains from (19.23) the equations

$$(N-1)R^m{}_{abp} = R_{ap}g^m_b - R_{ab}g^m_p$$

and

$$NR^m_p = Rg^m_p, \quad (19.26)$$

and hence the curvature tensor of such spaces of maximal symmetry has the form

$$R_{mabp} = \frac{R}{N(N-1)}(g_{ap}g_{mb} - g_{ab}g_{mp}). \quad (19.27)$$

The curvature scalar R must be constant, because of (19.24). Surfaces with these properties are called surfaces of constant curvature. The curvature $R/N(N-1)$ can be positive, zero or negative. The quantity

$$K = \sqrt{N(N-1)/|R|} \quad (19.28)$$

is known as the radius of curvature. In these spaces no point and no direction is preferred. They are isotropic and homogeneous. Flat spaces with vanishing curvature tensor are special instances of these spaces.

As one can easily show, a four-dimensional space of constant curvature is not a solution of the vacuum field equations, apart from the trivial case of a Minkowski space. The question of the maximum number of Killing vectors in spaces which correspond to vacuum gravitational fields can be answered in the following way. Vacuum solutions of type I or D have at most four Killing vectors (to this group belongs the Schwarzschild metric, for example, with one timelike Killing vector $\xi^i = (0, 0, 0, 1)$ and the three Killing vectors of the spherical symmetry); solutions of type N have at most six, and solutions of types II and III at most three Killing vectors.

19.5 Groups of motion

Translation in the direction of a Killing vector field can also be interpreted as a mapping of the space onto itself, or as a motion (for example, a rotation) of the space. Since we designate as motions precisely those transformations which do not alter the metric (for which the metric is the same, in a suitable coordinate system, at the initial-point and the end-point of the motion), these transformations form a group.

Groups of motion (Lie groups) are continuous groups whose elements are differentiable functions of a finite number of parameters r. One can imagine the entire group to be generated by repeated application of infinitesimal transformations (19.1) in the direction of the r Killing vectors of the space. These (linearly independent) Killing vectors thus serve as a basis for generating the group. Since every linear combination of Killing vectors is also a Killing vector, this basis is not uniquely determined.

One can characterize a group (and hence a space) by the number of linearly independent Killing vectors and their properties. An intuitive picture of the way in which the group acts is provided by the regions of transitivity, which are those regions of the space whose points can be carried into one another by the symmetry operations of the group. The surfaces of transitivity of the rotation group are spherical surfaces, for example, and the group is multiply transitive on them; that is, there exists more than one transformation which transforms one point into another.

The structure constants of a group of motion The structure of a group which is generated by r Killing vectors can be most clearly recognized if one examines the commutability of infinitesimal motions.

Two infinitesimal motions in the direction of the Killing vectors ξ^a_A and ξ^a_B, namely,

19 Killing vectors and groups of motion

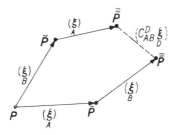

Fig. 19.1. Commuting of infinitesimal motions.

$$\bar{x}^a = x^a + \underset{A}{\xi^a}(x^i)\mathrm{d}\lambda_A + \underset{A}{F^a}(\mathrm{d}\lambda_A)^2 + \ldots \tag{19.29}$$

and

$$\tilde{x}^a = x^a + \underset{B}{\xi^a}(x^i)\mathrm{d}\lambda_B + \underset{B}{F^a}(\mathrm{d}\lambda_B)^2 + \ldots, \tag{19.30}$$

give

$$\begin{aligned}\tilde{\bar{x}}^a &= \bar{x}^a + \underset{B}{\xi^a}(\bar{x}^i)\mathrm{d}\lambda_B + \underset{B}{F^a}(\mathrm{d}\lambda_B)^2 + \ldots\\ &= x^a + \underset{A}{\xi^a}(x^i)\mathrm{d}\lambda_A + \underset{A}{F^a}(\mathrm{d}\lambda_A)^2 + \underset{B}{\xi^a}(x^i)\mathrm{d}\lambda_B\\ &\quad + \underset{B}{\xi^a}_{,n}(x^i)\underset{A}{\xi^n}(x^i)\mathrm{d}\lambda_A\mathrm{d}\lambda_B + \underset{B}{F^a}(\mathrm{d}\lambda_B)^2 + \ldots \end{aligned} \tag{19.31}$$

when performed one after the other. If one performs the transformations in reverse order and then takes the difference of the two results, then only that part of (19.31) antisymmetric in A and B remains:

$$\tilde{\bar{x}}^a - \bar{\tilde{x}}^a = \left(\underset{B}{\xi^a}_{,n}\underset{A}{\xi^n} - \underset{A}{\xi^a}_{,n}\underset{B}{\xi^n}\right)\mathrm{d}\lambda_A\mathrm{d}\lambda_B + \ldots. \tag{19.32}$$

Infinitesimal motions thus commute only to first order; in second order a difference term is left over, according to (19.32). We know, however, that just as the point $\tilde{\bar{P}}$ (coordinates $\tilde{\bar{x}}^a$) is equivalent to the initial point P, so also is the point $\bar{\tilde{P}}$ (coordinates $\bar{\tilde{x}}^a$) (cf. Fig. 19.1); because of the group property of the symmetry transformations it is thus also possible to construct a linear transformation from the Killing vectors which describes the transition from $\bar{\tilde{P}}$ to $\tilde{\bar{P}}$. Because of (19.32) we then have for this transformation

$$\left(\underset{A}{\xi^n}\frac{\partial}{\partial x^n}\underset{B}{\xi^a} - \underset{B}{\xi^n}\frac{\partial}{\partial x^n}\underset{A}{\xi^a}\right) = C^D_{AB}\underset{D}{\xi^a}, \quad A, B, D = 1, \ldots, r. \tag{19.33}$$

The quantities C^D_{AB} are called the *structure constants* of the group; they are independent of the choice of coordinate system, but do depend upon the choice of basis ξ^a_A and can be simplified (brought to certain normal forms) by suitable basis transformations.

Using the operators

$$X_A = \xi^n_A \frac{\partial}{\partial x^n}, \tag{19.34}$$

(19.33) can also be written in the form

$$X_A X_B - X_B X_A = \left[X_A, X_B\right] = C^D_{AB} X_D. \tag{19.35}$$

One can show that for arbitrarily specified structure constants a group always exists, if these constants have the antisymmetry property

$$C^D_{AB} = -C^D_{BA} \tag{19.36}$$

discernible in (19.33), and which satisfy the Lie identity

$$C^E_{AD} C^D_{BC} + C^E_{BD} C^D_{CA} + C^E_{CD} C^D_{AB} = 0, \tag{19.37}$$

which follows from the Jacobi identity

$$\left[X_A, \left[X_B, X_C\right]\right] + \left[X_B, \left[X_C, X_A\right]\right] + \left[X_C, \left[X_A, X_B\right]\right] = 0. \tag{19.38}$$

Examples The group of translations

$$\xi^a_A = \delta^a_A, \quad X_A = \delta^a_A \frac{\partial}{\partial x^a}, \quad a = 1, \ldots, n, \quad A = 1, \ldots, n, \tag{19.39}$$

of an *n*-dimensional flat space is an Abelian group. All its transformations commute, all the structure constants vanish.

The group of rotations of a three-dimensional flat space

$$\begin{aligned}
\xi^a_1 &= (y, -x, 0), & X_1 &= y\frac{\partial}{\partial x} - x\frac{\partial}{\partial y}, \\
\xi^a_2 &= (z, 0, -x), & X_2 &= z\frac{\partial}{\partial x} - x\frac{\partial}{\partial z}, \\
\xi^a_3 &= (0, z, -y), & X_3 &= z\frac{\partial}{\partial y} - y\frac{\partial}{\partial z}
\end{aligned} \tag{19.40}$$

has the commutators

$$\left[X_1, X_2\right] = X_3, \quad \left[X_2, X_3\right] = X_1, \quad \left[X_3, X_1\right] = X_2. \tag{19.41}$$

Since not all the structure constants vanish, but $C^3_{12} = C^1_{23} = C^2_{31} = 1$, rotations do not commute. The reader can verify that in spherical coordinates

19 Killing vectors and groups of motion

the basis (19.17) leads to the same commutators. The operators $\underset{A}{X}$ and their commutators correspond to the angular-momentum operators of quantum mechanics and their commutation rules.

Classification of spaces according to their groups of motion One characterizes the group of motion of a space by the number of its Killing vectors, the structure of the group and the regions of transitivity. Establishing all the non-isomorphic groups G_r of r Killing vectors, that is, of groups whose structure constants cannot be converted into one another by linear transformations of the basis, is a purely mathematical problem of group theory. It is in principle solved: in the literature one can find tables of all such possible groups for the cases of interest in relativity theory.

Thus every group with two elements is either an Abelian group

$$\left[\underset{1}{X},\underset{2}{X}\right]=0, \tag{19.42}$$

or else we have

$$\left[\underset{1}{X},\underset{2}{X}\right]=c_1\underset{1}{X}+c_2\underset{2}{X}, \tag{19.43}$$

with $c_1 \neq 0$. In the second case, however, we can always arrive at the normal form

$$\left[\underset{1}{\bar{X}},\underset{2}{\bar{X}}\right]=\underset{1}{\bar{X}} \tag{19.44}$$

by means of a basis transformation

$$\underset{1}{\bar{X}}=c_1\underset{1}{X}+c_2\underset{2}{X}, \qquad \underset{2}{\bar{X}}=\underset{2}{X}/c_1. \tag{19.45}$$

Equations (19.42) and (19.44) are just the two non-isomorphic groups G_2.

Of especial interest in cosmology are those groups whose regions of transitivity are three-dimensional spaces (homogeneous models of the Universe; all points of the three-dimensional Universe are equivalent). All simply transitive groups G_3 lead to such models. A list of all non-isomorphic groups G_3 can be obtained by using the relation

$$\tfrac{1}{2}\Delta^{ABE}C^D_{AB}=N^{DE}, \qquad A,B=1,2,3, \tag{19.46}$$

which, because of the antisymmetry condition (19.36), maps nine possible components of the matrix of the structure constants C^D_{AB} onto the 3×3 matrix N^{DE}, and then splitting this matrix further into its symmetrical part n^{DE} and the antisymmetric part, which in turn can be mapped onto a 'vector' a^A:

$$N^{DE}=n^{DE}+\Delta^{DEA}a_A. \tag{19.47}$$

Table 19.1: *The structure constants of the groups G_3*

Bianchi Type	I	II	VII	VI	IX	VIII	V	IV	III	VII	VI
a	0	0	0	0	0	0	1	1	1	a	a
n_1	0	0	0	0	1	1	0	0	0	0	0
n_2	0	0	1	1	1	1	0	0	1	1	1
n_3	0	1	1	-1	1	-1	0	1	-1	1	-1

If one substitutes the representation of the structure constants resulting from this decomposition,

$$C_{AB}^D = \Delta_{EAB} n^{DE} + \delta_B^D a_A - \delta_A^D a_B, \tag{19.48}$$

into the Jacobi–Lie identities (19.38), then these reduce to

$$n^{AB} a_A = 0. \tag{19.49}$$

One can always set $a_A = (a, 0, 0)$ by real linear transformations of the basis operators X_A and thus transform n^{AB} to principal axes, so that the diagonal elements have only the values $0, \pm 1$. One hence obtains the following normal form for the commutators and the structure constants of a group G_3:

$$\left.\begin{array}{l} \left[X_1, X_2\right] = n_3 X_3 + a X_2, \\[4pt] \left[X_2, X_3\right] = n_1 X_1, \qquad a n_1 = 0, \\[4pt] \left[X_3, X_1\right] = n_2 X_2 - a X_3, \qquad n_i = 0, \pm 1. \end{array}\right\} \tag{19.50}$$

As Table 19.1 shows, there are eleven types of groups G_3 altogether, which are distributed amongst the nine so-called Bianchi types *I* to *IX*. Notice that in the types *VII* and *VI* for $a \neq 0, 1$ one is always dealing with a whole family of non-isomorphic groups.

We should mention further that there are also groups G_4 and G_6 which are (multiply) transitive in three-dimensional space and can therefore likewise correspond to homogeneous models of the Universe.

19.6 Killing vectors and conservation laws

The conservation laws of physics are closely connected with the symmetry properties of physical systems. In the theory of gravitation, the properties

19 Killing vectors and groups of motion

of the four-dimensional space also have a physical significance. In this section we shall show how symmetry properties (that is, the existence of Killing vector fields) lead to conservation laws or other simple statements.

Mechanics of a point mass   The motion of a point mass on a surface or in Minkowski space in the absence of forces or the pure inertial motion in a Riemannian space (motion in the gravitational field) takes place along a geodesic:

$$\frac{D^2}{D\tau^2}x^a = \frac{D}{D\tau}u^a = u^a{}_{;b}u^b = 0. \tag{19.51}$$

Contraction of this equation with a Killing vector field $\underset{A}{\xi}{}^a$ leads to

$$\underset{A}{\xi}{}_a\frac{D}{D\tau}u^a = \frac{d}{d\tau}\left(\underset{A}{\xi}{}_a u^a\right) - u^a\underset{A}{\xi}{}_{a;b}u^b = 0, \tag{19.52}$$

and taking into account the Killing equation (19.5), that is, the antisymmetry of $\underset{A}{\xi}{}_{a;b}$, we have

$$\underset{A}{\xi}{}_a u^a = \text{constant}. \tag{19.53}$$

The quantities $\underset{A}{\xi}{}_a u^a$ do not change during the motion of the point mass. They are conserved quantities. Thus in mechanics a conservation law is associated with every Killing vector field. In Minkowski space with its ten Killing vectors (19.12) and (19.13) there are accordingly ten conservation laws: the four translational Killing vectors lead to the conservation law for the four-momentum, the three spatial rotations to the angular-momentum law and the three special Lorentz transformations to the centre-of-gravity law.

Interestingly enough, there exist Riemannian spaces in which there are more conservation laws than Killing vectors, that is to say, conservation laws which cannot be traced back to the presence of a symmetry. Thus consider the equation

$$\Xi_{ab;n} + \Xi_{bn;a} + \Xi_{na;b} = 0, \quad \Xi_{na} = \Xi_{an}, \tag{19.54}$$

which we shall take as the defining equations for a *Killing tensor* Ξ_{an}. If these equations possess a solution which is not a linear combination of products of Killing vectors, and which thus *cannot* be written in the form

$$\Xi_{ab} = c_0 g_{ab} + \sum_{A,B} c_{AB}\left(\underset{A}{\xi}{}_a\underset{B}{\xi}{}_b + \underset{A}{\xi}{}_b\underset{B}{\xi}{}_a\right), \quad c_0, c_{AB} = \text{constant}, \tag{19.55}$$

then the conservation laws

$$\frac{D}{D\tau}(\Xi_{ab}u^a u^b) = (\Xi_{ab}u^a u^b)_{;i}u^i = 0 \tag{19.56}$$

which follow from (19.51) and (19.54) are independent of the conservation laws (19.53). One can show that in Minkowski space there exist only the trivial Killing tensors (19.55). An example of a space with a non-trivial Killing tensor is the Kerr metric discussed in section 24.1. Killing tensors reflect symmetries of the (geodesic) differential equations in a space spanned by the variables (x^a, u^a) rather than those of space-time.

If forces are present and these have a potential,

$$\frac{D}{D\tau}u_a = -\Phi_{,a}, \tag{19.57}$$

then the conservation law (19.53) is still valid if the potential does not change under the symmetry operation of the space:

$$\underset{A}{\xi^a}\Phi_{,a} = \underset{A}{X}\Phi = 0. \tag{19.58}$$

The symmetry group of a mechanical (or general physical) system is thus always a subgroup of the symmetry group of the space in which the system is situated.

Scalar potentials in electrodynamics As the Killing equations (19.5) show, a space possesses a Killing vector field if and only if the Lie derivative of the metric in the direction of this vector field vanishes:

$$\underset{\xi}{£}g_{mn} = g_{nm,i}\xi^i + g_{im}\xi^i_{,n} + g_{ni}\xi^i_{,m} = 0. \tag{19.59}$$

We call a physical system in this space invariant under motion in the direction of the Killing vector field if the Lie derivatives of the physical variables vanish. This definition guarantees that the components of the field variables do not change under the motion when one introduces the old coordinate system again at the point reached by the motion (cf. the remarks on the intuitive interpretation of the Lie derivative in section 5.5).

Thus if in a Riemannian space there exists an electromagnetic field (a test field, or a field which acts gravitationally), then this field possesses a symmetry if and only if, in an appropriate gauge, the four-potential satisfies the condition

$$\underset{\xi}{£}A_m = A_{m,i}\xi^i + A_i\xi^i_{,m} = 0. \tag{19.60}$$

If this field tensor is contracted with the Killing vector, then the resulting vector E_m can be written as

19 Killing vectors and groups of motion

$$E_m = F_{mn}\xi^n = (A_{n,m} - A_{m,n})\xi^n = \xi^n A_{n,m} + A_n \xi^n{}_{,m}; \qquad (19.61)$$

that is, E_m can be represented as the gradient of a scalar function Φ:

$$E_m = (\xi^n A_n)_{,m} = -\Phi_{,m}. \qquad (19.62)$$

In the absence of charges and currents, or in a simply connected region outside the sources, or, when the current density vector j^a and the Killing vector ξ^a are parallel, then one can derive an analogous statement for the vector

$$H_m = \tilde{F}_{mn}\xi^n \qquad (19.63)$$

as well. From the Maxwell equations,

$$F^{mn}{}_{;n} = \frac{1}{2}\epsilon^{mn}{}_{ab}\tilde{F}^{ab}{}_{;n} = \frac{1}{c}j^m, \qquad (19.64)$$

we obtain, upon contracting with $\epsilon_{mrst}\xi^t$, the equation

$$(\tilde{F}_{rs,t} + \tilde{F}_{st,r} + \tilde{F}_{tr,s})\xi^t = 0, \qquad (19.65)$$

and since of course the Lie derivative of the dual field tensor vanishes, because of (19.60),

$$\underset{\xi}{\pounds}\tilde{F}_{mn} = \tilde{F}_{mn,a}\xi^a + \tilde{F}_{an}\xi^a{}_{,m} + \tilde{F}_{ma}\xi^a{}_{,n} = 0, \qquad (19.66)$$

H_m satisfies the condition $H_{m,a} = H_{a,m}$ and consequently can be written as the gradient of a potential Ψ:

$$H_m = -\Psi_{,m}. \qquad (19.67)$$

The six quantities E^m and H^m, which (for $\xi_a\xi^a \neq 0$) completely describe the Maxwell field, can thus be represented as gradients of two scalar potentials, if ξ^a is a Killing vector. These potentials are generalizations of the electrostatic and magnetic scalar potentials which one can introduce in Minkowski space when the fields are static; that is, when they do not change under motion in the direction of a timelike hypersurface-orthogonal Killing vector.

Equilibrium condition in thermostatics As we have shown in section 8.5, a system is in thermodynamic equilibrium only if the Lie derivative of the metric in the direction of u^a/T vanishes; that is, if this vector is a Killing vector. It is static if, further, the vector is hypersurface orthogonal. In the rest system $u^a = (0, 0, 0, c/\sqrt{-g_{44}})$ of the matter, the components $g_{4\alpha}$ then vanish, and the metric does not change under time reversal.

Substituting the vector $\xi^a = u^a/T$ into the Killing equations (19.59), we have (when $g_{4\alpha} = 0$)

$$g_{\alpha\beta,4} = 0, \qquad T_{,4} = 0, \qquad (\sqrt{-g_{44}}\, T)_{,\alpha} = 0. \tag{19.68}$$

By means of a transformation of time only, $dt' = \sqrt{-g_{44}}\, T\, dt$, $g_{4'4'} = -1/T^2$, one can convert these equations into

$$g_{mn,4} = 0 \tag{19.69}$$

and

$$(\sqrt{-g_{44}}\, T)_{,i} = 0 \tag{19.70}$$

(we have once again dropped the dash on the indices).

The equations (19.69) and (19.70) are the equilibrium conditions in the rest system of the matter. A system is thus in equilibrium not when the temperature gradient vanishes, but rather when the gradient of $\sqrt{-g_{44}}\, T$ is zero. This condition can be interpreted in the following way: in equilibrium, the change in termperature just compensates the energy which has to be fed in or carried away under (virtual) transport of a volume element in the gravitational field.

Observables in quantum mechanics In the usual coordinate representation of quantum mechanics the operators of momentum and angular momentum associated with the physical observables correspond to the operators X_A of the translations (19.39) and rotations (19.40) of the three-dimensional Euclidean space. There thus exists a close connection between those quantities which remain constant for a more extensive physical system (for example, an atom) because of the symmetry of the space, and those which can meaningfully be used to describe part of a system (for example, an electron). This also clarifies the difficulties involved in carrying over the quantum mechanics of Minkowski space to a general Riemannian space, which of course possesses no Killing vectors at all.

Conservation laws for general fields In a Riemannian space the local conservation law

$$T^{ik}{}_{;k} = 0 \tag{19.71}$$

holds for the energy–momentum tensor of an arbitrary field (an arbitrary matter distribution), but with which no genuine integral conservation law can be associated, because of the non-existence of a Gauss law for tensor fields of second or higher rank.

19 Killing vectors and groups of motion

If, however, a Killing vector field ξ^a_A exists in the space, then it follows from (19.59) and the Killing equation (19.5) that

$$\left(\xi_{A\,i} T^{ik}\right)_{;k} = \xi_{A\,i;k} T^{ik} + \xi_{A\,i} T^{ik}{}_{;k} = 0, \tag{19.72}$$

and the Gauss law can be applied to this local conservation law for a vector field (cf. section 7.5). Under certain mathematical assumptions a conservation law

$$T_A = \int_{x^4=\text{const}} T^{ia}\xi_{A\,i}\,\mathrm{d}f_a = \int_{x^4=\text{const}} T^{i4}\xi_{A\,i}\sqrt{-g}\,\mathrm{d}x^1\,\mathrm{d}x^2\,\mathrm{d}x^3 = \text{constant} \tag{19.73}$$

can then be associated with every Killing vector field of the space. If the Killing vector is timelike, the associated conserved quantity will be called energy. Whether for a spacelike Killing vector one uses the label 'momentum' or 'angular momentum' is sometimes only a matter of definition. In such a case one can be guided by the transitivity properties of the group of motion (the three translations in flat space yield a transitive group, the three spatial rotations are intransitive), by the commutators of the associated operators X_A or by the structure of the Killing vectors in the asymptotically flat far-field zone.

Starting from the identity

$$(\xi^{a;b} - \xi^{b;a})_{;a;b} = 0, \tag{19.74}$$

valid for all vectors, one can recast the conservation law (19.72) in a different form (Komar, 1959). If ξ^a is a Killing vector then this identity, (19.5) and (19.21) imply

$$(R^{mb}\xi_m)_{;b} = 0 = [(T^{mb} - \tfrac{1}{2}g^{mb}T)\xi_m]_{;b}, \tag{19.75}$$

which agrees with (19.72) since $T_{,m}\xi^m = 0$.

From the standpoint of the symmetry properties of a field and of the connection between symmetries and conservation laws, one would therefore answer the question, discussed in detail in section 14.5, of the validity of an energy law for and in a gravitational field in the following way. The energy of a gravitating system can only be defined if a timelike Killing vector exists. Furthermore, if it does exist, there is always an energy conservation law.

Bibliography to section 19

Textbooks   Eisenhart (1933), Weinberg (1972)

Monographs and collected works Abstracts of the 5th International Conference on Gravitation and the Theory of Relativity (1968), Neugebauer (1980), Petrov (1969)

Review and research articles Komar (1961)

20* The embedding of Riemannian spaces in flat spaces of higher dimension

In the history of mathematics a long process of development was needed before it was possible to advance from the idea of surfaces as structures in three-dimensional Euclidean space to the idea of the intrinsic geometry of a surface, and later of a space as well. In the Einstein theory of gravitation the object of physics is to investigate the *intrinsic* geometry of space. Our four-dimensional curved space-time (coordinates x^n) can also, however, be represented as a hypersurface in a flat space (coordinates y^A) of appropriate dimension: it can be embedded in such a higher-dimensional space. This space serves only to mediate study of the properties of the four-dimensional space; it has no inherent physical significance.

One speaks of a local isometric embedding of a Riemannian space V_4 with metric

$$ds^2 = g_{nm} dx^n dx^m \tag{20.1}$$

in a flat space

$$ds^2 = \sum_{A=1}^{N} e_A (dy^A)^2, \quad e_A = \pm 1, \tag{20.2}$$

of dimension N if a hypersurface can be defined by the $(N-4)$ independent equations

$$F_B(y^A) = 0, \tag{20.3}$$

or alternatively by

$$y^A = f^A(x^n), \tag{20.4}$$

in such a way that the flat-space metric (20.2) induces the metric of the Riemannian space on it. The necessary and sufficient condition is therefore the existence of N functions f^A, which for a given metric g_{mn} satisfy the equations

$$g_{nm} = \sum_A e_A \frac{\partial f^A}{\partial x^n} \frac{\partial f^A}{\partial x^m}. \tag{20.5}$$

The *minimum* number of additional dimensions $(N-4)$ necessary to embed the space is called the (embedding) *class* of a space. It can be shown that an n-dimensional Riemannian space has at most the class $\binom{n}{2}$. Our Universe can therefore always locally, that is, on leaving out of consideration the topological behaviour, be embedded in a flat space of at most ten dimensions.

The embedding permits an invariant characterization of solutions of the Einstein equations independent of the Petrov classification and of the classification according to groups of motion. It yields constructive methods for obtaining solutions of a certain class, just as do the other classifications. Because the immediate physical interpretation of the embedding class is lacking these methods have not, however, been so thoroughly investigated and elaborated as have the other procedures described above.

We mention the following results of the embedding classification:

(*a*) There is no vacuum solution of embedding class 1.

(*b*) Most solutions of embedding class 1 with perfect fluids are known; the interior Schwarzschild solution, the Friedman Universe and generalizations of these two solutions belong in this group.

(*c*) There are several theorems which relate embedding class on the one hand and Petrov class or structure of the group of motion on the other.

Bibliography to section 20

Monographs and collected works   Kramer, Stephani, MacCallum and Herlt (1981)

21 A survey of some selected classes of solutions

A compendium of all currently known solutions of the Einstein equations would fill a thick book. In spite of the complexity of the Einstein field equations many exact solutions are known, but most have little physical relevance, that is, it is most improbable that sources for such fields exist. On the other hand, exact solutions to many *real* problems, for example, the two-body problem, are unknown. Here we must restrict ourselves to a few brief references to, and remarks about, rather arbitrarily selected classes of solution. In sections 24, 26 and 28 we shall discuss at greater length several solutions which can be used as models for stars or the Universe.

21.1 Vacuum solutions

Several classes of degenerate vacuum solutions, that is, solutions with (at least) one shear-free, geodesic null congruence, have been systematically investigated. These classes include:

1 Type D solutions They are all known; their most important representative is the Kerr solution (Kerr, 1963), which in Boyer–Lindquist coordinates $(r, \vartheta, \varphi, t)$ has the form

$$\left.\begin{array}{l} ds^2 = \Sigma\left(\dfrac{dr^2}{\Delta} + d\vartheta^2\right) + (r^2+a^2)\sin^2\vartheta\, d\varphi^2 - c^2 dt^2 \\[4pt] \qquad + \dfrac{2Mr}{\Sigma}(a\sin^2\vartheta\, d\varphi - c\, dt)^2, \\[4pt] \Sigma \equiv r^2 + a^2\cos^2\vartheta, \quad \Delta \equiv r^2 - 2Mr + a^2, \end{array}\right\} \quad (21.1)$$

and which contains the Schwarzschild metric as a special case ($a=0$). The two degenerate, null eigenvectors of the curvature tensor, which according to the definition exist for every type D solution, are proportional to

and
$$\left.\begin{array}{l} k^i = (1, 0, a/\Delta, (r^2+a^2)/\Delta) \\ l^i = (-1, 0, a/\Delta, (r^2+a^2)/\Delta). \end{array}\right\} \quad (21.2)$$

An example of a type D solution which it has not (yet) been possible to interpret physically or apply is the metric

$$ds^2 = \frac{dz^2}{b/z-1} + \left(\frac{b}{z}-1\right)d\varphi^2 + z^2(dr^2 - \sinh^2 rc^2 dt^2), \quad (21.3)$$

which arises out of the Schwarzschild metric (10.19) via the transformation

$$\vartheta \to ir, \quad \varphi \to ict, \quad r \to z, \quad ct \to i\varphi. \quad (21.4)$$

Notice that the coordinate labels in (21.3) are completely arbitrary, for example, φ need not be an angular coordinate.

2 Degenerate solutions, whose null eigenvector field is rotation-free and divergence-free In section 17.3 it was explained that the most important physical and mathematical properties of a null vector field are contained in the three optical scalars σ (shear), ω (rotation) and θ (divergence). For degenerate solutions σ vanishes, in accordance with the definition, and the vanishing of the other scalars as well simplifies the field equations considerably. Besides the special type D solutions, which have this

21 A survey of some selected classes of solutions

property, the structure of all solutions of types *III* or *N* which fall in this category is known. Their metric has the form

$$ds^2 = |dx + i\,dy + B\,du|^2 + 2\,du\,dv + H\,du^2, \quad (21.5)$$

where the functions B and H must satisfy several differential equations. The most important representatives of this class are the plane-fronted waves with parallel rays (15.39) and (15.40) found in section (15.3).

3 Degenerate solutions, whose null eigenvector field is rotation-free, but has a non-vanishing divergence The metrics of this class of solutions are comprised in

$$\left. \begin{array}{l} ds^2 = \dfrac{v^2}{p^2}(dx^2 + dy^2) + 2\,du\,dv + c\,du^2, \\[6pt] \dfrac{\partial p}{\partial v} = 0, \quad c = -\dfrac{2m}{v} - p^2\left(\dfrac{\partial^2}{\partial x^2} + \dfrac{\partial^2}{\partial y^2}\right)\ln p - 2v\,\dfrac{\partial \ln p}{\partial u}, \quad m = 0, 1, \end{array} \right\} \quad (21.6)$$

where the function p is a solution of the equation (Robinson and Trautman, 1962)

$$p^2\left(\dfrac{\partial^2}{\partial x^2} + \dfrac{\partial^2}{\partial y^2}\right)\left[p^2\left(\dfrac{\partial^2}{\partial x^2} + \dfrac{\partial^2}{\partial y^2}\right)\ln p\right] = -12m\,\dfrac{\partial \ln p}{\partial u}. \quad (21.7)$$

The Schwarzschild metric is contained here too as a special case.

The hope of finding metrics amongst the solutions (21.5) and (21.6) which describe the radiation field of a physically meaningful matter distribution has not been realized.

21.2 Solutions with special symmetry properties

Such an extensive literature is available on solutions with symmetry properties (Killing vectors) that here we must limit ourselves to giving brief references to such classes of solutions that are used frequently or that have been especially thoroughly researched.

1 The Weyl class – axisymmetric, static vacuum fields In flat space physical configurations which are static and also spherically symmetric or axisymmetric (in cylindrical coordinates: φ-independent) are particularly simple. The analogue of spherical symmetry leads immediately for vacuum solutions to the Schwarzschild solution (Birkhoff theorem). The relativistic generalization of axially symmetric, static configurations is the Weyl class: in a suitable coordinate system the solution should not depend upon the time t nor the cyclic coordinate φ and should not change under the transformations $t \to -t$ and $\varphi \to -\varphi$ (should contain no terms $g_{\varphi r}$, $g_{\varphi \vartheta}$, g_{tr} or $g_{t\vartheta}$). The last condition means physically that a time-independent rotation of the source whose external field we are considering is forbidden.

To make an invariant formulation: all vacuum solutions with two commuting, hypersurface-orthogonal Killing vectors, of which one is timelike, whilst the world line congruence associated with the spacelike Killing vector consists of closed curves of finite length, belong to the Weyl class. (A metric is said to be *stationary* when it possesses a timelike Killing vector, and *static* if in addition the Killing vector is hypersurface orthogonal.) One can show that for this symmetry the metric can be transformed to the normal form (Weyl, 1917)

$$ds^2 = e^{-2U}[e^{2\nu}(d\rho^2 + dz^2) + \rho^2 d\varphi^2] - e^{2U}c^2 dt^2, \qquad (21.8)$$

where the functions $U(\rho, z)$ and $\nu(\rho, z)$ are to be determined from

$$\frac{\partial^2 U}{\partial \rho^2} + \frac{1}{\rho}\frac{\partial U}{\partial \rho} + \frac{\partial^2 U}{\partial z^2} = 0 \qquad (21.9)$$

and

$$\frac{1}{\rho}\frac{\partial \nu}{\partial \rho} = \left(\frac{\partial U}{\partial \rho}\right)^2 - \left(\frac{\partial U}{\partial z}\right)^2, \qquad \frac{1}{\rho}\frac{\partial \nu}{\partial z} = 2\frac{\partial U}{\partial \rho}\frac{\partial U}{\partial z}. \qquad (21.10)$$

Since (21.10) is always integrable when (21.9) holds, we have evidence which apparently suggests the astonishing fact that, from every (φ-independent) solution U of the potential equation (21.9) in flat space, that is, for every axisymmetric vacuum solution of the Newtonian gravitation theory, one can simply obtain a reasonable solution to the Einstein theory. The statement is however false in this oversimplified form. This is because we have not yet ensured that the singular line $\rho = 0$ of the coordinate system is not a singular line in the physical sense as well, with infinite mass density. The occurrence of such a singularity can be excluded by demanding that for every infinitesimal circle about the z-axis the ratio of circumference to radius is 2π (that is, that a Minkowski space exists locally); this is equivalent to the requirement that

$$\nu = 0 \quad \text{for} \quad \rho = 0, \ z \text{ arbitrary}. \qquad (21.11)$$

The differential equation (21.9) for U is of course linear, but the sum of two solutions U, which individually have a regular behaviour, will not in general satisfy the subsidiary conditions (21.11). The simple superposition of fields of two sources does not yield a field whose sources are in gravitational equilibrium.

2 Axisymmetric, stationary vacuum solutions In the Newtonian gravitation theory the gravitational field of an axisymmetric source distribution does not depend upon a possible rotation of the source about the symmetry axis. In the Einstein theory, on the other hand, the metric will be altered by the corresponding matter current, which enters the energy-momentum tensor. For uniform rotation about the symmetry axis, the metric will, of course, be independent of φ and t, but $g_{\varphi t}$ will not vanish. Formulated invariantly: axisymmetric, stationary vacuum fields possess two commuting Killing vectors (an Abelian group of motion G_2), of which one is timelike.

One can show that the metric of this class can be transformed into the canonical form

$$ds^2 = e^{-2U}[e^{2\nu}(d\rho^2 + dz^2) + \rho^2 d\varphi^2] - e^{2U}(c\,dt + a\,d\varphi)^2, \quad (21.12)$$

where the functions U, ν and a depend only upon ρ and z, and have to satisfy the differential equations

$$\left.\begin{aligned}\frac{\partial^2 U}{\partial z^2} + \frac{1}{\rho}\frac{\partial U}{\partial \rho} + \frac{\partial^2 U}{\partial \rho^2} &= -\frac{1}{2}\frac{e^{4U}}{\rho^2}\left[\left(\frac{\partial a}{\partial \rho}\right)^2 + \left(\frac{\partial a}{\partial z}\right)^2\right], \\ \frac{\partial}{\partial z}\left(\frac{e^{4U}}{\rho}\frac{\partial a}{\partial z}\right) + \frac{\partial}{\partial \rho}\left(\frac{e^{4U}}{\rho}\frac{\partial a}{\partial \rho}\right) &= 0,\end{aligned}\right\} \quad (21.13)$$

and

$$\left.\begin{aligned}\frac{1}{\rho}\frac{\partial \nu}{\partial \rho} &= \left(\frac{\partial U}{\partial \rho}\right)^2 - \left(\frac{\partial U}{\partial z}\right)^2 - \frac{e^{4U}}{4\rho^2}\left[\left(\frac{\partial a}{\partial \rho}\right)^2 - \left(\frac{\partial a}{\partial z}\right)^2\right], \\ \frac{1}{\rho}\frac{\partial \nu}{\partial z} &= 2\frac{\partial U}{\partial \rho}\frac{\partial U}{\partial z} - \frac{e^{4U}}{2\rho^2}\frac{\partial a}{\partial \rho}\frac{\partial a}{\partial z}.\end{aligned}\right\} \quad (21.14)$$

The system (21.14) is always integrable when (21.13) holds, so that ν can always be calculated by quadrature.

The system of field equations (21.13) has been extensively investigated with regard both to simple solutions and to the possibility of producing new solutions from those already known, for example, via Backlund transformations. The best known and most important representative of the class of axisymmetric, stationary vacuum solutions is the Kerr metric (21.1).

3 Solutions whose metric contains subspaces of constant curvature Important solutions of the Einstein field equations often contain a subspace

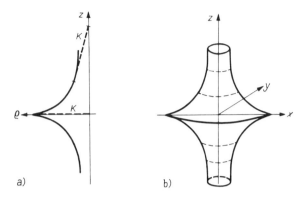

Fig. 21.1. Tractrix (a) and model (b) of a surface of constant negative curvature.

of higher symmetry, for example, a space of constant curvature, that is, a space with the greatest possible number of Killing vectors (cf. section 19.4). Because of the significance of the spaces of constant curvature we shall here describe their properties in greater detail.

Two-dimensional spaces of constant curvature

$$R_{EABF} = \epsilon K^{-2}(g_{AF}g_{EB} - g_{AB}g_{EF}), \quad A, B, \ldots = 1, 2, \\ \epsilon = \pm 1, 0, \quad K = \text{constant}, \quad (21.15)$$

can be regarded (at least to some extent) as surfaces in a flat three-dimensional space. Surfaces of positive curvature ($\epsilon = +1$) have the metric

$$ds^2 = K^2(d\vartheta^2 + \sin^2\vartheta\, d\varphi^2), \quad 0 \le \varphi \le 2\pi, \quad 0 \le \vartheta \le \pi; \quad (21.16)$$

they are spheres of radius K (cf. (1.35)–(1.38)). Surfaces of negative curvature ($\epsilon = -1$) are pseudo-spheres with the metric

$$ds^2 = K^2(d\vartheta^2 + \sinh^2\vartheta\, d\varphi^2), \quad 0 \le \varphi \le 2\pi, \quad \vartheta \ge 0. \quad (21.17)$$

They can be realized by surfaces of revolution generated by rotation of the tractrix

$$z = K \ln\left|\frac{K \pm \sqrt{K^2 - \rho^2}}{\rho}\right| \mp \sqrt{K^2 - \rho^2} \quad (21.18)$$

about the z-axis ($\rho^2 = x^2 + y^2$). The name tractrix is due to the fact that precisely this curve results if a man runs along the z-axis pulling behind him an object on the end of a rope of length K, the object not lying on the z-axis initially (cf. Fig. 21.1).

21 A survey of some selected classes of solutions 217

It is easy to convince oneself that the surface (21.18) really does have the metric (21.17), by inserting the differential equation of the tractrix

$$\left(\frac{dz}{d\rho}\right)^2 = \frac{\rho^2}{K^2 - \rho^2} \tag{21.19}$$

into the line element of flat space and finally making the substitution $\rho = K \sin \vartheta$.

While the line element (21.17) of the space of constant negative curvature is regular for all values of φ and ϑ, the surface has a singular line (for $z = 0$). One can prove quite generally that there exists no realization of the space of constant negative curvature by a surface in a flat three-dimensional space, which is regular everywhere. This two-dimensional Riemannian space is not *globally* embeddable in a flat three-dimensional space; such an embedding is only possible *locally*.

Finally, a surface of zero curvature is locally a plane,

$$ds^2 = d\varphi^2 + d\vartheta^2, \quad -\infty \le \varphi \le +\infty, \quad -\infty \le \vartheta \le +\infty \tag{21.20}$$

but can also be realized (after bending the plane appropriately), for example, by the surface of a cone or a cylinder. If one identifies the points on the baseline of the cylindrical surface with those at height H, that is, if one allows the variables φ and ϑ to occupy only the intervals $0 \le \varphi \le 2\pi R$ and $0 \le \vartheta \le H$ by identification of the end-points, then one has constructed a closed space of zero curvature.

The three types (21.16), (21.17) and (21.20) of a two-dimensional space of constant curvature can be summarized in the one metrical form

$$ds^2 = K^2 \frac{dx^2 + dy^2}{(1 + \epsilon r^2/4)^2}, \quad r^2 = x^2 + y^2, \tag{21.21}$$

which exhibits the isotropy of these spaces more clearly than the angular coordinates, but which singles out the point $r = 0$. Another frequently used version of the metric is

$$ds^2 = K^2 \left[\frac{d\bar{r}^2}{1 - \epsilon \bar{r}^2} + \bar{r}^2 d\varphi^2 \right], \quad \bar{r} = \frac{r}{1 + \epsilon r^2/4}. \tag{21.22}$$

If we drop the tacit assumption made up until now that the two-dimensional space has signature $+2$ (contains no timelike direction), then (21.21) can be generalized: the metric

$$ds^2 = K^2 \frac{e_1 (dx^1)^2 + e_2 (dx^2)^2}{[1 + \epsilon (e_1 x^{1\,2} + e_2 x^{2\,2})/4]^2}, \quad e_1 = \pm 1, \quad e_2 = \pm 1, \quad \epsilon = 0, \pm 1 \tag{21.23}$$

is the canonical metric of a two-dimensional space of constant curvature.

Subspaces of this kind occur with spherically symmetric solutions. Another interesting example is the metric

$$ds^2 = K^2[(dx^1)^2 + \sin^2 x^1 (dx^2)^2 + \sin^2 ct (dx^3)^2 - c^2 dt^2], \quad (21.24)$$

which is the product of two spaces of constant curvature and describes the gravitational field of a covariantly constant electrical field whose only non-vanishing component is

$$F_{34} = \frac{1}{K^3} \sin ct. \quad (21.25)$$

Three-dimensional spaces of constant curvature

$$\left. \begin{array}{l} R_{\mu\alpha\beta\nu} = \dfrac{\epsilon}{K^2} (g_{\alpha\nu} g_{\mu\beta} - g_{\alpha\beta} g_{\mu\nu}), \quad \alpha, \beta, \ldots = 1, 2, 3, \\[4pt] R = 6\epsilon/K^2, \quad \epsilon = 0, \pm 1, \quad K = \text{constant}, \end{array} \right\} \quad (21.26)$$

likewise split into three types if we restrict ourselves to purely positional spaces.

A space of positive constant curvature ($\epsilon = +1$) corresponds to a hypersphere

$$\left. \begin{array}{l} ds^2 = K^2[d\chi^2 + \sin^2 \chi (d\vartheta^2 + \sin^2 \vartheta\, d\varphi^2)], \\[4pt] 0 \le \chi \le \pi, \quad 0 \le \vartheta \le \pi, \quad 0 \le \varphi \le 2\pi, \end{array} \right\} \quad (21.27)$$

which can be embedded in a four-dimensional flat space according to

$$\left. \begin{array}{ll} y_1 = K \cos \chi, & y_3 = K \sin \chi \sin \vartheta \cos \varphi, \\[4pt] y_2 = K \sin \chi \cos \vartheta, & y_4 = K \sin \chi \sin \vartheta \sin \varphi. \end{array} \right\} \quad (21.28)$$

The volume of this sphere (the surface area, as regarded from the four-dimensional space) is

$$V = \int \sqrt{g}\, d\chi\, d\vartheta\, d\varphi = \int K^3 \sin^2 \chi \sin \vartheta\, d\chi\, d\vartheta\, d\varphi = 2\pi^2 K^3. \quad (21.29)$$

The 'radial' coordinate χ can take on the maximal value $\chi = \pi$; there is one point maximally distant from the null point $\chi = 0$, namely, the antipodal point $\chi = \pi$.

A space of constant negative curvature ($\epsilon = -1$) has the metric

$$ds^2 = K^2[d\chi^2 + \sinh^2 \chi (d\vartheta^2 + \sin^2 \vartheta\, d\varphi^2)]. \quad (21.30)$$

The 'radial' coordinate χ can vary arbitrarily, and the space can have infinite extent.

21 A survey of some selected classes of solutions

Finally, a space of zero curvature ($\epsilon = 0$) is (locally) a familiar flat space:

$$ds^2 = d\chi^2 + d\vartheta^2 + d\varphi^2. \tag{21.31}$$

The metric of all three types of space can be written in the form

$$ds^2 = K^2 \frac{dx^2 + dy^2 + dz^2}{(1 + \epsilon r^2/4)^2} = K^2 \left[\frac{d\bar{r}^2}{1 - \epsilon \bar{r}^2} + \bar{r}^2(d\vartheta^2 + \sin^2 \vartheta \, d\varphi^2) \right]. \tag{21.32}$$

Spaces with $\epsilon = 1$ are called *closed*, since although they are of course unbounded, they contain a finite volume, and the separation of two points is bounded. Spaces with $\epsilon = 0$ or -1 are frequently designated *open*. Since, however, amongst the spaces of negative curvature and the flat spaces, closed models which result from a suitable identification of points can readily be found (cf. the example discussed in connection with (21.20)), this designation is rather misleading. The problem of finding all possible realizations of a space of constant curvature is called the Caley–Klein space-structure problem.

Solutions which contain a three-dimensional spacelike surface of constant curvature are frequently used in cosmology (Robertson–Walker metrics). We shall describe their properties in detail later.

Four-dimensional spaces of constant curvature are known, as cosmological models, under the name of de Sitter universes; sometimes one uses this name only for the spaces of constant positive curvature. The de Sitter universes can all be represented as a hyperboloid

$$x^2 + y^2 + z^2 + \epsilon w^2 - v^2 = K^2 \tag{21.33}$$

in a flat space of metric

$$ds^2 = dx^2 + dy^2 + dz^2 + \epsilon \, dw^2 - dv^2. \tag{21.34}$$

For positive curvature they have the metric

$$ds^2 = K^2[\cosh^2 ct \{d\chi^2 + \sin^2 \chi (d\vartheta^2 + \sin^2 \vartheta \, d\varphi^2)\} - c^2 dt^2], \tag{21.35}$$

the space of zero curvature is Minkowski space, and spaces of negative curvature are represented locally by

$$ds^2 = K^2[\cos^2 ct \{d\chi^2 + \sinh^2 \chi (d\vartheta^2 + \sin^2 \vartheta \, d\varphi^2)\} - c^2 dt^2]. \tag{21.36}$$

Amongst the spaces of negative curvature there are some with closed, timelike curves.

De Sitter universes contain three-dimensional spaces of constant curvature and hence belong to the Robertson–Walker metrics.

Bibliography to section 21

Monographs and collected works Kramer, Stephani, MacCallum and Herlt (1981), Petrov (1969)

Review and research articles Ehlers and Kundt (1962), Godfrey (1972), Jordan, Ehlers and Kundt (1960), Kinnersley (1974), Robinson and Trautman (1962)

Gravitational collapse and black holes

In the examples and applications considered up until now we have always correctly taken into account the non-linearity of the Einstein equations, but most of the properties and effects discussed do not differ qualitatively from those of other classical (linear) fields. Now, in the discussion of black holes and of cosmological models, we are going to encounter properties of the gravitational field which deviate clearly from those of a linear field. The structure of the space-time is essentially changed by comparison with that of Minkowski space, and essentially new types of questions arise.

22 The Schwarzschild singularity

22.1 How does one examine the singular points of a metric?

A quick glance at the Schwarzschild metric,

$$ds^2 = \frac{dr^2}{1-2M/r} + r^2(d\vartheta^2 + \sin^2\vartheta \, d\varphi^2) - (1-2M/r)c^2 dt^2 \quad (22.1)$$

shows that a singularity of the metric tensor (of the component g_{rr}) is present at $r = 2M$. In our earlier discussion of the properties of the Schwarzschild metric in section 10 we had set this problem on one side with the remark that the radius $r = 2M$ lies far inside a celestial body, where the vacuum solution is of course no longer appropriate. Now, however, we shall turn to the question of whether and in what sense there is a singularity of the metric at $r = 2M$ and what the physical aspects of this are.

Places where a field is singular constitute a well-known phenomenon of classical physics. In electrostatics the spherically symmetric Coulomb field

$$U = \frac{e}{4\pi r} \quad (22.2)$$

is singular at $r = 0$, because an infinitely large charge density (point charge) is present there. In non-linear theories the situation is more complicated, because the singularity need not occur at the position of the source. Einstein hoped that the singularities of the gravitational field would represent elementary particles, that the general theory of relativity would thus to a certain degree automatically yield a (non-quantum-field-theoretical) theory of elementary particles. This hope has not been fulfilled. In the meanwhile, however, something has been learnt about the nature of singularities of the gravitational field and about the physical effects which occur there. Here we shall have to limit ourselves to the description of a few basic ideas, referring to the extensive specialist literature for the details.

A singular coordinate system can evidently give a false indication of a singularity of the space. For example, in flat three-dimensional space spherical coordinates are singular at $r = 0$ in the sense that $\sqrt{g}$ is zero there and $g^{\vartheta\vartheta}$ and $g^{\varphi\varphi}$ become infinite, without the space showing any peculiar properties there. Therefore if the metric is singular at a point, one investigates whether this singularity can be removed by introducing a new coordinate system. Or, appealing more to physical intuition, one asks whether a freely falling observer can reach this point and can use a local Minkowski system there. If both are possible, then the observer notices no peculiarities of the physical laws and phenomena locally, and hence there is no singularity present.

Singular points or lines can also arise if a hole has been cut in the universe by mistake, its edge appearing as a singularity. Of course one can repair such a defect by substituting a piece of universe back in, that is, one can complete the space by extension of a metric beyond its initially specified region of validity and by unbounded extension of geodesics.

In distinction to these two local types of investigation, one can also arrive at an understanding of the properties of a singularity by examining the topological properties of the space in its neighbourhood and, for example, asking what possibilities there are of interactions between the outside world and the neighbourhood of the singularity, that is, which points of the space can be linked to one another by test particles or by light rays.

We shall elucidate some of these questions by reference to the simple example of the Schwarzschild metric. But first a warning: the reader has perhaps assumed an exact definition of a singularity that includes all mathematical and physical properties, whereas research in this area has not yet advanced so far that such a definition can be given.

22.2 Radial geodesics in the neighbourhood of the Schwarzschild singularity

Soon after the Schwarzschild metric had been obtained as a solution of the field equations, it was established that both the determinant of the metric

22 The Schwarzschild singularity

$$-g = r^4 \sin^2 \vartheta \tag{22.3}$$

and also the invariant

$$R_{abcd}R^{abcd} = 48M^2/r^6 \tag{22.4}$$

associated with the curvature tensor are regular on the 'singular' surface $r = 2M$. This suggests that no genuine singularity is present there, but rather that only the coordinate system becomes singular.

In order better to understand the physical conditions in the neighbourhood of $r = 2M$, we investigate the radial geodesics, information about which is provided by the line element

$$ds^2 = \frac{dr^2}{1-2M/r} - (1-2M/r)c^2 dt^2. \tag{22.5}$$

From (22.5) or from (10.26) and (10.27) we obtain for the trajectories of test particles

$$\left.\begin{aligned}\frac{dr}{d\tau} &= \pm\sqrt{A^2 - c^2(1-2M/r)}, \\ \frac{dct}{d\tau} &= \frac{A}{1-2M/r},\end{aligned}\right\} \quad A = \text{constant} > 0, \tag{22.6}$$

while for photons $dx^2 = 0$; that is, we have

$$dr = \pm(1-2M/r)c\,dt. \tag{22.7}$$

For the test particle (for a freely falling observer) (22.5) and (22.6) tell us that an infinitely long time

$$\int dt = \int_{r_0}^{2M} \frac{A}{c} \frac{dr}{(1-2M/r)\sqrt{A^2-c^2(1-2M/r)}} \to \infty \tag{22.8}$$

is required to traverse the finite distance

$$L_0 = \int_{r_0}^{2M} \frac{dr}{\sqrt{1-2M/r}} \tag{22.9}$$

(the spatial displacement between $r = r_0$ and $r = 2M$), but that the destination is reached in the finite proper time

$$\tau_0 = \int_{r_0}^{2M} \frac{dr}{\sqrt{A^2-c^2(1-2M/r)}} \tag{22.10}$$

The freely falling observer would therefore probably not notice anything special at $r = 2M$; but the coordinates r and t are not suitable for describing his motion.

A photon would likewise require an infinitely long time, namely,

$$T_0 = \frac{1}{c}\int_{r_0}^{2M} \frac{dr}{1-2M/r} \qquad (22.11)$$

to cover the finite stretch L_0 (22.9) – and again the coordinate time t proves physically unsuitable for describing the process.

22.3 The Schwarzschild solution in other coordinate systems

We seek coordinate systems which are better adapted to the description of physical processes in the neighbourhood of $r = 2M$ than is the Schwarzschild metric, coordinate systems which may possibly even cover the spacetime completely. Notice that an extension of the Schwarzschild metric from the exterior space across the surface $r = 2M$ does not necessarily have to lead to the metric

$$ds^2 = (2M/r - 1)c^2 dt^2 + r^2(d\vartheta^2 + \sin^2\vartheta\, d\varphi^2) - \frac{dr^2}{2M/r - 1}, \quad r < 2M, \qquad (22.12)$$

which is also contained in (22.1), which one could of course regard as the metric 'inside' $r = 2M$ (where r is a timelike, and t a spacelike coordinate). For $r = 2M$ the metric (22.1) is completely undefined, and by extension of the metric of the exterior space into $r < 2M$ one could also arrive in a completely different region of the 'universal' Schwarzschild solution, just as by crossing a branch cut one can reach another branch of the Riemannian surface of an analytic function. We must therefore distinguish between the Schwarzschild *metric*, which is only valid for $r > 2M$, and the general Schwarzschild *solution*, which is the (yet to be discovered) maximal extension of the Schwarzschild metric, which contains (22.2) as one section, but which can also be described in completely different coordinates. We shall now meet three new coordinate systems (metrics) which all describe various sections of the Schwarzschild solution.

One can adapt the coordinate system to a freely falling observer by the transformation

$$\left. \begin{array}{l} dT = dt + \sqrt{\dfrac{2M}{r}}\, \dfrac{dr}{c(1-2M/r)}, \\[1em] cT(r,t) = ct + 2\sqrt{2Mr} + 2M \ln\left|\dfrac{\sqrt{r} - \sqrt{2M}}{\sqrt{r} + \sqrt{2M}}\right|, \end{array} \right\} \qquad (22.13)$$

and

22 The Schwarzschild singularity

$$dR = c\,dT + \frac{dr\sqrt{r}}{\sqrt{2M}} = c\,dt + \frac{\sqrt{r}}{\sqrt{2M}}\frac{dr}{1-2M/r},$$

$$r(R, cT) = \left[(R-cT)\frac{3\sqrt{2M}}{2}\right]^{2/3}. \quad (22.14)$$

In this way we pass from the Schwarzschild metric to the Lemaitre metric (Lemaitre, 1933)

$$ds^2 = \frac{2M}{r}dR^2 + r^2(d\vartheta^2 + \sin^2\vartheta\,d\varphi^2) - c^2 dT^2, \quad r = r(R, T). \quad (22.15)$$

T is clearly the proper time for particles, which are at rest in the coordinate system (22.15); and because of (22.14) and (22.6) $dR = 0$ holds exactly for those particles which are initially at rest at infinity ($A = c$) and then fall freely and radially. The line element (22.15) is regular at $r = 2M$, and a freely falling observer notices nothing peculiar there; only the point $r = 0$ is singular. A drawback of this metric is that the static Schwarzschild solution is described by time-dependent metric functions.

In another coordinate system null geodesics are preferred to the timelike geodesics used above. If one introduces a retarded time v by

$$\left.\begin{array}{l} dv = dt - dt^* = dt + \dfrac{dr}{c(1-2M/r)}, \\[6pt] cv = ct + r + 2M\ln(r-2M) - 2M\ln 2M \end{array}\right\} \quad (22.16)$$

(dt^* is the time needed by a radially falling photon to complete the distance dr), then from the Schwarzschild metric (22.1) one arrives at the Eddington–Finkelstein metric (Eddington, 1924; Finkelstein, 1958)

$$ds^2 = 2c\,dr\,dv + r^2(d\vartheta^2 + \sin^2\vartheta\,d\varphi^2) - (1-2M/r)c^2 dv^2, \quad (22.17)$$

in which light rays travelling inward radially are described by $dv = 0$. In these coordinates, too, the metric functions are only singular at $r = 0$ (the vanishing of g_{rr} at $r = 2M$ implies no loss of dimension, since the determinant of the metric (22.17) does not vanish there). The line element (22.17) is not invariant under time reversal $v \to -v$, which corresponds to a time reflection $t \to -t$ and a substitution of inward-travelling by outward-travelling photons. But from this time reversal we obtain another section of the universal Schwarzschild solution.

The maximal extension of the Schwarzschild metric is a metric which contains all the sections considered up until now as component spaces and which cannot be further extended. It is reached by introducing the advanced time u,

$$cu = ct - r - 2M \ln(r-2M) + 2M \ln 2M,$$
$$c(v-u) = 2r + 4M \ln(r-2M) - 4M \ln 2M, \quad (22.18)$$

into the metric (22.17), which hence (after elimination of dr) takes the form

$$ds^2 = r^2(d\vartheta^2 + \sin^2\vartheta\, d\varphi^2) - c^2 du\, dv(1 - 2M/r), \quad (22.19)$$

and by making the associated coordinate transformations

$$v' = e^{cv/4M}, \qquad u' = -e^{-uc/4M} \quad (22.20)$$

and

$$z = \tfrac{1}{2}(v' - u'), \qquad w = \tfrac{1}{2}(v' + u'). \quad (22.21)$$

The result is the Kruskal form of the metric representing the Schwarzschild solution

$$ds^2 = \frac{32M^3}{r} e^{-r/2M}(dz^2 - dw^2) + r^2(w,z)(d\vartheta^2 + \sin^2\vartheta\, d\varphi^2), \quad (22.22)$$

which is related to the original Schwarzschild metric by

$$z^2 - w^2 = \frac{1}{2M}(r - 2M)e^{r/2M},$$
$$\frac{w}{z} = \frac{e^{ct/4M} - e^{-ct/4M}}{e^{ct/4M} + e^{-ct/4M}} = \tanh\frac{ct}{4M}. \quad (22.23)$$

In the Kruskal metric (22.22) ϑ and φ are spherical coordinates in the sense that they are coordinates on the subspaces with spherical symmetry; the coordinates z and w are spacelike and timelike, respectively. They can take positive and negative values, but are restricted so that r is positive.

We now want to describe briefly how the Schwarzschild metric and its singularity appear from the standpoint of the Kruskal metric. The exterior space of the Schwarzschild metric ($r > 2M$, t finite) corresponds to region I of Fig. 22.1, where $z > |w|$. The rays $w = \pm z$, $z \geq 0$, form the boundary of this region, which is described in the r, t coordinates by $t = \pm\infty$ and/or $r = 2M$. If one crosses this boundary inwards into region II, that is, if one crosses $t = +\infty$, $r = 2M$, then one arrives in the 'interior' of the usual Schwarzschild metric; the Eddington–Finkelstein metric (22.17) covers precisely these two regions. The regions I' and II', which can be reached by further varying the coordinates w and z, are isometric to (metrically indistinguishable from) regions I and II: the maximal extension of the Schwarzschild solution contains both the exterior part ($r > 2M$) and the interior part ($r < 2M$) of the Schwarzschild metric twice. No

22 The Schwarzschild singularity

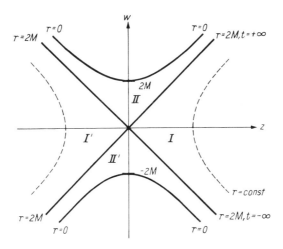

Fig. 22.1. The Kruskal diagram of the Schwarzschild solution (ϑ, φ suppressed).

boundaries or singularities occur here, with the exception of (what can be shown to be) the genuine singularity $r = 0$, which cannot be removed by coordinate transformations, and which is represented in the Kruskal diagram by two hyperbolae.

If we are inside the gravitational field of a spherically symmetric star, that is, within region I, then, because of the existence of the extended star (whose surface is described in Fig. 22.1 by the line $r =$ constant) the regions I', II and II'' are of course to be replaced by the interior Schwarzschild solution (11.24). Before turning to the question of whether and in what manner the complete Schwarzschild solution can be realized in nature, we want to discuss more closely the physical consequences of the maximal extension of the Schwarzschild metric.

22.4 The Schwarzschild solution as a black hole

From the mathematical description given in the previous section of the circumstances relating the different regions of the Schwarzschild solution, one might gain the impression that it would be possible for an observer to pass from our universe (region I) through the Schwarzschild singularity $r = 2M$ into the interior space (region II), and thus into another universe (region I'), which is again the exterior space of a Schwarzschild metric. Since, however, this observer requires an infinite time (as measured in proper time by the people left behind) just to reach $r = 2M$, he would have vanished forever to those remaining behind. Or, alternatively, while they believe him to be still on the way to $r = 2M$, he has long since (as measured in his proper time) been exploring in the new universe I'. To see whether such journeys are possible, we must examine more carefully the properties of the geodesics of the Schwarzschild solution. Our traveller need not necessarily fall freely (move on a geodesic), since he can of course use a rocket; but he can never travel faster than light. Hence with regard to possible journeys and to the physical relations between the different regions of the Schwarzschild solution, the route taken by null geodesics (light rays) is particularly important.

If we limit ourselves to purely radial motions ($\vartheta = $ constant, $\varphi = $ constant), then the line element

$$\mathrm{d}s^2 = \frac{32M^3}{r} e^{-r/2M}(\mathrm{d}z^2 - \mathrm{d}w^2) \tag{22.24}$$

determines the course of geodesics. This metric is conformally flat; because

$$\mathrm{d}z = \pm \mathrm{d}w \tag{22.25}$$

for the null geodesics ($\mathrm{d}s^2 = 0$), they are straight lines inclined at 45° (or 135°) in the zw-plane. This simple form for the null geodesics is not a coincidence; it follows from the choice of our coordinates, which were deliberately adapted to light propagation.

If one inserts into the Kruskal diagram all the null geodesics which in region I run radially inwards (t increasing, r decreasing) or radially outwards (t increasing, r increasing) and produces them across $r = 2M$, then one obtains the result sketched in Fig. 22.2. All light rays going radially inwards strike $r = 2M$ (for $t = \infty$), penetrate into region II and end up at the singularity $r = 0$; all light rays running outwards come from the region II' and the singularity there.

Thus one cannot send radially directed light rays from our universe (I) into the regions I' or II'; only region II is within reach, and once the photon is there it cannot avoid the singularity $r = 0$. One might think of escaping the singularity $r = 0$ by using non-radially directed light rays or by using observers with suitable rockets, but in fact once a photon or a point mass is in region II, it cannot avoid 'falling' to $r = 0$. Addition of new degrees of freedom to (22.24) implies for photons, for example, that, because

$$\mathrm{d}s^2 = 0 = \frac{32M^3}{r} e^{-r/2M}(\mathrm{d}z^2 - \mathrm{d}w^2) + \mathrm{d}\sigma^2, \quad \mathrm{d}\sigma^2 > 0, \tag{22.26}$$

and because of the term $\mathrm{d}\sigma^2$ which must also be compensated, then $\mathrm{d}w^2$ must be larger in relation to $\mathrm{d}z^2$ than for radial photons. In Fig. 22.2 the

22 The Schwarzschild singularity

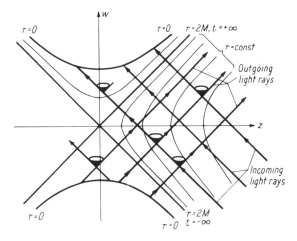

Fig. 22.2. Null geodesics and the local light cone in the Schwarzschild solution.

light rays bend up more steeply and reach $r = 0$ even earlier. The same thing holds for observers with rockets. Their fate can be described as follows. While in our part of the Universe it is always possible with the aid of rockets to remain at a fixed point, in spite of the gravitational field, in region II r cannot remain constant. As one can see from (22.12), it plays the rôle of a time coordinate there, and the observer cannot prevent the lapse of time. Moreover, in region II the Schwarzschild solution is no longer static, since the Killing vector which is timelike in region I becomes spacelike here.

Let us return to the observer whom we wanted to send into region I'. While we believe him to be on the way to $r = 2M$, he has long ago been lost to the singularity $r = 0$ in region II; he can never reach regions I' or II'.

We can thus establish the following causal structure for the Schwarzschild solution. From our universe I we can influence region II, but not regions I' or II', and we can be influenced by region II', but not by II or I'. While region I' is therefore rather uninteresting for us, region II is a 'black hole' (everything can go in, nothing can come out), and region II' is a 'white hole' (nothing can go in, things can only come out).

Is this nothing more than playing with mathematical constructions? The current opinion is that in our universe gravitational fields of this structure were not present from the beginning, but have possibly developed since. If they have thus only been in existence for a finite (coordinate) time, then region II' has no interest for us, since in a finite time nothing can reach us from there. Since, therefore, of the three regions

I', II and II' on the other side of the Schwarzschild singularity (it is probable that) only II has relevance for us, we often speak of a 'black hole' when we mean the full Schwarzschild solution.

The story of how a black hole comes into being, of the evolution of the gravitational field of a normal star to a field whose Schwarzschild radius $r = 2M$ can be crossed from outside (no longer lies within the star) – this will be the problem for the next section.

Bibliography to section 22

Textbooks   Rindler (1977)

Monographs and collected works   de Witt and de Witt (1972), de Witt-Morette (1974), Hawking and Ellis (1973), Stewart and Walker (1973), Zel'dovich and Novikov (1971)

Review and research articles   Geroch (1971), Penrose (1968)

23 Gravitational collapse – the possible life history of a spherically symmetric star

23.1 *The evolutionary phases of a spherically symmetric star*

In our universe a star whose temperature lies above that of its surroundings continuously gives up energy, and hence mass, mainly in the form of radiation, but also in explosive outbursts of matter. Here we want to sketch roughly the way in which it evolves. It is essentially characterized and determined by its innate properties (initial mass, initial density,...) and its behaviour in the critical catastrophic phases of its life.

According to observation, stars exist for a very long time after they have formed from hydrogen and dust. Therefore they can almost always settle down to a relatively stable state during the interplay between attractive gravitational force, repulsive (temperature-dependent) pressure and outgoing radiation.

The first stable state is reached when the gravitational attraction has compressed and heated the stellar matter to such a degree that the conversion of hydrogen into helium is a long-term source of energy sufficient to prevent the star cooling and to maintain the pressure (a sufficiently large thermal velocity of the stellar matter) necessary to compensate the gravitational force. The average density of such a star is of the order of magnitude 1 g cm^{-3}. A typical example of such a star is our Sun.

When the hydrogen of the star is used up, the star can switch over to other nuclear processes (possibly only after an unstable phase associated with explosions) and produce nuclei of higher atomic number. These processes always last a shorter time and follow one another more quickly. For sufficiently massive celestial bodies (the Earth would be too small!) the resulting pressure is then so great that the nuclei lose their electrons and a degenerate electron gas results. The density of this star is of order 10^7 g cm^{-3}; stars of such density have already been found as white dwarfs. The fact that white dwarfs with a mass above 1.2 solar masses cannot be stable is of great importance in the present context (their radius thus amounts at most to 5000 km). Stars of larger mass must therefore either lose a part of their mass after hydrogen burning or else evolve toward a different final state.

If the pressure (the mass of the star) is large enough, this can happen by the electrons and the protons of the nuclei (starting, for example, from ^{56}Fe) turning into neutrons, so that the whole star finally consists of the most closely packed nuclear matter. The density of such neutron stars is about 10^{14} g cm^{-3}. Although the details of the nuclear interactions are not known exactly, one can nevertheless show that neutron stars are only stable (the pressure can only then support the gravitational force in equilibrium) if their mass does not appreciably exceed the mass of the Sun. Neutron stars hence have radii of about 10 km. We are now convinced that the so-called pulsars are neutron stars. Pulsars are stars which send out optical or radio signals at regular intervals of 10^{-3} to 1 s; the period is kept so exactly that it can only be caused by the rotation of the star, and rotation times of this order are only possible for exceptionally small stars. However, the fact that more massive stars cannot end their lives as neutron stars is crucial here to the question of the final state of a star.

Before turning to the possible fate of more massive stars, we shall bring in the Einstein field equations and ask what they have to say about stable, spherically symmetric accumulations of matter.

23.2 The critical mass of a star

As we have shown in the discussion of the interior Schwarzschild solution in section 11, the gravitational field inside a static, spherically symmetric star is described by the metric

$$ds^1 = e^{\lambda(r)}dr^2 + r^2(d\vartheta^2 + \sin^2\vartheta \, d\varphi^2) - e^{\nu(r)}c^2dt^2. \tag{23.1}$$

A good approximation to stellar matter is given by the model of a perfect fluid medium with rest mass density $\mu(r)$ and pressure $p(r)$. The relation

$$e^{-\lambda(r)} = 1 - 2m(r)/r \tag{23.2}$$

follows simply from the field equations, where $m(r)$ is the mass function defined by

$$m(r) = \frac{\kappa c^2}{2}\int_0^r \mu(x)x^2 dx. \tag{23.3}$$

The remaining field equations can be converted, upon using (23.2), into

$$\nu' = -\frac{2p'}{p+\mu c^2} \tag{23.4}$$

and

$$\kappa p = \frac{\nu'}{r}\left(1 - \frac{2m}{r}\right) - \frac{2m}{r^3}. \tag{23.5}$$

While we integrated these field equations earlier for the special case $\mu = $ constant, we now want to derive a conclusion valid for arbitrary $\mu(r)$.

Suppose we have a star with finite (coordinate) radius r_0. The pressure p will vanish on the surface and will be greatest at the centre $r=0$ of the star; it cannot, however, be infinitely great there. The density $\mu(r)$ should likewise remain finite for $r=0$ and (on grounds of stability) decrease outwards:

$$\mu'(r) < 0. \tag{23.6}$$

Since e^ν and its derivative must be continuous on the surface, $m(r)$ takes the value

$$m(r_0) = M \tag{23.7}$$

there, where M is just the mass parameter occurring in the exterior, Schwarzschild solution; m/r^3 is finite at $r=0$ because of (23.3).

Our aim is to derive a condition for the maximum possible mass M for given r_0 from the condition that μ and p are finite.

From the field equation (23.5) one can see at once that ν'/r must be finite at $r=0$. If one introduces in place of e^ν the function $f(r)$, where

$$f(r) = e^{\nu(r)/2}, \tag{23.8}$$

then this requirement becomes

$$\frac{f'}{rf} \quad \text{finite at } r=0. \tag{23.9}$$

23 Possible life history of a spherically symmetric star

By eliminating p from (23.4) and (23.5) one obtains, after some transformations,

$$\frac{d}{dr}\left[\frac{1}{r}\sqrt{1-\frac{2m}{r}}\frac{df}{dr}\right] = \frac{f}{\sqrt{1-2m/r}}\frac{d}{dr}\left(\frac{m}{r^3}\right). \tag{23.10}$$

Since m/r^3 is the average mass density of the sphere of (coordinate) radius r, because of the definition (23.3), and since the average mass density cannot increase with r if μ decreases, then the right-hand side of (23.10) is negative or (and this only for $\mu = $ constant) zero:

$$\frac{d}{dr}\left[\frac{1}{r}\sqrt{1-\frac{2m}{r}}\frac{df}{dr}\right] \le 0. \tag{23.11}$$

On the surface of the star the metric must go over smoothly to the exterior Schwarzschild metric and the pressure must vanish, so that we must have

$$f^2(r_0) = 1 - \frac{2M}{r_0}, \quad \frac{df}{dr}\bigg|_{r=r_0} = \frac{M}{r_0^2}\frac{1}{\sqrt{1-2M/r_0}} \tag{23.12}$$

(cf. (23.3) and (23.8)). If one integrates (23.11) from r to r_0 using these relations, then one obtains

$$f'(r) \ge \frac{Mr}{r_0^3}\left(1-\frac{2m}{r}\right)^{-1/2}. \tag{23.13}$$

Since for finite f and m the right-hand side of (23.10) is finite, then $f'(r)/r$ will be bounded. The finiteness condition (23.9) for the pressure then reduces to the requirement that $f(0) > 0$. Integration of (23.13) between 0 and r_0, using (23.12), gives, however,

$$f(0) \le \left(1-\frac{2M}{r_0}\right)^{1/2} - \frac{M}{r_0^3}\int_0^{r_0}\frac{r\,dr}{(1-2m/r)^{1/2}}. \tag{23.14}$$

If we now express $\mu(r)$ in terms of the constant density $\mu_0 = 6M/\kappa c^2 r_0^3$ and the variable part $\rho(r)$,

$$\mu = \mu_0 + \rho, \quad \int_0^{r_0}\rho(x)x^2 dx = 0, \quad \rho' \le 0, \quad \rho(0) \ge 0, \tag{23.15}$$

then we see the integral in

$$m(r) = M\frac{r^3}{r_0^3} + \int_0^r \rho(x)x^2 dx \tag{23.16}$$

is always positive. The right-hand side of (23.14) can therefore be increased in magnitude by substituting Mr^3/r_0^3 for $m(r)$. We then obtain the final result of the analysis, namely,

$$f(0) \le \frac{3}{2}\left(1 - \frac{2M}{r_0}\right)^{1/2} - \frac{1}{2}. \tag{23.17}$$

As we have shown above, the central pressure $p(0)$ is only finite if $f(0)$ is greater than zero. Thus we can formulate the following important statement: a spherically symmetric star can only exist in a state of stable equilibrium (can only compensate its own gravitational attraction with a finite pressure) if its mass M and its radius r_0 satisfy the inequality

$$\frac{2M}{r_0} < \frac{8}{9}. \tag{23.18}$$

For the special case of the interior Schwarzschild solution with the equation of state $\mu = $ constant we have already derived this inequality in section 11. Now we know that it is valid for an arbitrary equation of state. In discussing this relation we must be careful about the definitions of M and r_0. M is (up to a factor) the integral of the mass density μ over the coordinate volume; it has the invariant significance of being the gravitating mass of the star as determined in the Newtonian far field. The stellar radius r_0 is defined so that the surface area of the star is $4\pi r_0^2$.

The inequality (23.18) expresses the fact that a star of fixed surface area is only stable as long as its mass lies below a critical mass. A star whose mass transgresses this limit must inevitably collapse into itself as a consequence of its gravitational attraction, which is now too strong. While in the linear Newtonian gravitational theory a predominance of the gravitational force can be compensated by a contraction and the associated finite increase in pressure, or by additional forces, in the non-linear Einstein theory a pressure increase or an extra force acts (via the energy–momentum tensor) to further increase the strength of the gravitational field above the critical mass (23.18).

An analysis of the maximum stable mass for given constant mass density μ, using the model of the interior Schwarzschild solution, leads from (23.3), (23.7) and (23.18) to the critical mass M_{crit} (measured in units of length)

$$M_{\text{crit}} = \frac{8}{9}\sqrt{\frac{2}{3\kappa c^2 \mu}}. \tag{23.19}$$

With $c^2 = 1.86 \times 10^{-27}$ cm g^{-1} one obtains for typical densities the following critical masses, which are compared with the mass of the Sun:

μ (in g cm^{-3})	1	10^6	10^{15}
M_{crit} (in cm)	1.685×10^{13}	1.685×10^{10}	0.532×10^6
$M_{\text{crit}}/M_\odot$	1.14×10^8	1.14×10^5	3.96

23 Possible life history of a spherically symmetric star

These very rough considerations already show that neutron stars can have only a few solar masses; more massive stars have no stable final state.

As an interesting side result we observe that, because of the general formula (10.53) and the inequality (23.18), the red shift of a light signal coming from the surface of a stable star has a maximum value of $z=2$.

23.3 Gravitational collapse

The considerations of the previous section have shown that if, during its evolution, a massive, spherically symmetric star does not succeed in ejecting or radiating away sufficient mass to become a neutron star, then there is no stable, final state available to it. At some time or other it will reach a state in which the pressure gradient can no longer balance the gravitational attraction. Consequently it will continue to contract further and its radius will pass the Schwarzschild radius $r = 2M$ and tend to $r = 0$: the star suffers a gravitational collapse.

Of course one would like to confirm these plausible intuitive ideas by making exact calculations on a stellar model with a physically reasonable equation of state (a reasonable relation between pressure and mass density). The only model for which this is possible without great mathematical complexity is that of dust ($p = 0$). Because the pressure vanishes it is to be expected here that once a star started to contract it would 'fall in' to a point. Nevertheless, this example is not trivial, since it yields an exact solution of the Einstein equations which is valid in the whole space, and which in a certain sense can serve as a model for all collapsing stars.

As the starting point for treating this collapsing stellar dust we take a form for the line element of a spherically symmetric star which deviates from the canonical form (10.2) used earlier. We obtain it by carrying out a transformation $r = r(\rho, c\tau)$, $t = t(\rho, c\tau)$ and hence bringing the metric (10.2) into the form

$$ds^2 = e^{\lambda(\rho,\, c\tau)} d\rho^2 + r^2(\rho, c\tau)(d\vartheta^2 + \sin^2\vartheta\, d\varphi^2) - c^2 d\tau^2. \qquad (23.20)$$

The coordinate τ is clearly the proper time of a particle at rest in the coordinate system (23.20), and the curves $\rho = \text{constant}$, $\vartheta = \text{constant}$, $\varphi = \text{constant}$ are geodesics. Now, because of (8.85), dust always moves along geodesics, and therefore (23.20) is a system co-moving with the dust; that is, the energy–momentum tensor has as its only non-vanishing component

$$T^4_4 = -c^2 \mu(\rho, c\tau). \qquad (23.21)$$

If we denote partial derivatives with respect to ρ and $c\tau$ by $'$ and $\dot{}$ respectively, then from (23.20) we obtain as the non-vanishing Christoffel symbols

$$\Gamma^1_{11} = \lambda'/2, \quad \Gamma^1_{22} = -e^{-\lambda}rr', \quad \Gamma^1_{33} = -e^{-\lambda}rr'\sin^2\vartheta, \quad \Gamma^1_{14} = \dot\lambda/2,$$
$$\Gamma^2_{12} = r'/r, \quad \Gamma^2_{24} = \dot r/r, \quad \Gamma^2_{33} = -\sin\vartheta\cos\vartheta,$$
$$\Gamma^3_{13} = r'/r, \quad \Gamma^3_{34} = \dot r/r, \quad \Gamma^3_{23} = \cot\vartheta,$$
$$\Gamma^4_{11} = \dot\lambda e^\lambda/2, \quad \Gamma^4_{22} = r\dot r, \quad \Gamma^4_{33} = r\dot r\sin^2\vartheta,$$
$$\tag{23.22}$$

with $x^1 = \rho$, $x^2 = \vartheta$, $x^3 = \varphi$ and $x^4 = c\tau$. The field equations finally take the form

$$R^1_1 - \frac{R}{2} = \frac{r'^2}{r^2}e^{-\lambda} - \frac{2\ddot r}{r} - \frac{\dot r^2}{r^2} - \frac{1}{r^2} = 0, \tag{23.23}$$

$$R^2_2 - \frac{R}{2} = R^3_3 - \frac{R}{2} = \left(\frac{r''}{r} - \frac{r\lambda''}{2r}\right)e^{-\lambda} - \frac{\dot r\dot\lambda}{2r} - \frac{\ddot\lambda}{2} - \frac{\dot\lambda^2}{4} - \frac{\ddot r}{r} = 0, \tag{23.24}$$

$$R^4_4 - \frac{R}{2} = \left(\frac{2r''}{r} - \frac{\lambda'r'}{r} + \frac{r'^2}{r^2}\right)e^{-\lambda} - \frac{\dot r\dot\lambda}{r} - \frac{\dot r^2}{r^2} - \frac{1}{r^2} = -\kappa\mu c^2, \tag{23.25}$$

and

$$R_{14} = \dot\lambda r'/r - 2\dot r'/r = 0. \tag{23.26}$$

First integrals of these equations can be obtained very easily. The first step is to write (23.26) as

$$\dot\lambda = (r'^2)^\cdot/r'^2 \tag{23.27}$$

and then integrate it to give

$$e^\lambda = \frac{r'^2}{1 - \epsilon f^2(\rho)}, \quad \epsilon = 0, \pm 1, \tag{23.28}$$

with $f(\rho)$ as an arbitrary function. Substitution into (23.23) leads to

$$2\ddot r r + \dot r^2 = -\epsilon f^2. \tag{23.29}$$

If one now chooses r as the independent variable and $u = (\dot r)^2$ as the new dependent variable, then one obtains the simple linear differential equation

$$\frac{d}{dr}(ru) = -\epsilon f^2, \tag{23.30}$$

whose solution is

$$\dot r^2 = -\epsilon f^2(\rho) + F(\rho)/r. \tag{23.31}$$

If one next eliminates f^2 in (23.28) with the aid of (23.31), then one finds that (23.24) is satisfied identically and that (23.23) leads to

$$\kappa\mu c^2 = \frac{F'}{r'r^2}. \tag{23.32}$$

23 Possible life history of a spherically symmetric star

The partial differential equation (23.31) can be integrated completely, since ρ only plays the part of a parameter. For $\epsilon \neq 0$ one can, through introducing

$$d\eta = f\, dc\tau/r, \tag{23.33}$$

bring the differential equation into the form

$$\left(\frac{\partial r}{\partial \eta}\right)^2 = \frac{F}{f^2} r - \epsilon r^2 \tag{23.34}$$

and solve it by

$$r = \frac{F(\rho)}{2f^2(\rho)} h'_\epsilon(\eta),$$

$$c\tau - c\tau_0(\rho) = \pm \frac{F(\rho)}{2f^3(\rho)} h_\epsilon(\eta), \qquad h_\epsilon(\eta) = \begin{cases} \eta - \sin\eta & \text{for } \epsilon = +1, \\ \sinh\eta - \eta & \text{for } \epsilon = -1, \end{cases} \tag{23.35}$$

while for $\epsilon = 0$ one immediately has from (23.31) that

$$c\tau - c\tau_0(\rho) = \pm \tfrac{2}{3} F^{-1/2}(\rho) r^{3/2}, \qquad \epsilon = 0. \tag{23.36}$$

The general spherically symmetric solution of the field equations for the case of dust (the Tolman solution, 1934) thus has the form in co-moving coordinates

$$ds^2 = \left(\frac{\partial r}{\partial \rho}\right)^2 \frac{d\rho^2}{1 - \epsilon f^2(\rho)} + r^2(\rho, c\tau)(d\vartheta^2 + \sin^2\vartheta\, d\varphi^2) - c^2 d\tau^2,$$

$$\kappa c^2 \mu(\rho, c\tau) = \frac{F'(\rho)}{r^2 \partial r/\partial \rho}, \tag{23.37}$$

where $r(\rho, ct)$ is to be taken from (23.35) and (23.36). Of the three free functions $F(\rho)$, $f(\rho)$ and $\tau_0(\rho)$, at most two have a physical significance, since the coordinate ρ is defined only up to scale transformation $\bar\rho = \bar\rho(\rho)$. Of course one cannot simply specify the matter distribution $\mu(\rho, c\tau)$ and then determine the metric, but rather through a suitable specification of f, F and τ_0 one can produce meaningful matter distributions. Since the layers of matter which move radially with different velocities can overtake and cross one another, one must expect the occurrence of coordinate singularities in the comoving coordinates used here.

We now want to apply the Tolman solution to the problem of a star of finite dimensions. To do this we have to obtain an interior ($\mu \neq 0$) solution and an exterior ($\mu = 0$) solution and join these two solutions smoothly at the surface of the star ($\rho = \rho_0$).

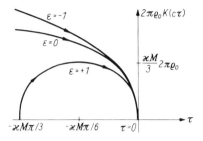

Fig. 23.1. Radius of a collapsing star as a function of time.

We obtain the simplest *interior solution* when μ does not depend upon position (upon ρ) and r has (for a suitable scale) the form $r = K(c\tau)\rho$. These restrictions lead to

$$f = \rho, \quad F = \frac{\kappa \hat{M}}{3}\rho^3, \quad \mu c^2 K^3(c\tau) = \hat{M} = \text{constant}, \quad \tau_0 = 0, \quad (23.38)$$

and the metric

$$\begin{aligned}
ds^2 &= K^2(c\tau)\left[\frac{d\rho^2}{1-\epsilon\rho^2} + \rho^2(d\vartheta^2 + \sin^2\vartheta\, d\varphi^2)\right] - c^2 d\tau^2, \\
K(\eta) &= \frac{\kappa \hat{M}}{6}h'_\epsilon(\eta), \quad c\tau = -\frac{\kappa \hat{M}}{6}h_\epsilon(\eta), \\
h_\epsilon(\eta) &= \begin{cases} \eta - \sin\eta & \text{for } \epsilon = +1, \\ \eta^3/6 & \text{for } \epsilon = 0, \\ \sinh\eta - \eta & \text{for } \epsilon = -1. \end{cases}
\end{aligned} \quad (23.39)$$

As comparison with (21.32) shows, the interior $\rho \leq \rho_0$ of the star is a three-dimensional space of constant curvature, whose radius K depends on time (in the language of cosmological models, it is a section of a Friedmann universe, cf. section 26.2). A great circle on the surface of the star has the radius $\rho_0 K(c\tau)$, and because of the time dependence of K the star either expands or contracts.

As Fig. 23.1 shows, models with $\epsilon = 0$ or -1 correspond to stars whose radius decreases continuously from arbitrarily large values until at the time $\tau = 0$ a collapse occurs, while models with $\epsilon = +1$ represent stars which first expand to a maximum radius and then contract.

The solution in the exterior space to the star is clearly a spherically symmetric, vacuum solution, and because of Birkhoff's law it can only be the Schwarzschild solution (cf. Fig. 23.2). Since the Tolman solution (23.37) holds for arbitrary mass density μ, it must contain the exterior

23 Possible life history of a spherically symmetric star

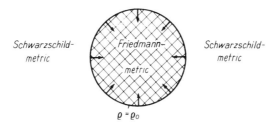

Fig. 23.2. Snapshot of a collapsing star.

Schwarzschild solution $\mu = 0$ as a special case (F = constant). In the Tolman solution the coordinates in the exterior space are chosen so that the surface of the star is at rest. In the usual Schwarzschild metric, on the other hand, the stellar surface is in motion. But in both cases the motion of a particle of the surface takes place on a geodesic. The equation (22.6) of the radial geodesics of the Schwarzschild metric, namely,

$$\left(\frac{dr}{d\tau}\right)^2 = A^2 - c^2 + 2Mc^2/r, \tag{23.40}$$

must therefore coincide with (23.31) for all times τ at $\rho = \rho_0$; that is, the relation

$$F = 2M \tag{23.41}$$

must hold. Since scale transformations $\bar{\rho} = \bar{\rho}(\rho)$ are still possible, $f(\rho)$ cannot be uniquely determined here; in the following we shall not need $f(\rho)$.

We must now ensure that the interior solution (23.39) and the exterior solution (23.35), (23.36), (23.37) and (23.41) match smoothly at the stellar surface $\rho = \rho_0$. The necessary condition for this is clearly

$$r(\rho_0, c\tau) = K(c\tau)\rho_0. \tag{23.42}$$

If we choose the origin of time in the exterior metric so that $\tau_0(\rho_0) = 0$ then for $\epsilon \neq 0$ the relation (23.42) can only be satisfied for all time τ if both sides have the same functional dependence on τ, that is, only if in (23.25) and (23.29) $h_\epsilon(\eta)$ has the same factor. This leads to the condition

$$6M/f^3(\rho_0) = \kappa\mu c^2 K^3. \tag{23.43}$$

From this and the above equations we obtain $f(\rho_0) = \rho_0$, and hence for condition (23.43) we have finally

$$\kappa\mu c^2 \rho_0^3 K^3 = 6M. \tag{23.44}$$

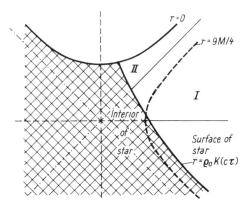

Fig. 23.3. Collapse of a star in the Kruskal diagram.

For $\epsilon = 0$ one immediately obtains the same condition from (23.36), (23.39) and (23.42). It is easy to convince oneself that when (23.44) is satisfied, then the metric is continuous on the surface of the star and the normal derivatives have the required continuity behaviour (16.34).

The condition (23.44) links the mass density μ of the star and its coordinate radius $K\rho_0$ with the externally acting Schwarzschild mass parameter M, in the same way as occurred in (11.23) when the interior and exterior Schwarzschild solutions were joined. If we recall the relation (10.21), that is, the relation $2M = \kappa mc^2/4\pi$ between the Schwarzschild radius $2M$ and the mass m which we would associate with the source of the Schwarzschild solution in the Newtonian gravitational theory, then we have

$$\frac{4\pi}{3}\mu\rho_0^3 K^3(c\tau) = \frac{4\pi}{3}\mu r_0^3 = m. \tag{23.45}$$

Notice that only for $\epsilon = 0$ is m the same as the integral over the mass density μ, calculated in the interior metric (23.39).

The solution found here for the gravitational field of a collapsing star clearly shows that in the interior of the star, too, no peculiarities occur when the stellar surface $\rho = \rho_0$ lies *inside* the Schwarzschild radius $r = K(c\tau)\rho_0 = 2M$; only at $K(c\tau) = 0$ does the interior field become singular.

To end this discussion, we shall follow the fate of a collapsing star in the Kruskal space-time diagram. To do this we draw a radial geodesic on which the points at the surface of the dust star move (Fig. 23.3). On its left is the stellar interior (hatched in the figure) with a metric which is regular up to the point $\tau = 0$ (from outside: $r = 0$). During the collapse a part of the region *II* is revealed to an observer in the exterior space; but the regions *I'* and *II'* (cf. Fig. 23.2) are not realized in the collapse.

23 Possible life history of a spherically symmetric star

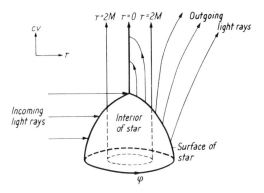

Fig. 23.4. Spherically symmetric collapse of a star in Eddington–Finkelstein coordinates.

For many purposes it is more convenient to describe the collapse in Eddington–Finkelstein coordinates (22.17), since one of the two rotational degrees of freedom can be more easily represented here (remember that the Eddington–Finkelstein coordinates describe just the regions I and II of the Kruskal diagram; that is, they include just those parts of the Schwarzschild solution essential to the collapse). One can see from Fig. 23.4 how the star contracts until it vanishes behind the Schwarzschild radius $r = 2M$, that is, until the emitted light rays

$$c\,dv = \frac{2\,dr}{1 - 2M/r} \qquad (23.46)$$

no longer succeed in reaching the exterior space ($dv < 0$ for $r < 2M$!). The radially ingoing light rays

$$dv = 0, \qquad (23.47)$$

however, can always reach the surface of the star (or the singularity $r = 0$). In order to interpret Fig. 23.4 or similar diagrams one has to remember that the metrical relations of a two-dimensional Riemannian surface are not correctly included in the plane of the paper. It is the possibilities of interaction represented by light rays or by geodesics that are essential.

Bibliography to section 23

Monographs and collected works   de Witt and de Witt (1972), de Witt-Morette (1974), Hawking and Ellis (1973), Hawking and Israel (1987), Stewart and Walker (1973), Zel'dovich and Novikov (1971)

Review and research articles   Miller and Sciama (1980)

242 Gravitational collapse and black holes

24 Rotating black holes

24.1 The Kerr solution

Most known stars are rotating relative to the local inertial system (relative to the fixed stars) and are consequently not spherically symmetric. Their gravitational field is therefore not described by the Schwarzschild solution. In Newtonian gravitational theory, although the field certainly changes because of the rotational flattening of the star, it still remains static, while in the Einstein theory, on the other hand, the flow of matter acts to produce fields. The metric will still be time-independent (for a time-independent rotation of the star), but not invariant under time reversal. We therefore expect that the gravitational field of a rotating star will be described by an axisymmetric, stationary, vacuum solution which goes over to a flat space at great distance from the source. According to the distribution of matter within the star there will be different types of vacuum fields which, in the language of the Newtonian gravitational theory, differ, for example, in the multipole moments of the matter distribution. One of these solutions is the Kerr solution found in 1963, almost fifty years after the discovery of the Schwarzschild metric. It proves to be especially important for understanding the gravitational collapse of a rotating star. To avoid misunderstanding we emphasize that the Kerr solution is not the gravitational field of an arbitrary axisymmetric rotating star, but rather only the exterior field of a very special source.

We shall now discuss the Kerr solution and its properties. Since its mathematical structure is rather complicated, we shall not construct a derivation from the Einstein field equations.

The line element of the Kerr metric has the form, in the so-called Boyer–Lindquist coordinates,

$$\begin{aligned} \mathrm{d}s^2 &= \Sigma\left(\frac{\mathrm{d}r^2}{\Delta}+\mathrm{d}\vartheta^2\right)+(r^2+a^2)\sin^2\vartheta\,\mathrm{d}\varphi^2-c^2\mathrm{d}t^2 \\ &\quad +\frac{2Mr}{\Sigma}(a\sin^2\vartheta\,\mathrm{d}\varphi-c\,\mathrm{d}t)^2, \\ \Sigma &\equiv r^2+a^2\cos^2\vartheta, \quad \Delta=r^2-2Mr+a^2. \end{aligned} \qquad (24.1)$$

For very large r it goes over to the line element of a flat space. From the structure of the far field one can disclose the meaning of the two parameters, M and a. Thus if *in the far field* one transforms the metric (24.1)

24 Rotating black holes

to 'Cartesian coordinates' by $r^2 = x^2 + y^2 + z^2$, $\vartheta = \arctan(\sqrt{x^2+y^2}/z)$, and $\varphi = \arctan(y/x)$, then one obtains

$$g_{4x} = 2May/r^3, \quad g_{4y} = -2Max/r^3, \quad g_{4z} = 0 \qquad (24.2)$$

and

$$g_{44} = 1 - 2M/r. \qquad (24.3)$$

By comparison with the representation (13.32) and (13.33), which is valid for every far field, we can deduce that M is the mass and Ma the z-component (the magnitude) of the angular momentum of the source of the Kerr field. This physical interpretation of the two constants of the Kerr metric is further consolidated by the facts that for $a = 0$ (absence of rotation) (24.1) reduces to the Schwarzschild metric and that the Kerr metric is invariant under the transformation $t \to -t$, $a \to -a$ (time reversal and simultaneous reversal of the sense of rotation).

The Boyer–Lindquist coordinates are generalized Schwarzschild coordinates and like these are not suitable for describing the solution over its full mathematical realm of validity. The coordinates (24.1) are clearly singular for $\Delta = 0$, provided that $0 < a^2 < M^2$, that is, for the two values

$$r_+ = M + \sqrt{M^2 - a^2}$$

and $\qquad (24.4)$

$$r_- = M - \sqrt{M^2 - a^2}.$$

For $a = 0$, r_+ goes over to the Schwarzschild radius $r = 2M$, while r_- goes over to $r = 0$. From now on we shall ignore the parameter region $M^2 < a^2$, which would correspond to very rapidly rotating bodies. It probably does not occur for the fields of real sources, as they would fly apart before rotating so rapidly.

In analogy to the transition from Schwarzschild coordinates to Eddington–Finkelstein coordinates, one can also transform the Kerr solution into a form which has no singularities at $r_\pm$. One introduces a new coordinate v adapted to light propagation by

$$c\,dv = c\,dt + \frac{r^2 + a^2}{\Delta} dr \qquad (24.5)$$

and a new 'angular coordinate' Φ by

$$d\Phi = d\varphi + \frac{a}{\Delta} dr, \qquad (24.6)$$

which takes into account the co-rotation of the local inertial system (compare with our discussion of the action of an angular momentum in section

13.5). The result of these transformations is the Kerr solution in Kerr coordinates,

$$ds^2 = \Sigma\, d\vartheta^2 - 2a \sin^2\vartheta\, dr\, d\Phi + 2c\, dr\, dv$$
$$+ \frac{\sin^2\vartheta}{\Sigma}[(r^2+a^2)^2 - \Delta a^2 \sin^2\vartheta]\, d\Phi^2 \qquad (24.7)$$
$$- \frac{4M}{\Sigma} ra \sin^2\vartheta\, d\Phi\, c\, dv - \left(1 - \frac{2Mr}{\Sigma}\right) c^2 dv^2.$$

The Kerr solution possesses (like every axially symmetric, stationary metric) two commuting Killing vectors, namely – in the coordinates $(r, \vartheta, \varphi, ct)$ or (r, ϑ, Φ, cv) – the vectors

$$\eta^i = (0, 0, 1, 0), \qquad \xi^i = (0, 0, 0, 1). \qquad (24.8)$$

The Killing vector ξ^i, which in the far field is associated with the stationarity (time independence), has an interesting property. Its magnitude

$$\xi_i \xi^i = -(1 - 2Mr/\Sigma) \qquad (24.9)$$

changes sign when one crosses the surface

$$\Sigma - 2Mr = r^2 - 2Mr + a^2 \cos^2\vartheta = 0. \qquad (24.10)$$

Inside this surface the Killing vector ξ^i is spacelike, the metric being no longer stationary there. This surface is therefore called the *limiting surface of stationarity (stationary limit)*.

The physical properties of the Kerr space-time are best brought out (as for the Schwarzschild solution) by studying the possible trajectories of test particles or photons. The details which one assembles in this way are, however, so complicated and confusing that we shall eschew an exhaustive description with proofs and merely give a qualitative discussion of the most important results.

When we approach (Fig. 24.1) the singularity of the Kerr solution, coming from the far field, we encounter first the stationary limit (24.10). Between it and the surface $r = r_+$ lies the so-called *ergosphere*. Particles and light rays can penetrate this region from outside and leave it again. Even the following physical process is in principle possible. A particle of positive energy E_0,

$$-m_0 u_i \xi^i = E_0 > 0, \qquad (24.11)$$

falls from outside along a geodesic into the ergosphere. The energy remains conserved (cf. (19.53)). Under conservation of four-momentum, the particle then splits into two parts:

24 Rotating black holes

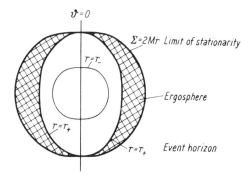

Fig. 24.1. The limiting surfaces of the Kerr solution.

$$m_0 u^i = m_1 u_1^i + m_2 u_2^i. \tag{24.12}$$

Since the Killing vector ξ^i is spacelike inside the ergosphere, because of (24.9), then the timelike vector u_1^i can be chosen so that $E_1 = -m_1 u_1^i \xi_i$ is negative. (In the exterior space this is impossible, because ξ^i is timelike there and the product of two timelike vectors is always negative.) We then have

$$-m_2 u_2^i \xi_i = E_2 = E_0 - E_1 > E_1, \tag{24.13}$$

and the particle leaves the ergosphere again with a greater energy than that of the particle shot in. The rotation of the source is what yields the energy for this process.

The surface $r = r_+$ can of course be crossed by particles or photons from outside, but it is impossible for photons or particles ever to leave the interior space: like the Schwarzschild solution, the Kerr metric describes the gravitational field of a black hole. Since all events which occur inside the radius $r = r_+$, such as the disintegration or radiation of test particles or real particles, are never recorded by an external observer (no photon can reach him from there), this surface is called the *event horizon*.

Further inside is the surface $r = r_-$, which has no particular physical significance, and finally one reaches $r = 0$ – a ring singularity, and not, as one might at first suppose, a point singularity.

The regions of the Kerr solution discussed up until now correspond to the regions *I* and *II* of the Kruskal diagram (Fig. 22.1) of the Schwarzschild solution. It is also possible to extend maximally the Kerr metric; that is, the points corresponding to the regions *I'* and *II'* can be made mathematically accessible.

24.2 Gravitational collapse – the possible life history of a rotating star

The life history of a rotating star differs from the life history sketched in section 23.1 of a spherically symmetric star not in the phases and the sequence of processes which yield up energy and the possible final stages of these processes, but rather by the influence of the rotation on the contraction phase of the star. If a rotating star contracts very strongly, then because of the conservation of angular momentum it will rotate more and more rapidly and possibly break up into separate fragments (only the nucleus carries on contracting). Or, put another way, a star can in general become extremely contracted (for example, to a neutron star) only if it gives up angular momentum to its surroundings. This can occur through ejection of matter or through gravitational interaction with other masses. If, however, at the end it still possesses sufficient mass at very high density, then the gravitational forces become so strong that gravitational collapse takes place, and then the Kerr solution remains as the external gravitational field. Although these ideas seem very plausible and are supported by a large number of facts and calculations, there are two gaps in the theory of the gravitational collapse of a rotating star which to date have not been closed.

The first gap is the complete absence of an internal Kerr solution. For the interior of the star there is no (stable or unstable) solution known with a reasonable equation of state which can be joined to the Kerr metric at the surface of the star; nor does one know any time-variable (interior or exterior) solution whose exterior part changes into a Kerr metric under collapse of the star. For these reasons one cannot say in detail exactly how the collapse proceeds.

The second gap in our present knowledge is our ignorance as to whether, under a gravitational collapse, a Kerr metric always results, or whether there are other, differently constituted, (singular) vacuum solutions, which describe the end stage of the gravitational field of a collapsed star. It is presumed that the star either does not collapse at all or just tends to a Kerr metric; but the proof of this has so far eluded us. The (supposed) uniqueness of the Kerr solution would be a typical property of the Einstein theory: the gravitational field of a collapsed star is characterized by only two parameters, namely, the mass M and the angular momentum Ma, in contrast to the infinitely many parameters (multipole moments) of a non-collapsed star. For the collapse itself this restriction to two parameters signifies that the star must lose all the higher moments

24 Rotating black holes

not appropriate in this scheme by ejection or radiation of mass before it disappears behind the even horizon.

24.3 Some properties of black holes

In this section we want to collect together some properties of black holes, taking particular account of those which are important for an observer in the exterior space. To some extent we shall repeat things said in sections 22-24, but we shall also bring out some new aspects and in particular take into regard the fact that a black hole can be electrically charged (the Kerr-Newman solution which is then appropriate contains the Kerr solution as a special case).

(a) Black holes are solutions of the field equations which describe the gravitational field of collapsed masses. This field is characterized by three parameters: by the mass M, the angular momentum Ma and the electrical charge Q. An external observer can determine these three parameters by observing the trajectories of uncharged and charged particles. Other possible physical properties that the source of the field had before collapse (baryon number, electrical dipole moment,...) are lost during the collapse. The relation between angular momentum and the magnetic moment produced by the rotation is, moreover, the usual 'anomalous' one for the electron.

(b) Black holes contain a closed event horizon. Within this surface the gravitational field is so strong that particles, light rays and time-dependent fields produced inside can no longer leave this region. Particles and light rays from outside can penetrate the horizon; for this they need (as seen by a distant observer) an infinite time. An observer can reach and pass the horizon in a finite time and inside can, it is to be hoped, convince himself of the correctness of the theory described; but he can never report back to the outside.

(c) Inside the event horizon there is a genuine singularity of the gravitational field, which forms during the collapse. Fortunately the universe is so constituted that (because of the event horizon) we cannot see this singularity ('cosmic censorship') and so it is without meaning for physics in the outside universe.

(d) Once it has formed, a black hole is (probably) stable and cannot be destroyed. Matter (mass, radiation) which reaches the black hole from outside can, however, change the charge Q, the mass M and the angular momentum $P = aM$ (the 'indigestible remains' of physical properties of the matter fed in will be emitted in the form of radiation, from outside the horizon of course). But during all these processes the quantity

$$A = 4\pi[2M^2 - Q^2 + 2(M^4 - M^2Q^2 - P^2)^{1/2}] \tag{24.14}$$

can only increase. In the Kerr metric (24.7) A can be visualized as the surface area of $r =$ constant $= r_+$, $v =$ constant (the event horizon). This law is also called (because of certain analogies to thermodynamics) the second law of black-hole dynamics.

The rearrangement

$$M^2 = \frac{A}{16\pi} + \frac{4\pi P^2}{A} + \frac{\pi Q^4}{A} + \frac{Q^2}{2} \tag{24.15}$$

of (24.14) clearly shows that it is indeed at the cost of charge and angular momentum that one obtains energy (mass) from a black hole, but that one cannot go below $M = (A/16\pi)^{1/2}$. These statements also hold for the possible union of two black holes into one. If, for example, two spherically symmetric black holes (of masses M_1 and M_2) coalesce to form one black hole, again spherically symmetric, then we must have

$$16\pi M^2 = A \geq A_1 + A_2 = 16\pi(M_1^2 + M_2^2), \tag{24.16}$$

and so at most the fraction

$$\eta = \frac{M_1 + M_2 - M}{M_1 + M_2} \leq 1 - \frac{\sqrt{M_1^2 + M_2^2}}{M_1 + M_2} \leq 1 - \frac{1}{\sqrt{2}} \tag{24.17}$$

of the mass can be given up in the form of gravitational radiation.

(e) The inclusion of quantum effects could alter this picture radically. We shall return to this point in section 30.4.

24.4 Can and do black holes exist?

The question as to whether these black holes with their remarkable properties really can exist and whether gravitational collapse of a star really does occur can be answered from various standpoints.

The simplest way of all is of course merely to refer to the Einstein theory and the fact that black holes are exact solutions of the field equations. Since the Einstein theory correctly describes the laws of nature, it is to be assumed, in accordance with all our experiences of those laws, that the theoretically possible properties of matter are also realized in nature. But from a purely logical point of view, by referring to the Einstein theory all one has really shown is the internal consistency of the mathematical theory of black holes.

24 Rotating black holes

The answer sketched above is, however, unsatisfactory. One can get rid of the sense of uneasiness which the existence of an event horizon may imply. The lack of a genuine interaction with the matter behind the horizon is only apparent; if the matter has disappeared behind the horizon, then it is left *only* with the properties of mass, angular momentum and charge, and these act outward, are determinable and (within limits) can even be changed from outside. With the physical ideas and concepts that we have developed up until now in weakly curved spaces, we can only formally describe the fate of a collapsing star which contracts to a point or of an observer sent behind the event horizon, but we cannot really understand these things. It is certain that sooner or later quantum effects must be taken into account; but perhaps this deep extrapolation of the realm of validity of the Einstein theory is inadmissible. But how, where and whether the Einstein theory must be modified can only be established by mathematical exploration of this theory and by comparison of the results with (astrophysical) observation.

Do black holes exist somewhere or other in our universe? Do stars suffer a gravitational collapse and is their final state then a black hole or a totally different kind of singularity?

Gravitational collapse is itself very probably associated with an explosive outburst of matter, so that the star would suddenly flare up, rather like a supernova. But this flaring up is not on its own very conclusive, because it could also indicate the formation of a neutron star. Since the far field of a black hole in no way differs from that of an ordinary star, only processes close to the horizon can provide a real proof.

Evidence for the existence of a black hole would become more convincing when one could be sure of the existence of a compact dense quiescent star, for example, the unobserved partner of the pulsar PSR1913+16; see section 10.7. One would then be able to examine the effect of this star on photons and particles which came close to the conjectured event horizon. Matter falling into a rotating black hole, via an accretion disc in the equatorial plane, may during the sharply accelerated terminal stages emit X-rays or gravitational waves of high intensity for short periods, and thus provide evidence for the existence of black holes. A number of such X-ray sources including CygnusX1 and the centre of the Galaxy are being considered seriously.

It is not yet possible to assert with absolute certainty whether or not black holes exist. Whatever the final answer turns out to be will improve our understanding of space-time, that is, gravity, significantly.

Bibliography to section 24

Monographs and collected works   Breuer (1975), Chandrasekhar (1983), de Witt and de Witt (1972), de Witt-Morette (1974), Hawking and Ellis (1973), Hawking and Israel (1987), Stewart and Walker (1973)

Review and research articles   Blandford and Thorne (1979), Carter (1979), Chandrasekhar (1979), Hawking (1975a, b)

Cosmology

Gravitational forces are the only forces presently known which are long range (in contrast to the nuclear forces, for example) and which cannot be compensated (there are no negative masses). It is therefore to be expected that, for large quantities of matter distributed over wide regions of space, they will be the decisive forces, and hence will determine the evolution and dynamics of the universe.

25 Robertson–Walker metrics and their properties

25.1 *The cosmological principle and Robertson–Walker metrics*

Cosmology makes statements about the whole universe; how do we really know that we are not in the situation of the fish in Fig. 25.1? Of course only a finite part of the universe is accessible to human observation and we know even this (spatial) part only over a very short time: a few thousand years for the planetary system, perhaps a hundred years for other galaxies. The answer to this question is that, as in many other areas of the natural sciences so also in cosmology, every new discovery can revolutionize the

Fig. 25.1. The cosmological principle.

structure of our knowledge, our present picture of the universe being in no way complete and secure. But up until now this picture has always proved compatible with assuming initially the universal validity of natural laws, making calculations with strongly simplified models of reality, then comparing with the observations, and thus in a stepwise manner approximating models and formulations of the laws of nature to reality.

The simplest model of the universe is obtained from the cosmological principle, that is, from the assumption that in the rest system of the matter there is no preferred point and no preferred direction, the three-dimensional universe being constituted in the same way everywhere. A glance at the sky (which of course ought to be uniformly bright or dark) shows us that this model is a very great simplification and that the universe is uniform at best only on the average. We do not know how large the spatial regions are over which the average should be taken – at any rate the galaxies are not uniformly distributed, but tend to be clustered. Nevertheless, this most simple of cosmological models can explain observations surprisingly well.

Translated into the language of Riemannian geometry, this cosmological principle clearly asserts that three-dimensional position space is a space of maximal symmetry, that is, a space of constant curvature whose curvature can, however, depend upon time:

$$\overset{(3)}{\mathrm{d}s^2} = g_{\alpha\beta}\mathrm{d}x^\alpha \mathrm{d}x^\beta = K^2(ct)\frac{\mathrm{d}x^2+\mathrm{d}y^2+\mathrm{d}z^2}{(1+\epsilon r^2/4)^2}$$

$$= K^2(ct)\left[\frac{\mathrm{d}\bar{r}^2}{1-\epsilon\bar{r}^2}+\bar{r}^2(\mathrm{d}\vartheta^2+\sin^2\vartheta\, \mathrm{d}\varphi^2)\right] = K^2(ct)\mathrm{d}\sigma^2, \quad \epsilon = 0, \pm 1. \tag{25.1}$$

Since the occurrence of terms $g_{4\alpha}$ in the full space-time metric picks out a spatial direction and $g_{44}(x^\alpha)$ signifies the dependence upon position of the proper time of a test particle at rest, then only the Robertson–Walker metrics (R–W metrics) (Robertson and Walker, 1936)

$$\mathrm{d}s^2 = K^2(ct)\mathrm{d}\sigma^2 - c^2\mathrm{d}t^2 \tag{25.2}$$

are in accord with the cosmological principle. The metric of this model is thus already substantially determined by symmetry requirements; the Einstein field equations can (if they are satisfied at all) now fix only the time behaviour of the universe – the function $K(ct)$ – and the type of the local space – the choice of ϵ. For the reasons explained in section 21.2, K

25 Robertson-Walker metrics and their properties

is called the radius of the universe, although K can only be visualized in this way for closed (three-dimensional) spaces ($\epsilon = +1$).

In the next section we shall discuss first of all some physical properties of the metrics (25.2); we shall not draw conclusions from the Einstein equations until section 26.

25.2 The motion of particles and photons in Robertson-Walker metrics

When, by introducing the coordinate system, we pick out a point in the universe (the origin of the coordinate system), we can also use the R-W metric in the form

$$ds^2 = K^2(ct)[d\chi^2 + f^2(\chi)(d\vartheta^2 + \sin^2\vartheta\, d\varphi^2)] - c^2 dt^2,$$

$$f(\chi) = \begin{cases} \sin\chi & \text{for } \epsilon = 1, \\ \chi & \text{for } \epsilon = 0, \\ \sinh\chi & \text{for } \epsilon = -1. \end{cases} \quad (25.3)$$

The coordinate χ is directly related to the radial displacement of the preferred point: the displacement D from the origin of a star at rest in the coordinate system (25.3) is

$$D = K(ct)\chi. \quad (25.4)$$

If the radius K of the universe changes with time, then the distances of the stars and galaxies also change among themselves, just as the separations of fixed points (fixed coordinates) on a balloon change when the balloon is blown up or deflated. The velocity $\dot{D}$ which thereby results is proportional to the displacement:

$$\dot{D} = \frac{\partial D}{\partial t} = \frac{\dot{K}}{K} cD. \quad (25.5)$$

A test particle or a photon which moves in the absence of forces describes, under suitable choice of the coordinate system, a purely 'radial' trajectory $\chi(\tau)$, $\vartheta = \text{constant}$, $\varphi = \text{constant}$, that is, a geodesic of the metric

$$ds^2 = K^2(ct)d\chi^2 - c^2 dt^2 = -c^2 d\tau^2. \quad (25.6)$$

For a test particle of mass m_0 the conservation law

$$K^2(ct)\frac{d\chi}{d\tau} = \frac{K^2}{\sqrt{1 - \frac{K^2}{c^2}\left(\frac{d\chi}{dt}\right)^2}}\frac{d\chi}{dt} = \text{constant} \quad (25.7)$$

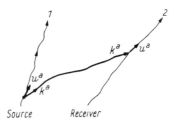

Fig. 25.2. The change in frequency of a photon in the gravitational field.

follows from the geodesic equation. If one denotes by v the speed $K\,d\chi/dt$ and by $p = mv = m_0 v/\sqrt{1-v^2/c^2}$ the momentum of the particle, then the conservation law (25.7) reads in three-dimensional form

$$pK = \text{constant}; \qquad (25.8)$$

that is, the product of the radius of the universe and the magnitude of the momentum is constant for force-free motion.

For photons one expects a similar result, that is, a dependence of the wavelength and the frequency of an emitted photon upon the radius of the universe K. We want now to derive the formula for the more general case that the source and the observer move arbitrarily with respect to the coordinate system (which we shall later identify as the rest system of the matter) (cf. Fig. 25.2).

The world line $x^a(v)$ (null geodesic) of a photon defines the null vector $k^a = dx^a/dv$ uniquely up to a factor which is constant along the world line if we use an affine parameter as the parameter v; that is, one which brings the geodesic equation to the form $k_{a;b}k^b = 0$ (cf. section 8.4). The frequency ν which an observer moving with the four-velocity u^a associates with this photon is proportional to the timelike component of k^a at the location, and in the rest system, of the observer, that is, proportional to $k^a u_a = g_{ab} u^a k^b$. The frequencies measured in the rest system of the source and by the observer are related according to

$$\frac{\nu_1}{\nu_2} = \frac{(k^a u_a)_1}{(k^a u_a)_2} = \frac{(g_{ab} k^a u^b)_1}{(g_{ab} k^a u^b)_2}. \qquad (25.9)$$

This formula describes not only the change in frequency which is a consequence of the relative motion (dependence upon u_1^a and u_2^a), that is, of the Doppler effect, but also the shift in frequency in the gravitational field (dependence upon the metric), and shows that the two effects can only be separated in an artificial manner depending upon the coordinate system.

25 Robertson-Walker metrics and their properties

Applying the formula (25.9) to sources and receivers, which are at rest in the coordinate system (25.6), we have to substitute

$$u^a = (0,0,0,c), \quad k_a = (1,0,0,-1/K), \qquad (25.10)$$

and obtain

$$\frac{\nu_1}{\nu_2} = \frac{K(ct_2)}{K(ct_1)}, \qquad (25.11)$$

or

$$\nu K = \text{constant}, \qquad (25.12)$$

in complete analogy with (25.8).

One usually expresses the change in frequency of the light received at two points through the *red shift* (relative change in wavelength)

$$z = \frac{\lambda_2 - \lambda_1}{\lambda_1}. \qquad (25.13)$$

The equation (25.11) thus yields the relation

$$z = \frac{K(ct_2)}{K(ct_1)} - 1 \qquad (25.14)$$

between the red shift z of the light received on the Earth, for example, at time t_2 and the radii of the universe at the times of emission and reception.

If on the Earth at the present time $t = t_2$ one examines the light emitted by a star at the time $t = t_1$, then, if the radius of the universe does not change too quickly and the light travel time $t_2 - t_1$ is not too large, one can replace $K(ct_1)$ by the first few terms of the Taylor series

$$K(ct) = K(ct_2)[1 + Hc(t - t_2) - \tfrac{1}{2}qH^2c^2(t - t_2)^2 + \ldots]. \qquad (25.15)$$

The parameters occurring in this expansion are the Hubble constant

$$H(ct_2) = \dot{K}(ct_2)/K(ct_2) \qquad (25.16)$$

and the acceleration parameter (retardation parameter)

$$q(ct_2) = -\ddot{K}(ct_2)K(ct_2)/\dot{K}^2(ct_2). \qquad (25.17)$$

Substitution of the series (25.15) into (25.14) gives the relation

$$z = Hc(t_2 - t_1) + \left(1 + \frac{q}{2}\right)H^2c^2(t_2 - t_1)^2 + \ldots \qquad (25.18)$$

between the red shift z and the light travel time $t_2 - t_1$.

The validity or applicability to our universe of the model of a Robertson–Walker metric is usually tested in the relation between the red shift and the distance of the source. Since $ds^2 = 0$ for light, from (25.6) and (25.15) it follows that, to first approximation,

$$\chi = \int_{t_1}^{t_2} \frac{c\,dt}{K(ct)} \approx \frac{c(t_2 - t_1)}{K(ct_2)} + \frac{Hc^2(t_2 - t_1)^2}{2K(ct_2)} + \ldots, \qquad (25.19)$$

and therefore, using (25.4) and (25.5),

$$z = HD + (q+1)H^2 \frac{D^2}{2} + \ldots = \frac{\dot{D}}{c} + \frac{1}{2c^2}(\dot{D}^2 - D\ddot{D}) + \ldots. \qquad (25.20)$$

The red shift is *to first approximation* proportional to the present distance D of the source and at the same time to the ratio of the (cosmological) escape velocity $\dot{D}$ of the source to the velocity of light.

25.3 Distance measurement and horizons in Robertson-Walker metrics

The determination of distance in astronomy is mostly done using the concepts and ideas of a three-dimensional Euclidean space. We therefore want to describe briefly how the laws of light propagation in R-W metrics influence the determinations of distance.

One possible way of determining the distance of an object is to compare its absolute brightness L, which is defined as the total radiated energy per unit time and is regarded as known, with the apparent brightness I of the energy reaching the receiver per unit time and per unit surface area. The *luminosity distance* D_L is defined by

$$D_L = \sqrt{L/4\pi I}, \qquad (25.21)$$

so that in the Euclidean space luminosity distance and geometrical distance coincide. In a Robertson-Walker metric there exists a complicated relationship between true distance $D = K\chi$ and brightness distance D_L (cf. Fig. 25.3). The photons streaming out from the source $\chi = 0$ are distributed, after a coordinate interval of χ, not of course over the surface $4\pi\chi^2$, but, in the metric (25.3), over the surface

$$F = 4\pi f^2(\chi) K^2(ct_2). \qquad (25.22)$$

Moreover, because $ds^2 = 0$, that is, because

$$K\,d\chi = c\,dt, \qquad (25.23)$$

the photons emitted during the time interval δt are distributed in the neighbourhood of the source over the interval $\delta_1 \chi = c\delta t / K(ct_1)$, while at the

25 Robertson–Walker metrics and their properties

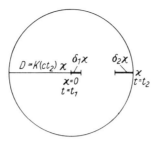

Fig. 25.3. The relation between apparent brightness and coordinate distance.

receiver in the time δt arrive all those photons which lie in an interval $\delta_2 \chi = c\delta t / K(ct_2)$. And third, the energy of an individual photon has also changed during its passage through the gravitational field by the factor $K(ct_1)/K(ct_2)$. We therefore finally obtain for the apparent brightness

$$l = \frac{L}{4\pi f^2(\chi) K^2(ct_2)} \frac{K^2(ct_1)}{K^2(ct_2)}, \tag{25.24}$$

that is, the relation

$$D_L = \frac{f(\chi) K^2(ct_2)}{K(ct_1)} = (1+z) D \frac{f(\chi)}{\chi} \tag{25.25}$$

between the luminosity distance D_L, the coordinate distance D (at time t_2) and the red shift z of a light source. Since one observes stars with $z > 2$, D and D_L can differ from one another considerably.

A second possible way of determining distance is to compare the true diameter Δ of a system with the angle δ which it subtends at the Earth. In Euclidean space we have of course

$$D_A \equiv D = \Delta / \delta \tag{25.26}$$

for the distance D_A determined by measurement of angle.

In an R–W metric, however, (25.3) implies, according to Fig. 25.4,

$$D_A = \frac{\Delta}{\delta} = f(\chi) K(ct_1) = \frac{D}{1+z} \frac{f(\chi)}{\chi}. \tag{25.27}$$

These two examples of how to determine distance show clearly how the space curvature comes into astronomical considerations concerning the law of light propagation. Unexpected effects can occur. If, for example, the function $K(ct_1)$ decreases with t, for increasing $f(\chi)$, then the more distant of two objects of identical dimensions will have the greater angular diameter.

Fig. 25.4. Measurement of distance by determination of angle.

Of course optical methods can only be used to determine the distances of objects whose light reaches us. In flat space we can in principle see every flash of light, however distant, if we wait sufficiently long to allow for the finite velocity of light. In a curved space, however, the situation is more complicated. Let us imagine, for the purposes of illustration, a fly which is crawling at constant velocity away from the south pole of a balloon. By blowing up the balloon (increasing the radius of curvature) sufficiently rapidly, can one prevent the fly (a photon) from reaching the north pole?

Light emitted at time t_1 at the origin $\chi = 0$ has, because of (25.3), reached the point

$$\chi = \int_{t_1}^{t_2} \frac{c\,dt}{K(ct)} \qquad (25.28)$$

by time t_2. If we want to know whether at the present time $t = t_2$ we can see all stars, then we must investigate whether the light sent out at the beginning of the universe t_B (the earliest possible time), from the furthest possible star, can reach us, or whether our signal sent out at the beginning of the universe and at the origin of the coordinate system has by now reached all stars. Depending on the cosmological model the beginning of the universe is here $t_B = -\infty$ or the first zero of $K(ct)$ (where the metric becomes singular) lying in the past; the terms 'beginning of the universe' and 'end of the universe' are discussed in section 27.2.

At the present time $t = t_2$ we can see stars up to a maximum coordinate distance

$$\chi_P = \int_{t_B}^{t_2} \frac{c\,dt}{K(ct)}. \qquad (25.29)$$

If this value χ_P is smaller than the maximum coordinate distance (which is π in closed universes and ∞ in open ones); that is, if not all stars are visible, then χ_P defines the horizon up to which we can see. It is called the *particle horizon*.

25 Robertson-Walker metrics and their properties

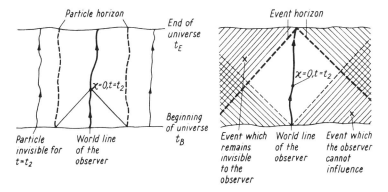

Fig. 25.5. Particle and event horizons.

If, for example, the radius of the universe changes according to the law $K(ct) = ct^2$ and we find ourselves in the contraction phase $t < 0$, then at the observer time $t_2 = -1$

$$\chi_P = \int_{-\infty}^{-1} \frac{dt}{t^2} = 1; \qquad (25.30)$$

that is, in this cosmological model there is a particle horizon.

Another physically interesting question is whether (by means of the photons emitted there) we can learn about every event occurring in the universe, no matter when or where, or whether the end of the universe t_E ($t_E = \infty$ or the next zero of $K(ct)$ lying in the future) coming prematurely prevents this. An equivalent question is whether our light signal sent out at the present time $t = t_2$ reaches all points of the universe before its end t_E. Since this light signal traverses a maximum coordinate distance

$$\chi_E = \int_{t_2}^{t_E} \frac{dct}{K(ct)} \qquad (25.31)$$

there exists an *event horizon* χ_E if χ_E is smaller than π or ∞: we shall never learn anything about events which at the present time $t = t_2$ are situated at distances greater than χ_E.

Figures 25.5 and 25.6 give a qualitative picture of how horizons work.

A possible misinterpretation of the significance of the horizons should also be dealt with. Should there be an event horizon in our universe, that is, events about which we can never learn anything, then that would not imply the absence of an interaction with that part of the universe or the establishing of something which is in principle not knowable. Our (very

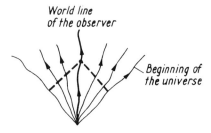

Fig. 25.6. A cosmological model without event horizon: the past light cone going out from the observer intersects the world lines of *all* particles.

poor) cosmological model presupposes from the beginning that the universe is everywhere the same, and we therefore know in advance that the same gravitational field and the same mass densities, and so forth, are present behind the horizon as close by, because without the presence of these masses as well the space in our neighbourhood would not be homogeneous and isotropic. The events which we cannot establish only affect test particles, that is, particles without a gravitational interaction, and it is only we who cannot detect these particles, which do act upon observers situated nearer to them.

25.4 *Physics in closed universes*

There exists a multitude of cosmological models (up until now we have encountered in the Robertson–Walker metrics only the most primitive), which do not always differ significantly from one another, since they form a continuous sequence. There are, however, some characteristics of spaces which can be expressed by integers; in this class belongs the property of whether a universe is open or closed. It is to be expected that closed universes also differ from open ones in a clear physical way and that this difference may even possibly lead to statements which can be tested on the Earth. We therefore want to describe in more detail some properties of closed universes with R–W metrics.

As can be shown by calculation of the Weyl tensor (18.1) or by explicitly carrying out a coordinate transformation, all R–W metrics are conformally flat. If we restrict ourselves to the closed space

$$\mathrm{d}s^2 = K^2(ct)[\mathrm{d}\chi^2 + \sin^2\chi(\mathrm{d}\vartheta^2 + \sin^2\vartheta\,\mathrm{d}\varphi^2)] - c^2\mathrm{d}t^2, \quad (25.32)$$

then after the transformations

$$T = \int \frac{dt}{K(ct)} \tag{25.33}$$

and

$$r = \frac{2 \sin \chi}{\cos \chi + \cos cT}, \quad c\eta = \frac{2 \sin cT}{\cos \chi + \cos cT} \tag{25.34}$$

the line element takes the form

$$ds^2 = \frac{K^2(ct)}{4}[\cos \chi + \cos cT]^2[dr^2 + r^2(d\vartheta^2 + \sin^2 \vartheta \, d\varphi^2) - c^2 d\eta^2], \tag{25.35}$$

which differs from that of a Minkowski space only by a conformal factor.

As one can immediately see from the transformation formula (25.34), this statement has only a local significance: the relations (25.34) map a section $(\cos \chi + \cos cT) \neq 0$ of the curved space-time onto the full Minkowski space, but a one-to-one mapping of the metrics (25.35) and (25.32) onto one another is impossible.

Diagrams, using the concept of a conformal factor mentioned above are very useful for the qualitative discussion of the behaviour near infinity, since a region of infinite extent can be portrayed as a finite one. The crucial point is that under a conformal transformation $ds^2 = 0$ implies $\Omega^2 ds^2 = 0$, so that light rays (null geodesics) remain light rays (see section 8.3). In such a picture the boundary of Minkowski space, given by $r, \eta \to \pm\infty$, becomes a light cone $\cos \chi + \cos cT = 0$, that is, $\chi = \pm cT \pm \pi$, in much the same way that after a stereographic projection the boundary of a plane becomes a point. These ideas are best illustrated in a *Penrose diagram* (Fig. 25.7) in which the spatial dimensions corresponding to ϑ and φ are suppressed.

The null geodesics of Minkowski space all start on the past light cone $\mathcal{J}^-$ (scri minus, from scri = script i) and end on the future light cone $\mathcal{J}^+$; timelike geodesics go from I^- to I^+, while spacelike geodesics end at I^0. (Because χ is cyclic I^0 is a point.) Asymptopia has therefore a structure; one must be precise in specifying the direction in which one goes to infinity.

It is possible to discuss infinity for other space-times using null, timelike and spacelike geodesics and introducing the corresponding boundaries $\mathcal{J}^\pm$, $I^\pm$ and I^0. However $\mathcal{J}^\pm$ is not necessarily a light cone, as, for example, in Fig. 25.5, where the boundaries labelled beginning of world and end of universe in Figs. 25.5 and 25.6 correspond to $\mathcal{J}^+$ and $\mathcal{J}^-$. Figure

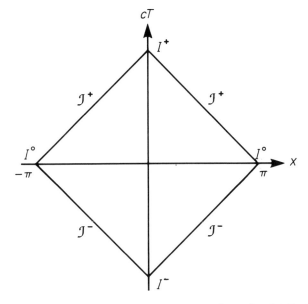

Fig. 25.7. The Penrose diagram of the section $ds^2 = dr^2 - c^2 d\eta^2$ of the Minkowski space $ds^2 = dr^2 + r^2(d\vartheta^2 + \sin^2\vartheta\, d\varphi^2) - c^2 d\eta^2$.

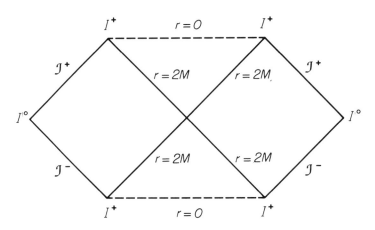

Fig. 25.8. The Penrose diagram of the Schwarzschild solution.

25.8 shows the Penrose diagram for the Schwarzschild space-time; see also section 22.3, especially Figs. 22.1 and 22.2.

We now return to the Robertson–Walker metrics.

For physical investigations the most important statement that one can deduce from the transformed line element

25 Robertson-Walker metrics and their properties

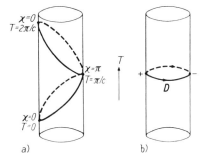

Fig. 25.9. The Einstein universe with (a) the light cone and (b) the field line of a point charge.

$$ds^2 = K^2(ct)[d\chi^2 + \sin^2\chi(d\vartheta^2 + \sin^2\vartheta\, d\varphi^2) - c^2 dT^2], \quad (25.36)$$

which involves only the new time coordinate T, is that all closed Robertson-Walker metrics are conformally equivalent to the spatially closed, static Einstein universe

$$\left.\begin{array}{l} ds^2_{\text{Einstein}} = d\chi^2 + \sin^2\chi(d\vartheta^2 + \sin^2\vartheta\, d\varphi^2) - c^2 dT^2, \\ 0 \leq \varphi \leq 2\pi, \quad 0 \leq \chi \leq \pi, \\ 0 \leq \vartheta \leq \pi, \quad -\infty < T < +\infty. \end{array}\right\} \quad (25.37)$$

This (special) Einstein universe is the simplest closed cosmological model. As we shall see later, it is not, however, a realistic model of our universe. By suppressing two dimensions (those corresponding to the coordinates ϑ and φ) one can explain some properties of this universe by a model of a cylindrical surface (Fig. 25.9).

Light rays emanating from a point (for example, $\chi = 0$) meet again at the antipodal point ($\chi = \pi$), and finally all return to the starting point at a correspondingly later time. Since light rays (null geodesics) of the Einstein universe coincide with those of the closed R-W models (the statement $ds^2 = 0$ is conformally invariant), light rays would also return to their source in the more general closed universe, if the proper time t available were sufficient to cover the interval $\Delta T = 2\pi/c$. Here one should compare (25.33) with the criteria (25.28) and (25.29) for the occurrence of horizons.

The source-free Maxwell equations are conformally invariant (cf. (8.41)); if we know their general solution in the Einstein universe, then we also have at hand the general solutions in every closed R-W universe. Since all spatial coordinates are periodic in the Einstein universe, the source-free

Maxwell equations have the character of eigenvalue equations for the frequency. Amongst the solutions one finds a generalized plane wave (eigenfunction with frequency c/λ), which in the neighbourhood of a point is practically a plane wave, but whose amplitude is noticeably different from zero only over a region

$$d \approx \sqrt{\lambda K}. \tag{25.38}$$

The influence of the space curvature 'localizes' the plane wave and makes it similar to a particle trajectory. For visible light and a radius of the universe of 2×10^{10} light years, we obtain the value $d \approx 10^7$ km.

If in an Einstein universe one draws the field lines of the D field emanating from a point charge at rest, then they all intersect at the antipodal point, but they arrive there with the opposite sign: in a closed universe, to every charge there corresponds a charge of opposite sign (which, however, is not necessarily situated at the antipodal point). This intuitively obtained statement can be derived from the Maxwell equations. For the charge density j^4/c we have

$$\frac{1}{c}j^4 = F^{4n}{}_{;n} = \frac{1}{\sqrt{-\overset{4}{g}}}[\sqrt{-\overset{4}{g}}\,F^{4n}]_{,n} = \frac{1}{\sqrt{\overset{3}{g}}}[\sqrt{\overset{3}{g}}\,F^{4\alpha}]_{,\alpha} \tag{25.39}$$

or, in three-dimensional form with $F^{4\alpha} = D^\alpha$,

$$D^\alpha{}_{;\alpha} = \frac{1}{c}j^4. \tag{25.40}$$

Since the closed three-dimensional space has no surface (a spherical surface has no boundary), application of the Gauss law yields

$$\int j^4 \mathrm{d}^3 V = 0; \tag{25.41}$$

that is, the Maxwell equations can only be integrated if the total charge vanishes. Our universe is uncharged on the average, and so in this respect a closed cosmological model would not stand in contradiction with experiment.

The conclusion deduced from (25.40) evidently concerns only the mathematical structure of this equation, not with its physical interpretation: the volume integral of any quantity which can be written as a three-dimensional divergence must vanish. Since there are only positive mass densities μ, a Newtonian gravitational theory, for example,

$$\Delta U = U^{,\alpha}{}_{;\alpha} = -\mu, \tag{25.42}$$

is not possible in a closed universe (25.36) and (25.37).

26 *Robertson–Walker metrics and Friedmann universe* 265

If there exists in a closed universe a Killing vector ξ^n proportional to a four velocity u^n,

$$\xi_{n;i} + \xi_{i;n} = 0, \qquad \xi^n = \alpha u^n, \qquad (25.43)$$

then because of the definition of the curvature tensor, the Einstein field equations and the general splitting (8.74) of the energy–momentum tensor of a fluid, we can write

$$(\xi^{i;n} - \xi^{n;i})_{;n} = -2\xi^{n;i}{}_{;n} = -2\xi^m R_m{}^{ni}{}_n = -2\kappa(\xi^m T_m{}^i - \tfrac{1}{2}\xi^i T^n{}_n)$$
$$= \alpha\kappa(3p + \mu c^2)u^i + \alpha\kappa c^2 q^i. \qquad (25.44)$$

Because of the formal similarity to the Maxwell equations one can conclude that for every closed universe the integral over the time component ($i = 4$) of the right-hand side of (25.44) – over the analogue of the charge density – must vanish. But this is clearly not possible for perfect fluids ($q^i = 0$, $p > 0$, $\mu > 0$): there exist no static or stationary, spatially closed cosmological model with perfect fluid media, whose Killing vector is parallel to the four-velocity. In the language of thermodynamics this can also be formulated as (cf. section 8.5): a cosmological model with closed three-dimensional space and perfect fluid ($q^i = 0$) cannot exist in complete thermodynamical equilibrium (the temperature vector u^a/T cannot be a Killing vector).

In these last considerations we have already made use of the Einstein equations. We shall now turn to the problem of determining the evolution of the R–W metrics from these field equations.

Bibliography to section 25

Textbooks Weinberg (1972)

Monographs and collected works Hawking and Ellis (1973), Neugebauer (1980), Peebles (1971, 1980)

Review and research articles Ellis (1971), Neugebauer (1974), Schrödinger (1940), Stephani (1974)

26 The dynamics of Robertson–Walker metrics and the Friedmann universe

26.1 *The Einstein field equations for Robertson–Walker metrics*

The Robertson–Walker metrics are completely determined by the behaviour with respect to time of the radius of the universe and by the sign of the curvature, that is, by $K(ct)$ and ϵ. We are thus confronted with the problem of calculating these parameters from the properties of the matter in our universe, and of seeing whether observational results and cosmological model can be brought into agreement.

The curvature tensor and the Ricci tensor of an R-W metric can be calculated relatively easily by applying the reduction formulae (16.20) to the line element

$$ds^2 = K^2(ct)d\sigma^2 - c^2dt^2 = g_{\alpha\beta}dx^\alpha dx^\beta - c^2dt^2; \qquad (26.1)$$

that is, by starting from

$$\begin{aligned} R_{\alpha\beta\mu\nu} &= \overset{3}{R}_{\alpha\beta\mu\nu} - \frac{\dot{K}^2}{K^2}(g_{\beta\mu}g_{\alpha\nu} - g_{\beta\nu}g_{\alpha\mu}), \\ R^4{}_{\beta\mu\nu} &= 0, \\ R^4{}_{\beta 4\nu} &= \frac{\ddot{K}}{K}g_{\beta\nu}, \\ K_{\alpha\beta} &= -\frac{\dot{K}}{K}g_{\alpha\beta}, \end{aligned} \qquad (26.2)$$

and in accordance with (21.26) substituting the relation

$$\overset{3}{R}_{\alpha\beta\mu\nu} = \epsilon K^{-2}(g_{\alpha\mu}g_{\beta\nu} - g_{\beta\mu}g_{\alpha\nu}) \qquad (26.3)$$

for the three-dimensional curvature tensor of the R-W metric, whose space is of course a space of constant curvature. For the non-vanishing components of the Ricci tensor we obtain

$$\begin{aligned} R_{\beta\nu} &= [\ddot{K}/K + 2(\dot{K}^2 + \epsilon)K^{-2}]g_{\beta\nu}, \\ R_{44} &= -3\ddot{K}/K. \end{aligned} \qquad (26.4)$$

Together with the Einstein field equations

$$R_{mn} - \tfrac{1}{2}Rg_{mn} = \kappa T_{mn}, \qquad (26.5)$$

the equations (26.4) show us that the energy–momentum tensor of the matter in the universe is spatially isotropic in the coordinates (26.1) and that no current of energy occurs ($T_{4\alpha} = 0$): in R-W metrics the energy–momentum tensor *must* be that of a perfect fluid,

$$T_{mn} = pg_{mn} + (\mu + p/c^2)u_m u_n, \qquad (26.6)$$

where the preferred coordinate system (26.1) is the rest system of the matter and μ and p depend only upon time.

26 Robertson-Walker metrics and Friedmann universe

As a consequence of (26.4) and (26.6) the field equations (26.5) reduce to

$$2\ddot{K}/K + (\dot{K}^2 + \epsilon)/K^2 = -\kappa p \tag{26.7}$$

and

$$3(\dot{K}^2 + \epsilon)/K^2 = \kappa\mu c^2. \tag{26.8}$$

These two equations are only mutually compatible if

$$\dot{\mu}/(\mu + p/c^2) = -3\dot{K}/K. \tag{26.9}$$

Since for $\dot{K} \neq 0$ and $\mu c^2 + p \neq 0$ (26.7) also follows from (26.9) and (26.8), the field equation (26.7) can be replaced by (26.9).

Equation (26.8) is called the Friedmann equation (Friedmann, 1922), and the special R-W metrics which satisfy it are called Friedmann universes. Occasionally, only the cosmological model arising from the special case $p = 0$ is designated the Friedmann universe. If one knows the equation of state $f(\mu, p) = 0$, then from (26.9) one can determine the radius K as a function of the mass density μ and hence calculate the behaviour of K and μ with respect to time from (26.8).

The Friedmann cosmological models can also be characterized invariantly by the fact that they are just those solutions of the Einstein field equations with a perfect fluid as source whose velocity fields $u^n(x^i)$ are free of rotation, shear and acceleration.

26.2 The most important Friedmann universes

The Einstein universe Soon after having set up his field equations, Einstein (1917) tried to apply them to cosmology. In accordance with the then state of knowledge, he started from a static cosmological model. Thus, for this Einstein universe, all the time derivatives in (26.7) and (26.8) vanish, so that we are left with the equations

$$\epsilon/K^2 = -\kappa p, \quad 3\epsilon/K^2 = \kappa\mu c^2. \tag{26.10}$$

These can only be brought into agreement with the observed data, which requires that $p \approx 0$, by rather artificial means, namely, by introduction of the cosmological constant Λ. According to this hypothesis, the energy-momentum tensor contains, in addition to the contribution due to the gravitating matter (here the dust), a contribution proportional to the metric tensor:

$$\kappa T_{mn} = -\Lambda g_{mn} + \bar{\mu} u_m u_n, \quad \bar{\mu} > 0, \quad \Lambda = \text{constant}. \tag{26.11}$$

Comparison of (26.11) with (26.10) gives us, upon use of

$$\kappa p = -\Lambda, \qquad \kappa \mu c^2 = \kappa \bar{\mu} c^2 + \Lambda, \qquad (26.12)$$

the relations

$$\epsilon = +1, \qquad \Lambda = +1/K^2, \qquad \kappa \bar{\mu} c^2 = 2/K^2; \qquad (26.13)$$

in an Einstein universe we are dealing with a closed universe of constant curvature:

$$ds^2 = K^2[d\chi^2 + \sin^2\chi(d\vartheta^2 + \sin^2\vartheta\, d\varphi^2)] - c^2 dt^2, \quad K = \text{constant}. \quad (26.14)$$

The de Sitter universes The introduction of the cosmological constant means that the space is curved even in the complete absence of matter. If one substitutes

$$\kappa T_{mn} = -\Lambda g_{mn}, \qquad \Lambda = \text{constant}, \qquad (26.15)$$

into the field equations (26.7) and (26.8), then one obtains

$$K\ddot{K} = \dot{K}^2 = \epsilon \qquad (26.16)$$

and

$$\Lambda = 3\frac{\dot{K}^2 + \epsilon}{K^2}. \qquad (26.17)$$

The best starting point for the integration of this system is equation (26.17) differentiated once, namely,

$$\ddot{K} - \frac{\Lambda}{3}K = 0. \qquad (26.18)$$

For positive Λ one obtains the proper de Sitter metrics (de Sitter, 1917)

$$\left.\begin{array}{ll} \epsilon = +1: & K = \dfrac{1}{B}\cosh Bct, \\[6pt] \epsilon = -1: & K = \dfrac{1}{B}\sinh Bct, \qquad \Lambda = 3B^2, \\[6pt] \epsilon = 0: & K = Ae^{Bct}, \end{array}\right\} \qquad (26.19)$$

for negative Λ

$$\epsilon = -1: \qquad K = \frac{1}{B}\cos cBt, \qquad \Lambda = -3B^2, \qquad (26.20)$$

and for $\Lambda = 0$ the flat space $\epsilon = 0$, $K = \text{constant}$.

The de Sitter universes have a higher symmetry than might be supposed from their description by Robertson–Walker metrics. If from (26.2), (26.3),

26 Robertson-Walker metrics and Friedmann universe

(26.16) and (26.17) one calculates the complete four-dimensional curvature tensor of these spaces, then one obtains

$$R_{abmn} = \frac{\Lambda}{3}(g_{am}g_{bn} - g_{an}g_{bm}). \tag{26.21}$$

Thus we are dealing with four-dimensional spaces of constant curvature (of positive curvature for $\Lambda > 0$) in which neither any space-direction nor any time-direction is singled out. In particular, the three metrics (26.19) are only three different sections of the same four-dimensional space of constant positive curvature. According to our present knowledge, the de Sitter models do not provide a realistic model of our universe except possibly for an inflationary period in the early stage (cf. section 27.2).

The radiation universe Incoherent, isotropic electromagnetic radiation can formally be described by the energy–momentum tensor (26.6) of a perfect fluid with

$$p = \tfrac{1}{3}\mu c^2. \tag{26.22}$$

With the aid of this equation of state we can at once integrate (26.9) and obtain

$$\mu c^2 K^4 = \text{constant} = A, \tag{26.23}$$

which says that when the universe expands or contracts the mass density (energy density) of the radiation is inversely proportional to the fourth power of the radius of the universe.

The behaviour of this universe with respect to time is determined by

$$\dot{K}^2 = \frac{\kappa A}{3K^2} - \epsilon; \tag{26.24}$$

upon introduction of $y = K^2$ the differential equation (26.24) becomes

$$\frac{\dot{y}^2}{4} = \frac{\kappa A}{3} - \epsilon y, \tag{26.25}$$

which can easily be integrated. If we choose the constant of integration so that $y(t_0) = 0$, then we obtain the solutions

$$\epsilon = 0: \quad K^2 = 2c\sqrt{\frac{\kappa A}{3}}(t - t_0),$$

$$\epsilon = -1: \quad K^2 = c^2(t-t_0)^2 + 2c\sqrt{\frac{\kappa A}{3}}(t-t_0), \tag{26.26}$$

$$\epsilon = +1: \quad K^2 = -c^2(t-t_0)^2 + 2c\sqrt{\frac{\kappa A}{3}}(t-t_0).$$

270 Cosmology

Although we certainly do not live in a radiation universe now, several properties of these solutions are worth noting. One such is the occurrence of a singularity of the metric at $t = t_0$. There K goes to zero, the separation of two arbitrary points in the universe becomes arbitrarily small, and in the neighbourhood of this singularity the radius K becomes independent of ϵ; that is, the same for open and closed universes. Another interesting statement is that electromagnetic radiation (light) alone can, by virtue of its own gravitational interaction, produce a closed universe whose radius K increases from zero to a maximum of $\kappa A/3$ and then after the time $\Delta T = 2\sqrt{\kappa A/3c^2}$ goes back to zero again.

The Friedmann universe By Friedmann universes in the strict sense one means cosmological models with dust:

$$T_{mn} = \mu u_m u_n. \tag{26.27}$$

For this special case one can immediately integrate (26.9) to give

$$\mu c^2 K^3 = \hat{M} = \text{constant}. \tag{26.28}$$

The integration constant $\hat{M}$ is evidently proportional to the total mass for closed universes. Notice the changed power-dependence upon K in comparison to the radiation universe (26.23)!

The remaining field equation (26.8) simplifies to the 'Friedmann differential equation'

$$\dot{K}^2 = \frac{\kappa \hat{M}}{3K} - \epsilon. \tag{26.29}$$

Introduction of the new variables

$$cT = \pm \int \frac{dct}{K(ct)} \tag{26.30}$$

brings it to the form

$$K'^2 = \kappa \hat{M} K/3 - \epsilon K^2, \tag{26.31}$$

in which it can easily be solved by separation of variables. If we again denote the time at which K vanishes by t_0, then the solutions of (26.29) are parametrically (cf. Fig. 26.1)

$\epsilon = 0$: $\quad K = \kappa \hat{M} c^2 T^2/12, \quad\quad c(t - t_0) = \pm \kappa \hat{M}(cT)^3/36,$ (26.32)

$\epsilon = -1$: $\quad K = \dfrac{\kappa \hat{M}}{6}(\cosh cT - 1), \quad c(t - t_0) = \pm \dfrac{\kappa \hat{M}}{6}(\sinh cT - cT),$

(26.33)

26 Robertson-Walker metrics and Friedmann universe

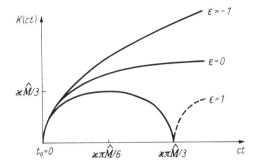

Fig. 26.1. The radius of the universe K as a function of time for the three Friedmann models.

and

$$\epsilon = +1: \quad K = \frac{\kappa \hat{M}}{6}(1 - \cos cT), \quad c(t - t_0) = \pm \frac{\kappa \hat{M}}{6}(cT - \sin cT). \tag{26.34}$$

All three types have a singularity at the 'beginning of the universe' $t = t_0$, where the radius K goes to zero. In the neighbourhood of this singularity the three types have the same dependence upon time, namely,

$$K(ct) \approx (3\kappa \hat{M}/4)^{1/3} t^{2/3}. \tag{26.35}$$

For the closed model ($\epsilon = +1$), $K(ct)$ reaches a maximum and then goes back again to zero, describing a cycloid. In the two open models $K(ct)$ increases continuously (if we take as the positive direction of time that in which T increases).

26.3 Consequences of the field equations for models with arbitrary equation of state having positive pressure and positive rest mass density

From the special theory of relativity we know that the equation of state of ordinary matter lies between that of dust ($p = 0$, $\mu > 0$) and that of incoherent radiation ($\mu c^2 = 3p$) in the sense that $\mu c^2 - 3p \geq 0$ (in the microscopic picture the pressure is caused by collisions of particles with at most the velocity of light). Some notable properties of the Friedmann model follow, however, from the field equations alone and the assumptions that $\mu > 0$ and $p > 0$, independently of the particular kind of equation of state (temporarily dependence!).

Thus from (26.7) and (26.8) one obtains the relation

$$6\ddot{K} = -\kappa(\mu c^2 + 3p)K, \qquad (26.36)$$

which can be interpreted in the following way. If $\dot{K}$ is positive at a time t, then, because of (26.36), $K(ct)$ is a curve concave downwards (like the curves of Fig. 26.1), which must have touched the axis $K = 0$ a finite time ago. If $\dot{K}$ is negative, this point $K = 0$ lies in the future. Since, as we shall describe in detail, we are currently observing a positive $\dot{K}$, the universes with Robertson-Walker metrics inevitably have an 'origin of the universe' $t = 0$ at which the metric becomes singular (K becomes zero), independently of the equation of state and the choice of ϵ. By comparison of the function $K(ct)$ with its tangent at the time t, the age of the universe can be estimated in terms of the Hubble parameter H according to

$$ct < K(ct)/\dot{K}(ct) = 1/H(ct). \qquad (26.37)$$

If one writes (26.9) in the form

$$\frac{\partial}{\partial K}(\mu c^2 K^3) = -3pK^2, \qquad (26.38)$$

then one can see that $\mu c^2 K^3$ increases into the past, possibly even becoming infinite: for $K \to 0$, μ increases at least as fast as K^{-3}. Hence one can ignore the term proportional to ϵ in equation (26.8) and near the origin of the universe calculate with

$$3\dot{K}^2 = \kappa\mu c^2 K^2. \qquad (26.39)$$

The expansion behaviour of the early universe does not depend upon ϵ; it is the same for open and closed models.

For the future behaviour of the universe, on the other hand, ϵ will have a real effect. Because of (26.38), for increasing K the rest mass density μ decreases at least like K^{-3}, and hence the term $\kappa\mu c^2 K^2$ goes at least like K^{-1}. Thus for $\epsilon = 1$, a maximum $\dot{K} = 0$ will be reached in a finite time, and since $K = $ constant is not a solution of (26.36), the radius function will decrease again and will necessarily reach $K = 0$ again: a closed universe with $\epsilon = 1$ executes a cycle (or several cycles). For $\epsilon = -1$, $\dot{K}^2$ can never become zero, and the universe expands continuously, $\dot{K}$ tending to the value 1 ($K(ct) \approx ct$ for $t \to \infty$). The universes with $\epsilon = 0$ also expand continuously, only now $\dot{K}$ and $\ddot{K}$ go to zero.

In all these universes there are *particle horizons;* an observer cannot always see the whole universe at time t. Because of (25.28), the existence of such horizons obviously depends crucially upon the behaviour of $K(ct)$ at the origin $t_B = 0$, and therefore we substitute the ansatz $K \sim (ct)^\alpha$, $\alpha > 0$, into (26.36). From the signs alone of both sides of the resulting equation it follows that $\alpha < 1$. For small times we have

$$\chi_P \sim (ct)^{1-\alpha}, \quad 0 < \alpha < 1, \qquad (26.40)$$

so that χ_P is finite near $t = 0$ and smaller than π, and also for arbitrary finite times χ_P is finite: in open models part of the universe is always invisible. In closed models, however, after a sufficient time χ_P can take the value π or even 2π. Thus, for example, in the Friedmann universe (26.34) $\chi_P = cT = \pi$ for the time of maximum expansion (the whole space is visible) and $\chi_P = 2\pi$ at the end of the universe (the observer sees his world line, that is, he himself, at the beginning of the universe).

The occurrence of *event horizons* depends upon the behaviour of $K(ct)$ at the end of the universe. Since for open models ($\epsilon = 0, -1$) the radius function K goes at most like t^{-1} for large t, then the integral (25.4) diverges for $t_E \to +\infty$: there is no event horizon; provided one waits long enough, one learns of every event. In closed models ($\epsilon = +1$), however, there do exist event horizons (the proof runs as with the above considerations regarding particle horizons): an observer will not necessarily learn anything before the universe comes to an end about events which take place after the stage of maximal expansion.

Bibliography to section 26

Textbooks Weinberg (1972)

27 Our universe as a Friedmann model

27.1 Red shift and mass density

It was one of the most important confirmations of the ideas of the theory of general relativity and its application to cosmology when the cosmological red shift was found by Hubble in the year 1929, about thirteen years after the basic equations had been set up and seven years after the publication of the Friedmann model. In the meantime there had of course been the detour and error of Einstein, who believed he could arrive at a cosmological model belonging to relativity theory only by the introduction of the cosmological constant, which led to the Einstein universe, a static model without red shift.

Not only does the red shift verify the cosmology of general relativity, and in particular the concept of the expanding universe, but its exact evaluation also gives us data to determine which of the homogeneous, isotropic cosmological models our universe most closely resembles. From the red shift (as a function of distance; cf. (25.20)) the Hubble parameter H and the acceleration parameter q can in principle be determined. If one substitutes them according to their definitions

$$H = \dot{K}/K, \quad q = -\ddot{K}K/\dot{K}^2 \tag{27.1}$$

into the field equations (26.7) and (26.8), then one obtains

$$\left.\begin{array}{l} 6qH^2 = \kappa(\mu c^2 + 3p), \\ 3H^2 = \kappa\mu c^2 - 3\epsilon/K^2. \end{array}\right\} \tag{27.2}$$

In general these two equations are of course not sufficient to determine the unknowns μ, p, ϵ and K from the red shift, that is, from a knowledge of q and H^2. But for our universe in its present state the predominant part of μ is contained in the masses of the galaxies and the pressure can consequently be ignored. For this dust we then have

$$6qH^2 = \kappa\mu c^2 \tag{27.3}$$

and

$$H^2(2q-1) = \epsilon/K^2. \tag{27.4}$$

Since ϵ can only take on the values $0, \pm 1$, it can be determined from the value of q alone: $q > \frac{1}{2}$ gives a closed universe, $q \leq \frac{1}{2}$ the two open models. If ϵ is fixed, then from H and q one can determine the radius function K and the mass density μ, and compare them with observations. That mass density $\dot{\mu}$ which corresponds precisely to the critical value $q = \frac{1}{2}$ (the transition from an open to a closed model of the universe) is called the critical mass density:

$$\mu_{\text{crit}} = 3H^2/\kappa c^2. \tag{27.5}$$

Unfortunately present measurements and analyses of the red shift–distance relation (25.20) are still so incomplete and inexact (essentially because of the large systematic error in the distance determination), that the relations (27.3) and (27.4) cannot yet be tested and evaluated. The following numerical values based on the red shift (H and q) and the analysis of galaxy counts (μ), are the most probable to date:

$$\left.\begin{array}{l} H = 6 \times 10^{-29}\,\text{cm}^{-1}, \quad cH = 55\,\text{km/s Mpc}, \\ 1/cH = 18 \times 10^9\,a, \end{array}\right\} \tag{27.6}$$

$$q = 1 \pm 1, \tag{27.7}$$

and
$$\mu = 3 \times 10^{-31} \text{ g cm}^{-3}. \tag{27.8}$$

If one compares the three numerical values with the relations (27.3) and (27.4) then one establishes that:

(a) The presently observed mass density lies below the critical density
$$\mu_{\text{crit}} = 6 \times 10^{-30} \text{ g cm}^{-3} \tag{27.9}$$
which means that we ought to be living in an open universe.

(b) Our universe has a radius of about $K \approx H^{-1} = 1.8 \times 10^{10}$ light years, is about 1.1×10^{10} years old and is in an expansion phase.

(c) Since q cannot yet be determined exactly enough from the red shift and also μ is not yet known with sufficient certainty, we cannot yet say whether our universe is open or closed.

Taking into consideration the surprisingly rapid change in the 'certain' numerical values of H, q and μ in the last few decades one can regard only the following as reliably established:

(a) The age of the universe, which follows from the Hubble parameter and from the age of rocks or of stellar systems, is of the order of magnitude 10^{10} years (the uncertainty is by a factor of 2 or 3).

(b) The average mass density in the universe is about $\mu = 10^{-30}$ g cm^{-3} (uncertain by a factor of 10).

(c) There is no doubt concerning the cosmological nature of the red shift, and hence the applicability of relativity theory to cosmology.

As we have shown above, in the early epoch of the universe the parameter ϵ played no essential part. Our ignorance of the exact value of the acceleration parameter q, that is, of the value of ϵ, thus does not put the value of the models for the earliest developmental stages of our universe in jeopardy. In the following we therefore want to sketch the ideas embodied in these models.

27.2 The earliest epochs of our universe and the cosmic background radiation

In direct optical observation of very distant objects we are really looking a considerable way back into the past of our universe. But the origin of the universe corresponds to an infinitely large red shift, and so the red shift is still very large indeed for times close to the origin, which is therefore invisible in practice. Thus if from our observations we want to obtain statements about the constitution of the universe in its early phase, then we must look at physical objects closer to us, and judge from their present condition, and the laws governing their (local) evolution, the state of the universe when they were formed.

How then did our universe appear at the beginning? In order to be able to answer this and similar questions at all meaningfully, we have to define more precisely the rather mysterious concept of the 'beginning of the universe.' The beginning of the universe does not mean that no matter was present before or that it was created at that instant; rather, this phrase should express the fact that on the basis of physical laws the state of the universe was essentially different from its present state. The concepts of the 'age of the universe' and the 'end of the universe' are to be understood analogously. In the framework of the Friedmann model, the beginning of the universe is that time in the past at which the radius of curvature K was zero, and the universe manifested a singular behaviour. But this statement requires amplification on two counts.

Firstly, both our knowledge of the laws of physics and the simplest cosmological model of a homogeneous isotropic space are insufficient to make well-founded statements about physical processes near the origin of the universe, that is, at extremely high desnities of matter. All observations and calculations point to the fact that about 10^{10} years ago the universe was essentially different from what it is like today, and was probably in a state of very high density. Cosmological models therefore begin with conditions in which interactions of elementary particles are the decisive processes.

Again, one must take into account the fact that time has no absolute meaning, that the measurement of time must always be seen in relation to the properties of the matter. The time coordinate (universal time) t of the Friedmann universe is the proper time for the mass element of the universe. The clocks which one uses for measuring proper time have zero dimensions in the abstract theory; in practice, this means that they are so small that the cosmological gravitational field does not change within the clocks and during the lapse of one period. While at the present time therefore the planetary system, for example, is a useful clock, in the early universe only elementary particles and their transformations are available. Measured by the number of characteristic, individual physical processes going on, the beginning of the universe is still very far away (possibly infinitely far away) even in the early phases (close to the singularity), and the unit of measurement derived from the planetary system, the year, does not correctly express this.

27 Our universe as a Friedmann model

Let us return to the model of the early stages of the Friedmann universe. Now the major contribution to the energy–momentum tensor comes from stars (pressure-free matter or "dust") and the contribution from radiation is negligibly small, but in the early stages of the universe a rather different balance must have occurred. On the one hand as the radius of the universe K decreases, the energy density of radiation increases faster than that for dust because of equations (26.23) and (26.28). On the other hand the energy density and temperature would rise so much that massive elementary particles and antiparticles, that would be unstable under terrestrial conditions, would be in thermal equilibrium with the high-energy photons. Thus a precise description of the earliest epoch is only possible if quantum theory (elementary particle theory) is taken into account, and so we can extrapolate into the past only as far as we know the laws of high-energy physics, *taking gravity into account*. This is a highly speculative area, but the following broad ideas are generally accepted.

The universe began in a state of extremely high temperature and density, which can only be described accurately through the projected unification of quantum theory and gravity. In the subsequent expansion an epoch could have occurred in which quantum effects produced an energy–momentum tensor proportional to the metric tensor, corresponding to a negative pressure! During this epoch the world could be described by a de Sitter universe, in which the radius K increased exponentially (see equation (26.19)); this is known as an *inflationary universe*.

This rapid expansion of the universe reduced the temperature, so that equilibrium preferred the stable particles, namely, the electrons, protons, atomic nuclei, the lighter chemical elements, and the neutrinos and photons generated in particular by pair annihilation. All these are particles whose physical behaviour we understand sufficiently to make confident predictions.

During further expansion and cooling the photons then decouple in the following senses. On the one hand, photons are not created to any great extent, they do not have sufficient energy for pair production, nor are they able to give their energy to the remaining matter (the universe is 'transparent'); and, on the other hand, the energy density of the photons decreases more rapidly than that of the rest of the matter, so that the subsequent behaviour of the expansion is not influenced by the photons.

From this time on the energy–momentum tensor of the photon gas alone thus obeys the conservation law

$$T^{ik}_{\text{Ph};k} = 0, \tag{27.10}$$

from which it follows, because of (26.9), that the energy density $\mu_{\text{Ph}} c^2$ again obeys the relation

$$\mu_{\text{Ph}} c^2 K^4 = A, \tag{27.11}$$

where now, however, in contrast to (26.23), the evolution of the radius function $K(ct)$ is dictated from outside (by the main component of the matter, that is, by the matter in atomic nuclei). Since we also have for the photon gas, according to the Planck radiation law,

$$\mu_{\text{Ph}} c^2 = \text{constant} \times T^4, \tag{27.12}$$

its temperature decreases with increasing radius function as

$$T \sim 1/K. \tag{27.13}$$

The experimental confirmation of these considerations (first made by Gamow as early as 1948), namely, the discovery of the incoherent cosmic background radiation by Penzias and Wilson in 1965 (Nobel Prize, 1978), was certainly the greatest success of relativity theory in cosmology since the interpretation of the Hubble red shift. Observations show that the Earth is bathed by an isotropic (incoherent) electromagnetic radiation, whose frequency spectrum corresponds to the radiation of a black body at temperature

$$T_0 \approx 2.7 \text{ K} \tag{27.14}$$

with a maximum intensity near the wavelength $\lambda_0 \approx 0.2$ cm. (Since the earliest measurements could be fitted by a slightly higher temperature, this radiation is still also called the 3 K radiation.) The energy density of this radiation today corresponds to a mass density of about

$$\mu_{\text{Ph}} \approx 4.4 \times 10^{-34} \text{ g cm}^{-3}. \tag{27.15}$$

If one assumes that the photons uncouple from the rest of the matter at about 4000 K, then the cooling which has taken place in the meantime corresponds, because of (27.14), (27.13) and (25.14), to a red shift of

$$z = \frac{4000}{2.7} - 1 \approx 1480. \tag{27.16}$$

The cosmic background radiation thus gives us immediate optical access to the early epoch of the Universe, back to much earlier times than are accessible to optical instruments by observation of distant objects (for which $z \leq 4$). The high degree of isotropy of this radiation shows that even at this time (if one assumes that initial anisotropies were dissipated) or up to this time (if one thinks of the inhomogeneities resulting later from the formation of galaxies) the universe was Friedmann-like and that the Earth moves at most with a small velocity relative to the rest system of the total matter.

A similar analysis of the neutrino radiation which is also present, according to the ideas contained in the model described above, is not yet possible because of the difficulties of obtaining experimental evidence for the neutrinos.

To end this section we make some brief remarks about the evolution of the universe *after* the formation of the electromagnetic background radiation. During the gradual cooling of the 'primaeval fire-ball' the chemical elements hydrogen and helium form in the preferred equilibrium ratio of about 73:27, almost no heavier elements being synthesized. Small disturbances to the homogeneity of the universe then lead to galaxy formation, and there the subsequent compression and heating of matter in the stars leads to nuclear processes, during which the heavier elements are produced. All these things are still the subject of research, however, and the details are not very well understood.

27.3 *A Schwarzschild cavity in the Friedmann universe*

The assumption of a position-independent mass-density in the universe leads, as we have seen, to useful cosmological models with properties which approximate to the observations, but they stand in flat contradiction to the mass distribution to be found in our neighbourhood. Here the mass is always concentrated into individual objects (planets, stars, galaxies), and the practically matter-free space in between exceeds the volume of these objects by several orders of magnitude.

This discrepancy can at least partially be taken into account, since the exact solution for the gravitational field of a spherically symmetric star which is situated in a special Friedmann universe ($p = 0$) and surrounded by a matter-free space is known (Einstein and Straus, 1945). We shall now state this solution and discuss its properties.

The details of our physical model are as follows (Fig. 27.1): A spherically symmetric star is surrounded by a space free of matter; the gravitational field inside the star can be described, for example, for a static star, by the interior Schwarzschild metric or for a collapsing or exploding star by the section of a Friedmann universe. A Schwarzschild solution

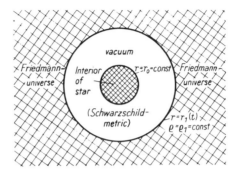

Fig. 27.1. The Schwarzschild vacuole in the Friedmann universe.

$$ds^2 = \frac{dr^2}{1-2M/r} + r^2(d\vartheta^2 + \sin^2\vartheta\, d\varphi^2) - (1-2M/r)c^2 dt^2,$$

$$r_0 \leq r \leq r_1, \tag{27.17}$$

can always be joined to this interior solution, and to this Schwarzschild solution a Friedmann universe can always be joined. This has been discussed in detail and proved in section 23.3, although always from the viewpoint of a connection 'inwards.' One can easily prove, however, that all the calculations are equally valid for the connection 'outwards' which is described here. This connection can be most simply achieved by introducing a new coordinate system into the Schwarzschild metric via the coordinate transformation

$$r = r(\rho, c\tau), \quad ct = ct(\rho, c\tau), \tag{27.18}$$

so that we have

$$\left.\begin{aligned}ds^2 &= \left(\frac{\partial r}{\partial \rho}\right)^2 \frac{d\rho^2}{1-\epsilon f^2(\rho)} + r^2(\rho, c\tau)(d\vartheta^2 + \sin^2\vartheta\, d\varphi^2) - c^2 d\tau^2, \\ \left(\frac{\partial r}{\partial c\tau}\right)^2 &= -\epsilon f^2(\rho) + 2M/r,\end{aligned}\right\} \tag{27.19}$$

where the boundary surface $\rho = \rho_1$ to the expanding or contracting universe

$$ds^2 = K^2(c\tau)\left[\frac{d\rho^2}{1-\epsilon\rho^2} + \rho^2(d\vartheta^2 + \sin^2\vartheta\, d\varphi^2)\right] - c^2 d\tau^2 \tag{27.20}$$

is at rest. The join between the Schwarzschild metric (27.17) or (27.19) and the Friedmann universe (27.20) is possible if and only if between the mass density $\mu(c\tau)$ and the curvature $K(c\tau)$ of the universe, on the one hand,

27 Our universe as a Friedmann model

and the gravitational radius $2M$, the Newtonian mass $m = 8M/\kappa c^2$ and the 'radius' ρ_1, or $r_1 = r_1(\rho_1, c\tau)$, on the other hand, the relations

$$\kappa \mu c^2 K^3 \rho_1^3 = 6M = 3\kappa mc^2/4\pi, \quad r_1 = \rho_1 K(c\tau) \tag{27.21}$$

hold. These relations ensure that the dust particles of the boundary surface between the vacuum and the cosmic matter move on geodesics both of the interior Schwarzschild metric and of the exterior universe; that is, the system exists in a dynamical equilibrium.

In a Friedmann universe one can thus construct one or several spherically symmetric cavities which gather up the originally dusty matter within each of these spaces and put it back into the middle of the cavities as stars. Notice, however, that the total mass extracted

$$m_h = \int_0^{\rho_1} \sqrt{g}\, \mu\, d\vartheta\, d\varphi\, d\rho = 4\pi K^3 \mu \int_0^{\rho_1} \frac{\rho^2\, d\rho}{\sqrt{1-\epsilon\rho^2}} \tag{27.22}$$

differs, for $\epsilon \neq 0$, from the stellar mass m (only insignificantly for small radii ρ_1).

The most interesting property of the vacuole solution is that inside the vacuole the field is static. The expansion or contraction of the universe has no influence on the physical processes inside the vacuole, except that the radius r of the vacuole is time-dependent. An observer inside is only made aware of the cosmic expansion through the red shift of objects lying beyond the boundary surface. Stars inside show no red shift.

How then is the vacuole in which we live constructed? The relation

$$r_1 = \sqrt[3]{6M/\kappa \mu c^2} \tag{27.23}$$

between the coordinate radius r_1 of the vacuole, the Schwarzschild radius $2M$ of the central body and the cosmic mass density μ is crucial for the size of the vacuole. If we measure r_1 and $2M$ in centimetres and substitute the value 3×10^{-31} g cm^{-3} for μ, then we have

$$r_1 \approx 1.75 \times 10^{19} \sqrt[3]{2M}. \tag{27.24}$$

The vacuole radius of the Earth ($2M = 0.88$) would thus extend out far beyond the Sun, and many of the nearest fixed stars would in fact be contained within the vacuole of the Sun ($2M = 2.95 \times 10^5$). But we obtain a realistic model if we identify the central body with the local group of galaxies ($2M \approx 5 \times 10^{17}$) to which belongs not only, for example, our Galaxy, but also the Andromeda nebula; there is then no other galaxy in the associated vacuole. Inside this system the expansion of the universe is not effective, the gravitational field being determined exclusively by the masses contained within the vacuole (in so far as the model is applicable, that is, as the mass distribution is to good approximation spherically symmetric). The radius of the Earth or of the Earth's orbit thus cannot change because of the cosmic expansion. But of course the central body, that is, our Galaxy, could also be in a state of general expansion which is independent of the cosmic expansion.

Of more theoretical interest is the fact that measurements of time in the universe and in the vacuole do not exactly agree. As a consequence of the method of joining, the universal time τ of the Friedmann universe is identical with the proper time τ of the particles on the boundary layer $r = r_1$, which move on radial geodesics. Because of (22.6) and (23.42)–(23.46), this proper time differs from the coordinate time t of the Schwarzschild metric by the factor

$$\frac{dt}{d\tau} = \frac{c\sqrt{1-\epsilon \rho_1^2}}{1 - 2M/\rho_1 K(c\tau)}. \tag{27.25}$$

In general this factor is ignorably small.

Vacuoles cannot be arbitrarily large. Trivially, the vacuole radius $\rho_1 = \sin \chi_1$ may not exceed the maximum value π of the coordinate χ in a closed universe ($\epsilon = +1$). (The coordinate system used here covers only the half universe with $0 \leq \rho \leq 1$.) A second, more important, condition follows from the requirement that the vacuole radius r_1 must be outside the Schwarzschild radius $r = 2M$. Because of the relation (27.21) between vacuole radius and mass and the special form

$$K(c\tau) = \frac{\kappa \mu c^2 K^3}{6} k(c\tau), \tag{27.26}$$

of the time dependence of the Friedmann model (cf. (26.28) and (26.32)–(26.34)), the condition $r_1 > 2M$ just mentioned is only satisfied for

$$\rho_1^2 < \tfrac{1}{2} k(c\tau). \tag{27.27}$$

The maximum allowed vacuole radius $K\rho_1$ depends upon the age of the universe. In a closed Friedmann universe the vacuole is always smaller than the semi-universe ($\rho_1 < 1$). A galaxy (a group of galaxies) of mass M can thus only occur in a universe if the age of the universe τ is greater than that required by

$$M^2 = k(c\tau) K^2(c\tau)/8. \tag{27.28}$$

In a closed model ($\epsilon = +1$) with a period of about 6×10^9 years, that is, $\kappa M/6 \approx 9 \times 10^{27}$ s, a vacuole of mass $M \approx 2.5 \times 10^{17}$ cm, for example, can form at the earliest 128 days after the beginning of the universe.

Bibliography to section 27

Textbooks Weinberg (1972)

Monographs and collected works Dautcourt (1972), Goldberg *et al.* (1973), Hawking and Israel (1987), Kundt (1971), Longair (1974), Sciama (1971), Weinberg (1977)

Review and research articles Einstein and Straus (1945), Gamow (1948), Gott *et al.* (1974), Kramer (1969), Sandage and Tamman (1971, 1974*a, b, c, d,* 1975*a, b*), Schücking (1954), Wesson (1974), Zel'dovich (1979)

28 General cosmological models

28.1 *What is a cosmological model?*

A cosmological model is a model of our universe which, taking into account and using all known physical laws, predicts (approximately) correctly the observed properties of the universe, and in particular explains in detail the phenomena in the early universe. Such a model must also explain *inter alia* why the universe was so homogeneous and isotropic at the epoch of last scattering of the cosmic microwave background, and how and when inhomogeneities (galaxies and stars) arose. It should also explain whether the special properties of our universe (compared with other conceivable universes) depend on particular initial conditions, or whether the laws of nature ensure that a stable and quasi-permanent universe, similar to our own, must have occurred independently of the initial conditions.

In a more restricted sense cosmological models are exact solutions of the Einstein field equations for a perfect fluid that reproduce the important features of our universe. Because there is only one actual universe the large number of known or possible cosmological models may at first seem surprising. There are, however, two reasons for this multiplicity.

Firstly, only a section of our universe is known, both in space and in time, and even in this section the observational results are often very uncertain (for example, the value of the acceleration parameter). All cosmological models which differ only near the origin of the universe must be accepted as equally valid. In fact a series of solutions is known which are initially inhomogeneous or anisotropic to a high degree, and which then increasingly come to approximate a Friedmann universe. All cosmological models which yield a red shift and a cosmic background radiation can hardly be refuted. The possibility cannot be excluded that our universe is *not* homogeneous and isotropic, but has those properties only approximately in our neighbourhood. An expanding 'dust star,' that is, a section of a Friedmann universe which is surrounded externally by a static Schwarzschild metric (cf. the model of the collapsing star in section 23.3), may also perhaps be an excellent model of the universe.

Secondly, one also examines solutions of the field equations where it is clear in advance that they do not correctly reproduce the properties of our universe. Every model is of course a great simplification of reality, and only by the study of many solutions can one establish which simplifications are allowed and which assumptions are essential. Exaggerating somewhat, one can say that there is almost no exact solution of the field equations to which one could not attribute the name 'cosmological model.'

A special rôle is played now as before by cosmological models which satisfy the cosmological principle to such an extent that the universe (the three-dimensional position space) is homogeneous, that is, that the points on a section $t = $ constant are physically indistinguishable. Besides the Robertson–Walker metrics, these models include all spaces which possess a simply transitive group of motion G_3 and are accordingly to be associated with one of the Bianchi types *I* to *IX* (cf. section 19.5) or which permit a transitive group G_4 which possesses no transitive subgroup G_3 (Kantowski–Sachs model). We want to go briefly into two examples of such cosmological models.

28.2 *Solutions of Bianchi type I with dust*

If the three-dimensional space is the rest space of the matter and possesses three commuting Killing vectors, then we are dealing with homogeneous cosmological models of Bianchi type *I*. Since one can simultaneously transform the three Killing vectors to the normal forms

$$\underset{1}{\xi^a} = (1,0,0,0), \quad \underset{2}{\xi^a} = (0,1,0,0), \quad \underset{3}{\xi^a} = (0,0,1,0) \quad (28.1)$$

in suitably chosen coordinates, the metric depends only upon the time coordinate $x^4 = ct$. By the transformation $x^{4\prime} = x^{4\prime}(x^4)$, $x^{\alpha\prime} = x^\alpha + f^\alpha(x^4)$ one can, without destroying (28.1), always bring the metric to the normal form

$$\mathrm{d}s^2 = -c^2\mathrm{d}t^2 + g_{\alpha\beta}(ct)\mathrm{d}x^\alpha \mathrm{d}x^\beta. \quad (28.2)$$

28 General cosmological models

As one can see, the subspaces $t = $ constant are flat, three-dimensional spaces in which Cartesian coordinates can always be introduced, for a fixed t.

To calculate the curvature tensor of this metric we need the reduction formulae (16.20). They give

$$
\begin{aligned}
R_{\alpha\beta\mu\nu} &= -\tfrac{1}{4}(\dot{g}_{\beta\mu}\dot{g}_{\alpha\nu} - \dot{g}_{\beta\nu}\dot{g}_{\alpha\mu}), \\
R^4{}_{\beta\mu\nu} &= 0, \\
R^4{}_{\beta 4\nu} &= \tfrac{1}{2}\ddot{g}_{\beta\nu} - \tfrac{1}{4}\dot{g}_{\beta\alpha}\dot{g}_{\mu\nu}g^{\alpha\mu}.
\end{aligned} \quad (28.3)
$$

The field equations

$$R_{ab} - \tfrac{1}{2} R g_{ab} = \kappa \mu u_a u_b \qquad (28.4)$$

then read, if we remember the relations

$$\dot{g}/g = g^{\alpha\beta}\dot{g}_{\alpha\beta} \quad \text{and} \quad \dot{g}^{\beta\nu} = -g^{\beta\alpha}g^{\nu\mu}\dot{g}_{\alpha\mu},$$

$$R_4^4 - \frac{R}{2} = -\frac{1}{8}\dot{g}_{\beta\nu}\dot{g}^{\beta\nu} - \frac{1}{8}\left(\frac{\dot{g}}{g}\right)^2 = -\kappa\mu c^2, \qquad (28.5)$$

and

$$R_\beta^\alpha - \delta_\beta^\alpha \frac{R}{2} - \frac{1}{2\sqrt{-g}}(\sqrt{-g}\,g^{\alpha\rho}\dot{g}_{\rho\beta})^\cdot - \delta_\beta^\alpha \frac{\kappa\mu c^2}{2} = 0 \qquad (28.6)$$

(the equations $R_\alpha^4 = 0$ are satisfied identically). Because of the equation of conservation of rest mass (8.89), which always holds for dust, the system of field equations is only integrable if

$$\kappa\mu c^2 \sqrt{-g} = \hat{M} = \text{constant}. \qquad (28.7)$$

In order to integrate the field equations we take the trace of (28.6), which gives the differential equation

$$(\sqrt{-g})^{\cdot\cdot} = \tfrac{3}{2}\hat{M}, \qquad (28.8)$$

which we can solve as

$$\sqrt{-g} = \frac{3ct}{4}(\hat{M}ct + A). \qquad (28.9)$$

The complete system (28.6) can be integrated once, using (28.7), with the result

$$\dot{g}_{\beta\alpha} = \frac{\hat{M}ct}{\sqrt{-g}} g_{\beta\alpha} + \frac{a_\alpha^\mu}{\sqrt{-g}} g_{\mu\beta}. \qquad (28.10)$$

If one introduces a Cartesian coordinate system at an arbitrary fixed point and arranges its axes so that the constant matrix a_α^μ is diagonal, then because of (28.10) the diagonal form of the metric remains preserved for all time. Hence from (28.9) and (28.10) follows

$$\dot{g}_{11} = \left[\frac{4}{3} \frac{\hat{M}}{\hat{M}ct + A} + 2 \frac{p_1 A}{ct(\hat{M}ct + A)} \right] g_{11}, \quad p_1 A = \tfrac{2}{3} a_1^1 \qquad (28.11)$$

with the solution

$$g_{11} = \text{constant}\,(\hat{M}ct + A)^{4/3} \left(\frac{ct}{\hat{M}ct + A} \right)^{2p_1}. \qquad (28.12)$$

Analogous results can be obtained for g_{22} and g_{33}. Thus we have finally the solution

$$\begin{aligned}
&ds^2 = -c^2 dt^2 + g_{11} dx^2 + g_{22} dy^2 + g_{33} dz^2, \\
&g_{11} = (-g)^{1/3} \left(\frac{ct}{\hat{M}ct + A} \right)^{2p_1 - 2/3}, \\
&g_{22} = (-g)^{1/3} \left(\frac{ct}{\hat{M}ct + A} \right)^{2p_2 - 2/3}, \quad \kappa\mu c^2 \sqrt{-g} = \hat{M}, \\
&g_{33} = (-g)^{1/3} \left(\frac{ct}{\hat{M}ct + A} \right)^{2p_3 - 2/3}, \quad \sqrt{-g} = 3ct(\hat{M}ct + A)/4,
\end{aligned} \qquad (28.13)$$

in which, because of (28.6) and (28.5), the three coefficients p_i must satisfy the conditions

$$p_1 + p_2 + p_3 = 1, \quad p_1^2 + p_2^2 + p_3^2 = 1 \qquad (28.14)$$

which is guaranteed by, for example,

$$\begin{aligned}
&2p_1 - \frac{2}{3} = \frac{4}{3} \sin \alpha, \quad 2p_2 - \frac{2}{3} = \frac{4}{3} \sin\left(\alpha + \frac{2\pi}{3}\right), \\
&2p_3 - \frac{2}{3} = \frac{4}{3} \sin\left(\alpha + \frac{4\pi}{3}\right), \quad -\frac{\pi}{6} < \alpha \le \frac{\pi}{2}.
\end{aligned} \qquad (28.15)$$

For the four-velocity $u^a = (0, 0, 0, c)$ of the field-producing matter we have

$$u_{a;b} = \frac{c}{2} g_{ab,4}. \qquad (28.16)$$

Thus we are dealing (compare with the definitions (17.11) of the kinematic quantities) with a geodesic ($\dot{u}_a = 0$), rotation-free ($\omega_{ab} = 0$) flow, whose expansion velocity is

$$\Theta = \frac{2\hat{M}ct + A}{t(\hat{M}ct + A)}, \qquad (28.17)$$

28 General cosmological models

and the components of whose shear velocity are

$$\sigma_{ii} = \frac{cg_{ii}}{\sqrt{-g}} \frac{A}{4}(3p_i - 1) \quad \text{(no summation over } i\text{).} \tag{28.18}$$

The integration constant A is therefore a measure of the shear, while the p_i characterize its dependence upon direction.

The particular case $A = 0$ which is contained in (28.13) leads at once to

$$\dot{g}_{\beta\alpha} = \frac{4}{3ct} g_{\beta\alpha} - \frac{a_\beta^\mu g_{\mu\alpha}}{c^2 t^2}. \tag{28.19}$$

The field equations, however, can only be satisfied for $a_\beta^\mu \equiv 0$: when $A = 0$ we are dealing with an (isotropic) Friedmann universe with $\epsilon = 0$.

The metric (28.13) describes an anisotropic, homogeneous universe, which is expanding or contracting. The distances between the dust particles (which of course are at rest in these coordinates) change in a direction-dependent fashion, as the isotropic case $p_1 = p_2 = p_3$ stands in contradiction to (28.14). For $A > 0$ (which can always be achieved by choice of the time direction) the metric becomes singular at $t = 0$, if we approach the origin from the positive t side.

In the general case $\alpha \neq \pi/2$ ($p_3 \neq 0$) precisely one of the p_i, namely, p_3, is negative. Because

$$\frac{\dot{g}_{33}}{g_{33}} = \frac{1}{ct(\hat{M}ct + A)}\left(\frac{4}{3}\hat{M}ct + 2p_3 A\right), \quad p_3 < 0, \tag{28.20}$$

then the relative change in distances in the z-direction is very strongly negative at very small times. This collapse comes to a halt for $ct = -3p_3 A/2\hat{M}$ and it is followed by an expansion. In the x-direction and the y-direction, on the other hand, the universe expands continuously. If we follow its history backwards from positive t, then from an initial sphere we find a very long, thin, elongated ellipsoid, and in the limiting case $t \to +0$ a straight line – a 'cigar' singularity – occurs.

It is worth noting that the mass $\hat{M}$ does not affect the behaviour as $t \to 0$; the metric (28.13) can be approximately replaced by the vacuum solution (Kasner metric)

$$\left.\begin{array}{l} ds^2 = (ct)^{2p_1}dx^2 + (ct)^{2p_2}dy^2 + (ct)^{2p_3}dz^2 - c^2 dt^2, \\ p_1 + p_2 + p_3 = 1, \quad p_1^2 + p_2^2 + p_3^2 = 1. \end{array}\right\} \tag{28.21}$$

In the exceptional case $\alpha = \pi/2$, that is, $p_1 = 1, p_2 = p_3 = 0$, we have

$$\frac{\dot{g}_{11}}{g_{11}} = \frac{\frac{4}{3}\hat{M}ct + 2A}{ct(\hat{M}ct + A)}, \quad \frac{\dot{g}_{22}}{g_{22}} = \frac{\dot{g}_{33}}{g_{33}} = \frac{4\hat{M}}{3(\hat{M}ct + A)}. \tag{28.22}$$

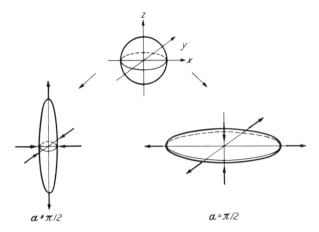

Fig. 28.1. The two types of singularity of a Bianchi type I universe.

A singular behaviour occurs for $t \to +0$ only in the x-direction, and then in such a way that (followed backwards in time) out of a sphere first a strongly flattened, rotating ellipsoid is formed and finally a 'pancake' singularity (cf. Fig. 28.1).

For large times the metric approaches (independently of α) that of a homogeneous and isotropic Friedmann universe with $\epsilon = 0$.

28.3 The Gödel universe

The Gödel universe (Gödel, 1949), is a homogeneous, but anisotropic, four-dimensional space with the metric

$$ds^2 = a^2[dx^2 + \tfrac{1}{2}e^{2x}dy^2 + dz^2 - (e^x dy + c\,dt)^2], \quad a = \text{constant}, \quad (28.23)$$

or

$$ds^2 = 4a^2[dr^2 + dz^2 + (\sinh^2 r - \sinh^4 r)d\varphi^2$$
$$- 2\sqrt{2}\sinh^2 r\, d\varphi\, c\,dt - c^2 dt^2]. \quad (28.24)$$

It possesses five Killing vectors, which in the coordinates (28.23) have the form

$$\underset{1}{\xi^a} = (0,1,0,0), \quad \underset{2}{\xi^a} = (0,0,1,0), \quad \underset{3}{\xi^a} = (0,0,0,1),$$
$$\underset{4}{\xi^a} = (1,-y,0,0), \quad \underset{5}{\xi^a} = (y, e^{-2x} - \tfrac{1}{2}y^2, 0, -2e^{-x}). \quad (28.25)$$

Its gravitational field is produced by the energy–momentum tensor

$$T^{mn} = \frac{1}{2\kappa a^2} g^{mn} + \frac{u^m u^n}{\kappa c^2 a^2}, \quad u^m = (0, 0, 0, c/a), \tag{28.26}$$

which we can interpret either as the energy–momentum tensor of a perfect fluid with

$$p = \mu c^2 = 1/2\kappa a^2, \tag{28.27}$$

or, since for physically reasonable media or for application in cosmology this pressure is too large in relation to the rest-mass density, as an energy-momentum tensor which besides the contribution from the dust also contains the cosmological term Λg_{mn} (cf. (9.4)):

$$\mu = 1/\kappa c^2 a^3, \quad \Lambda = -1/2a^2. \tag{28.28}$$

Since only the components

$$u_{1;2} = -u_{2;1} = ac\,e^x/2 \tag{28.29}$$

of the derivative $u_{a;b}$ of the four-velocity are non-zero, the matter current is geodesic, shear-free and expansion-free, but rotates with the constant velocity

$$\omega = \sqrt{\omega_{ab}\omega^{ab}/2} = c/a\sqrt{2}. \tag{28.30}$$

The Gödel universe is certainly not a realistic model of the universe, but it does possess a series of interesting properties. It is one of the few cosmological models which contains rotating matter, and it also contains closed timelike lines; that is, an observer can influence his own past.

28.4 Singularity theorems

Of the cosmological models which we have so far discussed, the physically reasonable ones (Friedmann model, Bianchi type *I* universes) have a singularity in their evolutionary history, that is, a beginning to the universe or a 'primaeval bang' ('big bang'), while the physically less realistic ones (Einstein universe, de Sitter universe, Gödel universe) certainly do not possess such a singularity, but they involve the cosmological constant, or possess a rather implausible equation of state for the matter, show no red shift or else contradict our ideas about causality.

Since a singularity at the beginning of the universe is, however, a rather unwelcome property of cosmological models, one would very much like to know whether this singularity is unavoidable, if the model is to be physically reasonable. Do singularities perhaps occur only in cosmological models of high symmetry and vanish under the small deviations from symmetry which are always present in reality; or were we unlucky in our selection of the model: are singularities absent in other universes of high symmetry (for other Bianchi types)?

In answer to the last of the questions raised here we shall now show that in gravitational fields which are produced by perfect fluids whose elements move without rotation along geodesics, then under certain plausible assumptions singularities must occur. Our starting point is the decomposition (17.11) of the covariant derivative of the velocity field u_m of the fluid, that is, the representation

$$u_{m;i} = \omega_{mi} + \sigma_{mi} + \tfrac{1}{3}\Theta(g_{mi} + u_m u_i/c^2) - \dot{u}_m u_i/c^2. \quad (28.31)$$

If we substitute this into the equation

$$(u_{m;i;n} - u_{m;n;i})g^{mi}u^n = -R_{an}u^n u^a \quad (28.32)$$

which is valid for every vector u_m, then, using the field equations, we obtain

$$\frac{d\Theta}{d\tau} = -\Theta^2/3 - \sigma_{in}\sigma^{in} - \kappa c^2(3p + \mu c^2)/2 + \omega_{in}\omega^{in} + \dot{u}^n{}_{;n}. \quad (28.33)$$

If we also assume that

$$\mu c^2 + 3p \geq 0 \quad (28.34)$$

(physically we would of course expect further that $\mu \geq 0$ and $p \geq 0$), then all terms on the right-hand side of (28.33) except $\omega_{in}\omega^{in}$ and possibly $\dot{u}^n{}_{;n}$ are negative. Therefore if the rotation and the acceleration vanish we have

$$\frac{d}{d\tau}\left(\frac{1}{\Theta}\right) \geq \frac{1}{3}. \quad (28.35)$$

Accordingly Θ^{-1} was either ($\Theta > 0$) zero at a finite proper time in the past, or ($\Theta < 0$) will take the value zero after a finite proper time. Since the expansion Θ is a measure of the relative change in volume, then singularities (with $\Theta = \infty$) are always present in such models. Because $\Theta = 3c\dot{K}/K$ in the Friedmann universe, these singularities just correspond to the zero points of the radius function K.

A similar conclusion can be drawn if the matter itself is not necessarily non-rotating, but if there does exist in the space a rotation-free, geodesic congruence of timelike world lines (cluster of test particles). Since for two timelike, future-directed, unit vectors u^i/c and V^i/c we always have $u^i V_i \leq -c^2$, and hence the field equations yield

$$R_{an}V^a V^n \geq \kappa c^2(3p + \mu c^2)/2, \quad (28.36)$$

the inequality (28.35) follows also for these geodesics, and the family of test particles shows a singular behaviour. The space-time is therefore singular in the mathematical sense. Whether the physical quantities (pressure, rest-mass density) behave singularly must be investigated separately.

As a generalization of these laws one can show that in every universe which is at some time homogeneous (which possesses a transitive spatial group of motion), for which the associated initial value problem can be solved uniquely on this initial surface and in which the condition $R_{ab}V^aV^b < 0$ is satisfied for all timelike or null vectors V^a, then there exists a singularity. This singularity is characterized by the occurrence of geodesics which although of finite length cannot be extended. The type of physical singularity must in every case be clarified separately.

The existence of singularities can be proved under still weaker assumptions; singularities occur, for example, in every spatially closed universe which at some time or other expands or contracts.

Bibliography to section 28

Textbooks   Landau and Lifschitz (1975)

Monographs and collected works   Hawking and Ellis (1973), Hawking and Israel (1987), Ryan and Shepley (1975)

Review and research articles   Ellis and King (1974), Ellis and Sciama (1966), Gibbons, Hawking and Siklos (1983), Gödel (1949), Kundt (1956), Liang and Sachs (1980), MacCallum (1979)

Non-Einsteinian theories of gravitation

29 Classical field theories

29.1 Why and how can one generalize the Einstein theory?

Notwithstanding the great success of the Einstein gravitational theory, one is still justified in asking whether this theory correctly reproduces the properties of matter. Here a distinction must be made between the question of whether the theory is in fact valid or not within certain narrow limits, for example, in the celestial mechanics of the planetary system or the equation of state of the stars, and the question, which should in principle be asked in every theory, of its realm of validity and the inclusion of other kinds of interaction.

The motives for modifying the Einstein theory cover a broad range. They include not only the idea already pursued by Einstein of 'geometrizing' other fields as well (perhaps the electromagnetic field), but also attempts at modification on the grounds of doubt about the correctness of the theory under extreme conditions (gravitational collapse, cosmological singularity). They also contain the efforts at throwing doubt on the Einstein theory by comparing experimental findings with the predictions of the Einstein or other theories, as well as rejection based on prejudice. At the moment there is certainly no indication of false predictions of the Einstein theory, and therefore no compulsion to change it.

Most of the non-Einsteinian gravitational theories follow the path of Einstein in 'geometrizing' the gravitational force; that is, they employ a geometry – which is thus not necessarily Riemannian – to describe the gravitational interaction of matter. The deeper physical reason for this is that the equality of (passive) gravitational and inertial mass can then be taken into account most easily. Without going into details, we want to sketch the basic ideas of several theories, which we have selected quite arbitrarily.

In the Einstein theory the matter and its interactions are described by two different types of physical quantities (which indeed can be distinguished

29 Classical field theories

just by inspection of the field equations, when these are written in the usual way), namely, by metrical quantities, which describe the gravitational field, and by non-metrical quantities, which represent the other fields. In order to have a 'unified' field theory, and also to describe the Maxwell field, for example, by purely geometrical quantities, then one abandons the framework of Riemannian geometry.

In the Weyl theory of 1918 (Weyl, 1918) measurements of length and angle are given by a symmetrical metric tensor g_{mn} or the fundamental metric form

$$ds^2 = g_{mn} dx^m dx^n, \quad m, n = 1, \ldots, 4, \quad g_{mn} = g_{nm}. \tag{29.1}$$

The affine connection is different from that in Riemannian geometry: the connection components occurring in the definition

$$\frac{dT^a}{d\lambda} + \Gamma^a_{nm} T^m \frac{dx^n}{d\lambda} = 0 \tag{29.2}$$

of the parallel shift of a vector along a curve are defined by

$$\Gamma^a_{nm} = \tfrac{1}{2} g^{ab}(g_{bn,m} + g_{bm,n} - g_{nm,b})$$
$$+ \tfrac{1}{2} g^{ab}(g_{bn} Q_m + g_{bm} Q_n - g_{nm} Q_b), \tag{29.3}$$

and so differ from the Christoffel symbols through the last terms. The new degrees of freedom can be used to describe the Maxwell field by identifying the vector Q_m with the four-potential.

In the asymmetric Einstein theory (1946, 1950), the geometry is determined by a sixteen-component (generalized metric) asymmetric tensor g_{mn}, whose additional antisymmetric part is associated with the six field components of the Maxwell field. The field equations are then derived (as also in the Weyl theory) from a variational principle with the generalized curvature scalar as Lagrangian density.

In the different variants of the projective field theory (Kaluza, 1921; Klein, 1926; and many others) one also tries to carry through a unified description of electromagnetism and gravity on a geometrical basis. Here the starting point is the property of a five-dimensional space, that the group of transformations, under which the new coordinates are homogeneous functions of first degree of the old coordinates, is isomorphic to the group generated by coordinate transformations of a four-dimensional space and by gauge transformations of the four-potential, that is, isomorphic to the invariance group of the Einstein–Maxwell theory. One hence expects that the Einstein–Maxwell theory can be described especially simply in a five-dimensional space; and of course all the quantities of the abstract five-dimensional space can be mapped uniquely onto the physical quantities of the space-time. It is therefore at once possible to make a description in a five-dimensional space completely equivalent to the usual theory, where of course the fifteen components (degrees of freedom) of a symmetric metric tensor in five dimensions have to be rather artificially restricted (a four-dimensional metric plus a four-potential gives only fourteen components). If one drops this restriction, then a new scalar field appears in the theory, and one can interpret this, for example, as the field of a position-dependent or time-dependent gravitational number (it is dependent upon the distribution of charges and masses via the field equations). This variability of the gravitational 'constant' is of course particularly significant for cosmology, but the possibility of demonstrating it currently lies beyond the limits of accuracy in measurement.

One much discussed projective field theory besides Jordan's theory (1955) is the scalar–tensor gravitational theory of Brans and Dicke (1961). In this theory the gravitational field is described not only by the metric tensor, but also by a scalar field $\Phi(x^i)$, which is determined by the trace $T = T^a_a$ of the energy–momentum tensor according to

$$\Phi^{,n}{}_{;n} = T/(3 + 2\omega) \tag{29.4}$$

and enters the generalized field equations

$$R_{mn} - \frac{R}{2} g_{mn} = \frac{1}{\Phi} T_{mn} + \frac{\omega}{\Phi^2} \left(\Phi_{,n} \Phi_{,m} - \frac{1}{2} \Phi^{,a} \Phi_{,a} g_{nm} \right)$$
$$+ \frac{1}{\Phi} (\Phi_{;nm} - g_{nm} \Phi^{,a}{}_{;a}) \tag{29.5}$$

in such a way that the motion of the field-producing masses depends only indirectly (via the metric) upon the reciprocal of the gravitational number Φ, that is, we now have, as before,

$$T^m{}_{n;m} = 0. \tag{29.6}$$

These field equations can be derived from the variational principle

$$\delta \int \left[\Phi R - \frac{\omega}{\Phi} \Phi_{,n} \Phi^{,n} + \frac{1}{2} \overset{M}{L} \right] \sqrt{-g} \, d^4x = 0. \tag{29.7}$$

The constant ω is a dimensionless coupling parameter which is to be determined from experiment; for suitable boundary conditions, $\omega \to \infty$ ($\Phi =$ constant) corresponds to the Einstein theory.

The aim of the Cartan theory (Cartan, 1922), and to some extent of the so-called tetrad theories too, is quite different from that of including the

Maxwell field or introducing a variable gravitational number. In the classical theory of fields and in quantum field theory the energy–momentum tensor (associated with space-time translations of the Minkowski space) and the spin tensor or the angular momentum tensor (associated with rotations in space-time) of course stand side by side and are formally equivalent. In the Einstein gravitational theory, however, only the energy–momentum tensor (the mass) corresponds to a geometrical property of the space. One can now associate a geometrical image with the spin as well by using, in addition to the metric tensor g_{mn}, connection components

$$\Gamma^m_{ab} = \tfrac{1}{2} g^{mn}(g_{na,b} + g_{nb,a} - g_{ab,n}) + g^{mn}(S_{abn} + S_{nab} - S_{bna}) \quad (29.8)$$

which are asymmetric and contain the torsion tensor $S_{abn} = -S_{ban}$ with twenty-four independent components as new degrees of freedom. One can see how this torsion tensor acts by introducing four orthonormal unit vectors at each point of the space-time and transporting them parallel with the connection. During this process they are rotated (six degrees of freedom), depending on the direction of the parallel transport (four degrees of freedom).

29.2 Possible tests of gravitational theories and the PPN formalism

The correctness of a theory of gravitation is decided not by the opinion of its discoverer and his possible followers, but rather only by the agreement of the predictions of the theory with the observational results. Logical and formal simplicity and beauty – arguments which are particularly often brought up in favour of the Einstein theory – are certainly important in the setting up of a theory, but they are not in themselves sufficient.

Like every new physical theory, a theory of gravitation must also satisfy a series of more or less 'natural' conditions. Thus, it must be free of internal contradiction, it must be compatible with the system of theories already at hand and it must fit in with these theories (for example, it must give the orbits not only of neutral, but also of charged, point masses), and it must contain its predecessor, that is, the Newtonian theory of gravitation, as a limiting case or a special case.

A gravitational theory must also correctly reproduce at least qualitatively the effects which are today established, that is, the equality of inertial and gravitational mass, the Principle of Equivalence, the cosmological red shift, the deflection of light by the Sun, the lengthening of travel times in the solar field, and the perihelion precession.

Metric theories can most easily be tested formally. Metric theories are theories in which, as in the Einstein theory, the action of gravitation upon the matter is due exclusively to the metric tensor g_{mn}; they differ from one another in the field equations from which the metric is to be determined. Thus the Brans-Dicke theory is metric, the Cartan not. Metric theories automatically take into account the Principle of Equivalence and the equality of inertial and gravitational mass, which has been verified by measurements in the field of the Earth or the Sun up to an error of at most 10^{-12}. They moreover contain the Newtonian gravitational theory, if in a coordinate system in which the centre of gravity is at rest (i.e. T_{44} is the most significant component of the energy–momentum tensor) to first approximation the expressions

$$\overset{1}{g}_{44} = -\left(1 + \frac{2U}{c^2}\right), \quad U(x^\alpha, t) = -\frac{\kappa c^2}{8\pi} \int \frac{T_{44}(\bar{x}^\alpha, t)}{|x^\alpha - \bar{x}^\alpha|} \, d^3\bar{x},$$

$$\overset{1}{g}_{4\alpha} = 0, \quad \overset{1}{g}_{\alpha\beta} = \eta_{\alpha\beta}$$

(29.9)

give the correct equations of motion for a test particle.

As the measurements of gravitational effects in our solar system are far more exact than those of cosmological effects, one could take just these solar–terrestrial effects to test gravitational theories; sufficient accuracy is obtained by carrying the approximation one step further than in (29.9); that is, by taking the post-Newtonian approximation. In this approximation one can distinguish the different metric theories by the numerical values of certain dimensionless parameters, which specify how the matter distribution (the components of the energy–momentum tensor) enter the metric. One therefore speaks of a parametrized post-Newtonian (PPN) formalism. The numerical values of these parameters can be determined from observations and experiments (cf. section 10.7). At the present there is no serious reason to doubt the Einstein values of the paramaters; rather, the rôle of the PPN formalism is to offer a testbed on which general relativity can be compared with other theories, and hence why, on experimental grounds, it is to be preferred.

Bibliography to section 29

Textbooks   Schmutzer (1968), Weyl (1970)

Monographs and collected works   Hlavaty (1958), Jordan (1955, 1966), Ludwig (1951)

Review and research articles   Hehl (1973, 1974), Thorne and Will (1971 a, b, c), Will (1979)

30 Relativity theory and quantum theory

30.1 *The problem*

The general theory of relativity is completely compatible with all other classical theories. Even if the details of the coupling of a classical field (Maxwell field, Dirac field, neutrino field, Klein–Gordon field) to the metric field are not always free of arbitrariness and cannot yet be experimentally tested with sufficient accuracy, no doubt exists as to the inner consistency of the procedure.

This optimistic picture becomes somewhat clouded when one appreciates that besides the gravitational field the only observable *classical* field in our universe is the Maxwell field, while the many other interactions between the building blocks of matter can only be described with the aid of quantum theory. A unification of relativity theory and quantum theory has not yet been achieved, however.

One of the main postulates of relativity theory is that a locally geodesic coordinate system can be introduced at every point of space-time, so that the action of the gravitational force becomes locally ineffective and the space is approximately a Minkowski space. Hence it is easily understandable why in our neighbourhood, with its relatively small space curvature, the space is flat, to very good approximation, as it is assumed to be in quantum theory. But it also shows us the limits of this more or less undisturbed coexistence of quantum theory and relativity theory: in regions of strong curvature (close to singularities) and in questions which concern the behaviour of far-extended physical systems, the two theories are no longer compatible, since they start out from different space structures. Quantum theory presupposes a Minkowski space of infinite extent both in its fundamental commutation rules, which are formulated explicitly using the group of motion of the space (the Lorentz group), and in detailed technical issues like expansion in plane waves, asymptotic behaviour at infinity or the formulation of conservation laws. Relativity theory shows, however, that the space is a Riemannian space.

On the other hand, the idea of relativity theory that the properties of the space are properties of the interaction of the matter and can be measured out by material test bodies leads to contradictions with the definition or measurement of very small distances, and thus with the metric in very small regions of space. If the dimensions are so small that atoms or elementary particles should be taken as test objects, then their location is no longer so precisely defined that one can really speak of a measurement, even be it only in a gedanken experiment.

In nature, however, there exist stars, which consist of elementary particles and whose motion obeys the gravitational laws, and therefore a self-consistent synthesis of relativity theory and quantum theory must be possible, where it is to be expected that at least one of these two theories must be modified. Theoretical physicists are certainly in a difficult situation: in contrast to the physics of elementary particles, which provide a large amount of experimental data seeking interpretation, here there are no experimental findings (or at least none recognized as such) which could give an indication of the course to be followed.

Three distinct possibilities appear as likely candidates for the unification of gravity and quantum theory; these will be outlined in the following sections.

30.2 Unified quantum field theory and quantization of the gravitational field

After the successful unification of the weak and electromagnetic interactions in a unified quantum field theory (Salam and Weinberg, 1967) theoretical physicists have set themselves the ambitious task not only to include the strong interactions in the scheme (Grand Unified Theories or GUTs), but to describe all four known interactions in a unified field theory, for example, in a supergravity or superstring theory. As yet there have been no resounding successes. But perhaps this 'theory of everything' sought after both by relativists and elementary particle physicists is an illusion; the unity of our world need not be reflected, even at the most basic physical level, by a simple comprehensive formula.

A less demanding approach is to quantize only the gravitational field. A range of physical and formal mathematical grounds suggest that all fields and interactions should be handled in a uniform manner; thus the gravitational field also should be quantized. The first steps have already been made. In chapter 16 an analysis of the Cauchy problem for the vacuum Einstein field equations was made, and this can be used to isolate the true dynamical degrees of freedom of the gravitational field, that is, those which do not arise from pure coordinate transformations. This is the starting point for canonical quantization. Although a great deal of effort has been invested not only to construct a formal theory, but also to understand and interpret it physically, the task is still in its infancy. The picture of the (four-dimensional) world would look quite different in a quantized theory of gravity. At each event the world is a mixture of states, each with an *a priori* probability. Each state corresponds to a possible three-geometry, including its topological properties, and can be described by a point in *superspace*. How one couples in non-metric fields, how man is to interpret the wave function of the universe and how measurement processes and observers are to be described is unclear.

There is no problem in quantizing the *linearized* Einstein field equations, that is, the classical field $\bar{f}_{mn}$ described by the equations

$$\bar{f}_{mn} = 0, \quad \bar{f}^{mn}{}_{,n} = 0 \tag{30.1}$$

(see section 13.2). It shows that the massless particles of this field, analogous to the photons of the electromagnetic field, have spin 2. Of course by restricting consideration to source-free weak fields the real problems have been swept under the carpet.

30.3 Semiclassical gravity

A possible resolution of the problems caused by attempting to quantize gravity is to treat the gravitational field classically, but quantize all other fields. This school of thought is supported especially by those who regard the gravitational field as playing a privileged rôle, which should not and cannot be quantized. An extreme standpoint of this conservative-relativistic view was taken by Einstein himself. For a time he believed that quantum theory could be encompassed in a (possibly generalized) theory of relativity that would link space-time singularities to elementary particles. This hope has not been realized.

In a semiclassical theory the coupling of gravity to the quantized fields depends, on the one hand, on the fact that the field equations of the latter can be formulated covariantly, and thus can be made to depend on the gravitational field. On the other hand the gravitational field can be seen to have the quantum fields as its source. They occur, however, in the 'source' of the Einstein field equations, the energy–momentum tensor, not as operators but as expectation values:

$$R_{mn} - \tfrac{1}{2} R g_{mn} = \kappa \langle T_{mn} \rangle. \tag{30.2}$$

In order for the field equations (30.2) to be integrable the expectation values for the components of the energy–momentum tensor must be divergence-free,

$$\langle T^{mn} \rangle_{;n} = 0. \tag{30.3}$$

However, as a deeper analysis reveals, (30.3) is not a simple consequence of the equations governing the quantum fields (which have not been given explicitly), but rather a constraint on those quantities, for example, the states, which are used to form the expectation values. One sees immediately that the main problem in this form of unification of quantum and relativity theory is the choice, meaning and interpretation of states, even the 'vacuum state.' In addition there are the difficulties in carrying over to a non-linear theory, in which the superposition principle is invalid, the usual interpretation of measurement processes. It is not possible to decide now whether such a semiclassical theory is self-consistent, and to what extent it is a good approximation or even consistent with observations.

30.4* Quantization in a given classical gravitational field. The thermodynamics of black holes

One can obtain a valuable insight into the problems and consequences of the as yet unknown unified theory by considering the influence of a given gravitational field on the quantum field and ignoring the back-reaction, that is, the inertia field produced by the quantum field. As an example we outline the typical procedure, some results and some problems by considering a real massless scalar field $\varphi(x^i)$,

$$\Box \varphi = \varphi^{,n}{}_{;n} = 0. \tag{30.4}$$

In order to quantize classical fields φ satisfying the wave equation (30.4) in *Minkowski space* one can proceed as follows. One first represents the general (classical) solution of (30.4) by its Fourier transform with respect to time t and splits the inversion integral into waves of positive ($e^{-i\omega t}$) and negative ($e^{+i\omega t}$) frequency:

$$\varphi(x^a) = \int_0^\infty [\varphi_\omega(x^\alpha) e^{-i\omega t} + \bar\varphi_\omega(x^\alpha) e^{i\omega t}] \, d\omega, \quad \alpha = 1, 2, 3. \tag{30.5}$$

On the surfaces $t =$ constant one constructs a complete orthonormal system, $f_p(x^\alpha)$ and $\bar f_p(x^\alpha)$, of solutions of the transformed wave equation which can be used to represent φ_ω and $\bar\varphi_\omega$. Here the norm is defined by

$$(\Psi_1, \Psi_2) = -i \int (\Psi_1 \dot{\bar\Psi}_2 - \dot\Psi_1 \bar\Psi_2) \, d^3x = \overline{(\Psi_2, \Psi_1)}. \tag{30.6}$$

Every solution of the wave equation can be represented as a superposition of partial waves g_n of the form $f_p(x^\alpha) e^{-i\omega t}$ and their complex conjugates $\bar g_n$; the index n represents symbolically the possible values of p, which are often discrete, and the continuous frequency parameter ω. Because of the structure of the norm (30.6) g_n and $\bar g_n$ satisfy the equations

$$(g_n, \bar g_m) = 0, \quad (g_n, g_m) = -(\bar g_n, \bar g_m). \tag{30.7}$$

30 Relativity theory and quantum theory

The general Hermitian field operator $\varphi(x^\alpha, t)$ can then be written in the form

$$\varphi(x^\alpha, t) = \sum_n (a_n g_n(x^\alpha, t) + a_n^\dagger \bar{g}_n(x^\alpha, t)), \tag{30.8}$$

where the operators a^n and $a_n^\dagger$ satisfy the commutator rules

$$[a_n, a_{n'}] = 0 = [a_n^\dagger, a_{n'}^\dagger], \quad [a_n, a_{n'}^\dagger] = \delta_{nn'}. \tag{30.9}$$

The set of states which can be constructed by single or multiple application of the creation operator $a_n^\dagger$ to the vacuum state $|0\rangle$ forms the Hilbert space of the system. Here the vacuum state is defined as that state in which no particles can be annihilated

$$a_n|0\rangle = 0. \tag{30.10}$$

A single particle state (of type n) $|1_n\rangle$ is then constructed via

$$a_n^\dagger|0\rangle = |1_n\rangle. \tag{30.11}$$

The total number of particles in a given state can be found by using the particle-counting operator

$$N = \sum_n N_n = \sum_n a_n^\dagger a_n. \tag{30.12}$$

It can be shown that this quantization procedure is Lorentz invariant. In particular the vacuum state is independent of the (arbitrary) choice of surfaces $t = \text{constant}$.

However, the attempt to carry over the procedure sketched above to a *curved space-time* leads to a series of difficulties, which occur essentially because of the non-existence of a preferred foliation of space-time by three-dimensional surfaces and because the topology of space-time may differ from the Minkowski one. Two different foliations of space-time lead in general to different systems g_n and $\hat{g}_n$ of partial waves, that is, to different definitions of particles.

Consider the representation of a general field operator with respect to two such systems

$$\varphi = \sum_n (a_n g_n + a_n^\dagger \bar{g}_n) = \sum_m (\hat{a}_m \hat{g}_m + \hat{a}_m^\dagger \hat{\bar{g}}_m), \tag{30.13}$$

with corresponding vacuum states

$$a_n|0\rangle = 0, \quad \hat{a}_n|\hat{0}\rangle = 0. \tag{30.14}$$

Because of the completeness of both systems, the functions and operators of each system can be represented in terms of the other. In particular, there exist relations of the form

$$\hat{g}_n = \sum_m (\alpha_{nm} g_m + \beta_{nm} \bar{g}_m),$$
$$g_n = \sum_m (\bar{\alpha}_{mn} \hat{g}_m - \beta_{mn} \hat{\bar{g}}_m),$$
(30.15)

with constant (complex) coefficients α_{nm} and β_{nm}, a 'Bogoliubov transformation.' On inserting (30.15) in (30.13) one obtains the transformation law for the operators

$$a_m = \sum_n (\alpha_{nm} \hat{a}_n + \bar{\beta}_{nm} \hat{a}_n^\dagger).$$
(30.16)

Thus not only are the particle (partial wave) definitions in the two systems different, but also, if $\beta_{nm} \neq 0$, what one observer regards as a vacuum state $|\hat{0}\rangle$ is seen by the other to be a mixture of particles

$$a_m |\hat{0}\rangle = \bar{\beta}_{nm} \hat{a}_n^\dagger |\hat{0}\rangle = \bar{\beta}_{nm} |\hat{1}_n\rangle.$$
(30.17)

This surprising result shows clearly that within general relativity the concept of particle is more problematical than one might have expected. Proper Lorentz transformations in Minkowski space-time have $\beta_{nm} = 0$ and so do not alter the vacuum state. However, an accelerated observer in the 'usual' Minkowski space-time vacuum state would detect particles (with a thermal spectrum).

An immediate consequence of this property of a quantum field is the possibility that a gravitational field can create particles. Suppose, for example, that initially (as $t \to -\infty$) there is a flat space with vacuum state $|\hat{0}\rangle$, then a gravitational field is switched on and off, and finally (as $t \to +\infty$), the space is again Minkowski. However, the final vacuum state $|0\rangle$ will not always agree with the initial one $|\hat{0}\rangle$; particles have been produced.

The most spectacular example for the creation of particles by a gravitational field is produced by the gravitational field of a collapsing star, that is, the creation of a black hole. We shall sketch the basic ideas in the case of spherically symmetrical collapse. We shall use the Eddington–Finkelstein coordinates introduced in chapter 22, in which the Schwarzschild line element has the form

$$ds^2 = 2c\,dr\,dv + r^2(d\vartheta^2 + \sin^2\vartheta\,d\varphi^2) - (1 - 2M/r)c^2 dv^2,$$
$$cv = ct + r + 2M \ln(r/2M - 1) = ct + r^*.$$
(30.18)

Of course this metric represents only the exterior of the star, whose boundary is given by

$$f(r, v) = 0;$$
(30.19)

30 Relativity theory and quantum theory

see Figs. 23.4 and 30.1. The metric in the interior could be, for example, a part of the Friedmann universe (23.39). What matters is that it is regular and shows no peculiarities even when the surface of the star disappears behind the horizon.

We consider a state of the system in which incoming waves do not occur, especially as $t \to -\infty$. This corresponds to the choice of partial waves

$$\hat{g}_\omega = \hat{h}_\omega(r, \vartheta, \varphi) e^{-i\omega v} = \hat{h}_\omega(r, \vartheta, \varphi) e^{-i\omega(t+r^*/c)} \qquad (30.20)$$

as waves of positive frequency (with respect to v). Thus we write the solution of the wave equation $\Box \varphi = 0$ as

$$\varphi = \int_0^\infty (\hat{a}_\omega \hat{g}_\omega + \hat{a}_\omega^\dagger \hat{\bar{g}}_\omega)\, d\omega, \qquad (30.21)$$

and require the system to be in the corresponding vacuum state $|\hat{0}\rangle$. If the gravitational field creates particles, outgoing particles must be present, although there are no ingoing ones. However, outgoing waves are best described in terms of retarded time u given by

$$cu = ct - r^* = cv - 2r^* = cv - 2r - 4M \ln(r/2M - 1), \qquad (30.22)$$

and the corresponding preferred system of partial waves with positive frequency with respect to u is

$$g_{\omega'} = h_{\omega'}(r, \vartheta, \varphi) e^{-i\omega' u} = h_{\omega'}(r, \vartheta, \varphi) e^{-i\omega'(v - 2r^*/c)}. \qquad (30.23)$$

Unlike (30.20), this system is not complete, for in a general state incoming waves will occur which will be absorbed by the black hole, and will not propagate to infinity as in (30.23). However, in spite of this one can represent the functions $g_{\omega'}$ in terms of the complete system $\hat{g}_\omega$ and $\hat{\bar{g}}_\omega$,

$$g_{\omega'} = \int_0^\infty (\bar{\alpha}_{\omega\omega'} \hat{h}_\omega e^{-i\omega v} - \beta_{\omega\omega'} \hat{\bar{h}}_\omega e^{i\omega v})\, dv. \qquad (30.24)$$

Outgoing particles occur if and only if the Bogoliubov coefficient $\beta_{\omega\omega'}$ does not vanish.

It is not possible to carry out exactly the transformation between the two systems of functions $g_{\omega'}$ and $\hat{g}_\omega$; in particular, the radial dependence of the functions $h_{\omega'}$ and $\hat{h}_\omega$ cannot be given as simple analytic expressions. However, as often in the discussion of wave propagation, a geometrical optics (eikonal) approximation (see section 8.4) allows further progress. Both systems of functions have the form

$$\varphi = A(x^i) e^{-iW(r,v)}, \qquad (30.25)$$

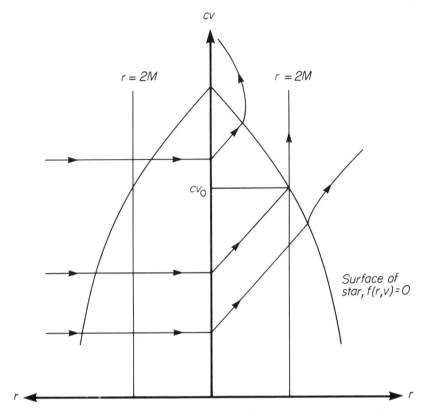

Fig. 30.1. A collapsing star and light rays in Eddington–Finkelstein coordinates.

where, because of the wave equation (30.4), the eikonal $W(r, v)$ is to satisfy

$$0 = W_{,a}W^{,a} = W_{,r}[2W_{,v} + c(1 - 2M/r)W_{,r}] = W_{,r}[2W_{,v} + cW_{,r^*}]. \quad (30.26)$$

Clearly the solutions of this equation form two classes: ingoing waves having $W = W(v)$, and outgoing waves having $W = W(v - 2r^*/c) = W(u)$.

The approximation that we shall now make is to convert only the eikonal function W of the outgoing wave to the ingoing form, thus neglecting the factors $\hat{h}_\omega$ and $h_{\omega'}$, and to include only those terms in the eikonal whose derivative is especially large. (In the geometrical optics ansatz (30.25) it is implicitly assumed that derivatives of W are large in comparison with those of A.)

How does the conversion of the eikonal of the outgoing wave, given by (30.23),

30 Relativity theory and quantum theory

$$W = \omega'u = \omega'\left[v - \frac{2r}{c} - \frac{4M}{c}\ln(r/2M - 1)\right], \quad (30.27)$$

that is, the determination of the eikonal $\hat{W} = \hat{W}(v)$ of the corresponding ingoing wave, occur? Because $\hat{W}$ does not depend on r it is obviously sufficient to know $\hat{W}$ on the surface of the star. In order to extract from W, the eikonal of the outgoing wave, the eikonal of the ingoing wave on the surface of the star we should trace the outgoing wave back to the surface of the star, thence to the centre and further back to the surface again, and sum up all changes in phase occurring along this path. Although we cannot compute the eikonal or phase change within the star, we can at least estimate the required quantities.

If the surface of the star is at rest or almost so (e.g. at the start of the collapse, $t \to -\infty$), the forward and backward directions within the star are equivalent and in particular are v-independent. Therefore the eikonal $\hat{W}(v)$ coincides with $W(v)$ (up to an additive constant) on the surface of the star,

$$\hat{W}(v) = W(u)|_{r=r_0} + \text{constant} = W(v, r)|_{r=r_0} + \text{constant}$$
$$= \omega'[v - 2r^*(r_0)/c] + \text{constant} = \omega'v + \text{constant} \quad (30.28)$$

A constant contribution to the eikonal is inessential and can be incorporated in the amplitude A (see (30.25)). Thus an outgoing wave $e^{-iW} = e^{-i\omega'u}$ is associated with an ingoing wave $e^{-i\hat{W}} = e^{-i\omega'v}$ of the same frequency; if there is no ingoing wave then outgoing waves will not exist and particles are not produced.

If the surface of the star moves, and especially when it approaches the horizon $r = 2M$, the forward and backward directions in the stellar interior are no longer equivalent; the eikonal $\hat{W}(v)$ on the surface of the star (and hence for all values of v) differs from $W(u)$ on the surface $f(r, v) = 0$ by an additive v-dependent function $F(v)$

$$\hat{W}(v) = W(u)|_{f(r,v)=0} + F(v). \quad (30.29)$$

Near the horizon the equation $f(r, v) = 0$ for the surface of the star has the approximate form

$$r = 2M + Bc(v_0 - v) + \ldots, \quad B = \text{constant}, \quad v < v_0. \quad (30.30)$$

From (30.27) and (30.30) follows

$$W(u)\Big|_{f=0} = \omega'\left[v - \frac{4M}{c} - 2B(v_0 - v) - \frac{4M}{c}\ln\frac{Bc(v_0 - v)}{2M} + \ldots\right].$$
$$(30.31)$$

If we are only interested in the dominant part of $\hat{W}(v)$ near $v = v_0$ we need only retain the ln term in (30.31), and we can also discard $F(v)$ in (30.29) which has a finite derivative at $v = v_0$ because of the regularity of the metric in the stellar interior. Thus for small positive $v_0 - v$

$$\hat{W}(v) \approx -\omega' \frac{4M}{c} \ln(v_0 - v). \tag{30.32}$$

Ingoing waves cannot produce outgoing waves with $v > v_0$ because the former must pass within the horizon $r = 2M$ and can never return.

Let us summarize. The outgoing partial wave (30.23) has the following representation in terms of incoming waves:

$$\left.\begin{array}{ll} g_{\omega'} = 0, & v > v_0, \\ g_{\omega'} \sim e^{+i\omega'(4M/c)} \ln(v_0 - v) & \dfrac{c(v_0 - v)}{2M} \ll 1, \\ g_{\omega'} \sim e^{-i\omega' v}, & v \to -\infty. \end{array}\right\} \tag{30.33}$$

Because the Fourier transform of the middle line contains $\Gamma(1 - (i\omega 4m/c))$, $g_{\omega'}$ contains all frequencies ω (and not just the positive ones), and the $\beta_{\omega\omega'}$ of (30.24) are non-zero; particles are produced! The same result (30.33) would have been obtained when instead of a collapsing star we had considered a shrinking reflective spherical surface. Here too the essential condition is that on the surface the phase (eikonal) of the incoming wave agrees with that of the outgoing one.

The important parameter for particle production and the frequency spectrum is the quantity $4M/c$. A more precise analysis (Hawking, 1975) shows that the particles have a thermal spectrum, that is, a black hole with (Newtonian) mass m radiates like a black body of temperature

$$T = \frac{\hbar c}{8\pi M k} = \frac{\hbar}{km\kappa c^2} \approx \frac{1.2 \times 10^{26} \text{ K}}{m[g]}, \tag{30.34}$$

where in the last equation the mass m is to be given in grammes and the temperature is obtained in degrees Kelvin. Black holes are therefore not black but emit radiation continuously; smaller (lighter) ones are hotter.

Even before Hawking had found this astonishing relationship between thermodynamics, quantum theory and gravitation, Bekenstein (1973) had suggested that a temperature and entropy could be associated with a black hole. Just as the total entropy of a process involving several thermodynamical systems can never decrease, so the sum of all surface areas A_i of a system of (rotating or non-rotating) black holes cannot decrease (see

30 Relativity theory and quantum theory

section 24.3). In fact the Hawking's discussion can be generalized to rotating black holes, whose temperature T and entropy S are given by

$$T = \frac{2(r_+ - M)\hbar c}{Ak}, \quad S = \frac{2\pi k}{\kappa c \hbar} A. \tag{30.35}$$

The first law of thermodynamics then gives for spherically symmetric black holes

$$T\,\mathrm{d}S = \mathrm{d}(8\pi M/\kappa) = \mathrm{d}mc^2. \tag{30.36}$$

Let us return to the derivation and discussion of particle production in the gravitational field of a spherically symmetric collapsing star. The derivation of the effect outlined above may appear to include somewhat arbitrary approximation procedures. However the main equation (30.32) furnishes all the important details about outgoing waves (particles) that would be observed by a distant observer at late times, since all those come from a neighbourhood of the horizon. One sees immediately that the basic idea can easily be carried over to other massless fields (e.g. the Maxwell field) because the eikonal equation (30.26) is the same for all such fields. It can also be shown that massive particles are produced.

Where precisely do these particles originate? The analytic structure of the eikonal suggests that the creation can be localized in a close neighbourhood of the horizon; however, the global nature of the particle concept in quantum field theory suggests caution before accepting so simple an interpretation.

If a collapsing star emits particles continuously, its energy (mass) must of course decrease. Because a solar mass black hole has a temperature $T = 6 \times 10^{-8}$ K this mass loss for conventional celestial objects undergoing collapse is totally negligible. However, very low mass black holes can have only a short life; because of the energy loss the temperature rises rapidly, more is radiated, and in a self-accelerating process the black hole disappears.

In order to decide whether these considerations are correct one needs a theory which correctly describes the back-reaction of the quantum field on the gravitational field, and we do not yet have one. Therefore it is not clear whether in gravitational collapse a black hole must occur, or whether particle production (which will have started before the star disappears within the horizon) decreases the mass so quickly, and so forces the horizon $r = 2M$ to shrink rapidly, that the outer surface of the star always remains outside the horizon. It is highly plausible that a horizon is created, but as yet we have no detailed ideas or theory as to how it might subsequently disappear.

Bibliography to section 30

Textbooks   Straumann (1984)

Monographs and collected works   Birell and Davies (1984), de Witt and Graham (1973), Gibbons, Hawking and Siklos (1983), Hawking and Israel (1987), Isham, Penrose and Sciama (1975, 1981)

Review and research articles   Ashtekhar and Geroch (1974), Bekenstein (1973), de Witt (1980), Gibbons (1979), Hawking (1975, 1976, 1979), Israel (1985), Kuchar (1973), Sciama, Candelas and Deutsch (1981)

Bibliography

Textbooks

Anderson, J. L. (1967). *Principles of Relativity Physics.* New York, London: Academic Press.
Bergmann, P. G. (1958). *Introduction to the Theory of Relativity.* Englewood Cliffs, NJ: Prentice-Hall.
Birkhoff, G. (1923). *Relativity and Modern Physics.* Cambridge, MA: Harvard University Press.
Eddington, A. A. (1923). *The Mathematical Theory of Relativity.* Cambridge: Cambridge University Press.
Einstein, A. (1950). *The Meaning of Relativity.* Princeton, NJ: Princeton University Press.
Einstein, A. (1969). *Über spezielle und allgemeine Relitivitätstheorie,* 2nd edn. Berlin: Akademie-Verlag.
Einstein, A. (1970). *Grundzüge der Relativitätstheorie,* 2nd edn. Berlin: Akademie-Verlag.
Eisenhart, L. Pf. (1933). *Continuous Groups of Transformations.* Princeton, NJ: Princeton University Press.
Eisenhart, L. Pf. (1949). *Riemannian Geometry.* Princeton, NJ: Princeton University Press.
Fock, V. (1964). *The Theory of Space, Time and Gravitation,* 2nd edn. Oxford: Pergamon Press.
Hicks, N. J. (1971). *Notes on Differential Geometry.* New York: Van Nostrand.
Landau, L. D. and Lifschitz, E. M. (1975). *Course of Theoretical Physics,* vol. 2. *The Classical Theory of Fields,* 4th edn. Oxford: Pergamon Press.
v. Laue, M. (1965). *Die Relativitätstheorie.* Braunschweig: Vieweg.
Lichnerowicz, A. (1955). *Théories Relativistes de la Gravitation et de l'Electromagnétisme.* Paris: Masson et Cie.
Misner, C. W., Thorne, K. S. and Wheeler, J. A. (1973). *Gravitation.* San Francisco: Freeman.
Møller, C. (1972). *The Theory of Relativity.* Oxford: Clarendon Press.
Peebles, P. J. E. (1980). *The Large-Scale Structure of the Universe.* Princeton, NJ: Princeton University Press.
Rindler, W. (1977). *Essential Relativity,* 2nd edn. New York: Springer-Verlag.
Schmutzer, E. (1968). *Relativistische Physik.* Leipzig: B. G. Teubner Verlagsgesellschaft.
Schouten, J. A. (1954). *Ricci-Calculus (Die Grundlagen der Mathematischen Wissenschaften, Bd. 10).* Berlin, Göttingen, Heidelberg: Springer-Verlag.

Schutz, B. (1980). *Geometric Methods of Mathematical Physics.* Cambridge: Cambridge University Press.
Spivak, M. (1979). *A Comprehensive Introduction to Differential Geometry,* vols. 1–5. Publish or Perish.
Straumann, N. (1984). *General Relativity and Relativistic Astrophysics.* Berlin, Heidelberg, New York: Springer-Verlag.
Synge, J. L. (1960). *Relativity – The General Theory.* Amsterdam: North-Holland.
Synge, J. L. (1965). *Relativity – The Special Theory,* 2nd edn. Amsterdam: North-Holland.
Wald, R. (1984). *General Relativity.* Chicago: Chicago University Press.
Weinberg, S. (1972). *Gravitation and Cosmology.* New York: Wiley.
Weyl, H. (1970). *Space, Time, Matter,* 6th edn. Berlin, Heidelberg, New York: Springer-Verlag.

Monographs and collected works

Abstracts of the 5th International Conference on Gravitation and the Theory of Relativity. (1968). Tbilisi: Publishing House of Tbilisi University.
Bertotti, B., de Felice, F. and Pascolini, A. (eds). (1984). *General Relativity and Gravitation.* Dordrecht: Reidel.
Birrell, N. D. and Davies, P. C. W. (1984). *Quantum Fields in Curved Space.* Cambridge: Cambridge University Press.
Breuer, R. A. (1975). *Gravitational Perturbation Theory and Synchrotron Radiation,* Lecture Notes In Physics, vol. 44. Berlin, Heidelberg, New York: Springer-Verlag.
Carmeli, M., Fickler, S. J. and Witten, L. (eds). (1970). *Relativity.* New York: Plenum.
Chandrasekhar, S. (1983). *The Mathematical Theory of Black Holes.* Oxford: Oxford University Press.
Chiu, H.-Y. and Hoffmann, W. F. (eds). (1964). *Gravitation and Relativity.* New York: Benjamin.
Clarke, R. W. (1974). *Albert Einstein.* Esslingen: Bechtle-Verlag.
Dautcourt, G. (1972). *Relativisitische Astrophysik.* Berlin: Akademie-Verlag.
de Sabbata, V. and Weber, J. (eds). (1977). *Topics in Theoretical and Experimental Gravitation Physics.* New York: Plenum.
de Witt, C. and de Witt, B. (eds). (1963). *Relativity, Groups and Topology.* Les Houches. (1964, New York: Gordon & Breach.)
de Witt, C. and de Witt, B. (eds). (1972). *Black Holes.* Les Houches. (1973, New York: Gordon & Breach.)
de Witt-Morette, C. (ed). (1974). *Gravitational Radiation and Gravitational Collapse,* Proceedings of IAU Symposium No. 64 Poland, 5–8 September, 1973. Dordrecht: Reidel.
de Witt, B. S. and Graham, N. (eds). (1973). *The Many-Worlds Interpretation of Quantum Mechanics.* Princeton, NJ: Princeton University Press.
de Witt, C. and Wheeler, J. A. (eds). (1968). *Batelle Rencontres. 1967 Lectures in Mathematics and Physics.* New York: Benjamin.

Farnsworth, D. *et al.* (eds). (1972). *Methods of Local and Global Differential Geometry in General Relativity,* Lecture Notes in Physics, vol. 14. Berlin, Heidelberg, New York: Springer-Verlag.

Gibbons, G. W., Hawking, S. W. and Siklos, S. T. C. (eds). (1983). *The Very Early Universe.* Cambridge: Cambridge University Press.

Goldberg, L. *et al.* (1973). *Annual Review of Astronomy and Astrophysics,* **11.** Palo Alto, CA: Annual Review Inc.

Hawking, S. W. and Ellis, G. F. R. (1973). *The Large-Scale Structure of Space-Time.* Cambridge: Cambridge University Press.

Hawking, S. W. and Israel, W. (eds). *General Relativity.* Cambridge: Cambridge University Press.

Hawking, S. W. and Israel, W. (eds). (1987). *300 Years of Gravitation.* Cambridge: Cambridge University Press.

Held, A. (ed). (1980). *General Relativity and Gravitation,* vols I and II. New York: Plenum.

Hlavatý, V. (1958). *Geometry of Einstein's Unified Field Theory.* Groningen: Noordhoff.

Hoffmann, B. (ed). (1966). *Perspectives in Geometry and Relativity.* Bloomington, IN: Indiana University Press.

Infeld, I. (ed). (1965). *Relativistic Theories of Gravitation.* Oxford: Pergamon Press.

International Conference on Relativistic Theories of Gravitation. (1965). London.

Isham, C., Penrose, R. and Sciama, D. (1975, 1981). *Quantum Gravity* 1, 2. Oxford: Clarendon Press.

Jordan, P. (1955). *Schwerkraft und Weltall.* Braunschweig: Vieweg.

Jordan, P. (1966). *Die Expansion der Erde.* Braunschweig: Vieweg.

Klauder, J. R. (ed). (1972). *Magic without Magic: John Archibald Wheeler.* San Francisco: Freeman.

Kramer, D., Stephani, H., MacCallum, M. and Herlt, E. (1981). *Exact Solutions of Einstein's Field Equations.* Cambridge: Cambridge University Press.

Kundt, W. (1971). *Survey of Cosmology,* Springer Tracts in Modern Physics, vol. 58. Berlin, Heidelberg, New York: Springer-Verlag.

Longair, M. S. (ed). (1974). *Confrontation of Cosmological Theories with Observational Data,* Proceedings of IAU Symposium No. 63, Cracow, 10–12 September 1973. Dordrecht: Reidel.

Ludwig, G. (1951). *Fortschritte der projektiven Relativitätstheorie.* Braunschweig: Vieweg.

Neugebauer, G. (1980). *Relativistische Thermodynamik.* Berlin: Akademie-Verlag.

O'Raifeartaigh, L. (ed). (1972). *General Relativity.* Oxford: Clarendon Press.

Peebles, P. J. E. (1971). *Physical Cosmology.* Princeton, NJ: Princeton University Press.

Penrose, R. and Rindler, W. (1984), *Spinors and Space-Time.* 1 *Two-Spinor Field Calculus and Relativistic Fields.* Cambridge: Cambridge University Press.

Petrov, A. Z. (1969). *Einstein Spaces.* Oxford: Pergamon Press.

Ryan, M. P. and Shepley, L. C. (1975). *Homogeneous Relativistic Cosmologies.* Princeton, NJ: Princeton University Press.

Sachs, R. K. (ed). (1971). *General Relativity and Cosmology.* New York, London: Academic.
Schmutzer, E. (ed). (1983). *Proceedings of the 9th International Conference on Relativity and Gravitation.* Berlin: VEB Deutsche Verlag der Wissenschaften.
Sciama, D. W. (1971). *Modern Cosmology.* Cambridge: Cambridge University Press.
Smarr, L. (ed). (1979). *Sources of Gravitational Radiation.* Cambridge: Cambridge University Press.
Stewart, J. and Walker, M. (1973). *Black Holes: The Outside Story,* Springer Tracts in Modern Physics, Vol. 69. Berlin: Springer-Verlag.
Symposia Mathematica. (1973). vol. 12. New York: Academic.
Tolman, R. C. (1934). *Relativity, Thermodynamics and Cosmology.* Oxford: Clarendon Press.
Trautman, A., Pirani, F. A. E. and Bondi, H. (1964). *Lectures on General Relativity,* Brandeis Summer Institute, 1964. Englewood Cliffs, NJ: Prentice-Hall.
Treder, H. J. (1962). *Gravitative Stoßwellen – Nichtanalytische Wellenlösungen der Einsteinschen Gravitationsgleichungen.* Berlin: Akademie-Verlag.
Weber, J. (1961). *General Relativity and Gravitational Waves.* New York: Wiley-Interscience.
Weinberg, S. (1977). *The First Three Minutes.* London: Fontana/Collins.
Wheeler, J. A. (1968). *Einstein's Vision.* Berlin: Springer-Verlag.
Will, C. M. (1981). *Theory and Experiment in Gravitational Physics.* Cambridge: Cambridge University Press.
Witten, L. (ed). (1962). *Gravitation.* New York: Wiley.
Yano, K. (1955). *The Theory of Lie Derivatives and Its Applications.* Amsterdam: North-Holland.
Zakharov, V. D. (1973). *Gravitational Waves in Einstein's Theory.* New York: Wiley.
Zel'dovich, Ya. B. and Novikov, I. D. (1971). *Relativistic Astrophysics,* vol. 1. Chicago: University of Chicago Press.

Review and research articles

Ashtekar, A. and Geroch, R. (1974). Quantum theory of gravitation, *Rep. Progr. Phys.,* **37**, 1211.
Bekenstein, J. D. (1973). Black holes and entropy, *Phys. Rev.,* **D7**, 2333.
Blandford, R. D. and Thorne, K. S. (1979). Black hole astrophysics. In *General Relativity,* eds. S. W. Hawking and W. Israel. Cambridge: Cambridge University Press.
Bondi, H., Pirani, F. and Robinson, I. (1959). Exact plane waves, *Proc. Roy. Soc.,* **A251**, 519.
Bondi, H. *et al.* (1962). Waves from axi-symmetric isolated systems, *Proc. Roy. Soc.,* **A269**, 21.
Bonnor, W. B. (1966). Multipole fields in linearised general relativity. In *Perspectives in Geometry and Relativity,* ed. B. Hoffmann. Bloomington, IN: Indiana University Press.

Boulware, D. G. and Deser, S. (1975). Classical general relativity derived from quantum gravity, *Ann Phys.*, **89**, 193.
Braginski, B. G. and Manukin, A. B. (1974). *Die Messung Kleiner Kräfte in physikalischen Experimenten.* Moscow: Nauka.
Braginski, V. B. and Thorne, K. S. (1983). Present status of gravitational-wave experiments. In *Proceedings of the 9th International Conference on General Relativity and Gravitation,* ed. E. Schmutzer. Berlin: VEB Deutsche Verlag der Wissenschaften.
Brans, C. and Dicke, R. H. (1961). Mach's principle and a relativistic theory of gravitation, *Phys. Rev.*, **124**, 925.
Bruhat, Y. (1962). The Cauchy problem. In *Gravitation,* ed. L. Witten. New York: Wiley.
Campbell, W. B. (1973). The linear theory of gravitation in the radiation gauge, *Gen. Rel. Grav.*, **4**, 137.
Cartan, É. (1922). Sur une généralisation de la notion de courbure de Riemann et les espaces a torsion, *Compt. Rend. Acad. Sci. (Paris)*, **174**, 593.
Carter, B. (1979). The general properties of the mechanical, electromagnetic and thermodynamic properties of black holes. In *General Relativity,* eds. S. W. Hawking and W. Israel. Cambridge: Cambridge University Press.
Chandrasekhar, S. (1979). An introduction to the theory of the Kerr metric and its perturbations. In *General Relativity,* eds. S. W. Hawking and W. Israel. Cambridge: Cambridge University Press.
Choquet-Bruhat, Y. and York, J. W. (1980). The Cauchy problem. In *General Relativity and Gravitation I,* ed. A. Held. New York: Plenum.
Darwin, C. (1959). The gravity field of a particle, *Proc. Roy. Soc.*, **A249**, 181.
Dautcourt, G. (1963). Zum charakteristischen Anfangswertproblem der Einsteinschen Feldgleichungen, *Ann. Phys.*, **12**, 302.
de Sitter, W. (1917). On the curvature of space, *Proc. Kon. Ned. Akad. Wet,* **20**, 229.
de Witt, B. S. (1980). Quantum gravity: the new synthesis. In *General Relativity,* eds. S. W. Hawking and W. Israel. Cambridge: Cambridge University Press.
de Witt, B. and Brehme, R. (1960). Radiation damping in a gravitational field, *Ann. Phys.*, **9**, 220.
Douglas, D. H. and Braginski, V. B. (1980). Gravitational radiation experiments. In *General Relativity,* eds. S. W. Hawking and W. Israel. Cambridge: Cambridge University Press.
Dreven, R. W. P. (1984). Laser interferometer gravitational wave experiments. In *General Relativity and Gravitation,* eds. B. Bertotti, F. de Felice and A. Pascolini. Dordrecht: Reidel.
Duff, M. J. (1974). On the significance of perihelion shift calculations, *Gen. Rel. Grav.*, **5**, 441.
Eddington, A. S. (1924). A comparison of Whitehead's and Einstein's formulae, *Nature,* **113**, 192.
Ehlers, J. (1961). Beiträge zur relativistischen Mechanik Kontinuierlicher Medien. *Abh. Mainzer Akad. Wiss., Math.-naturwiss,* No. 11.
Ehlers, J. (1966). Generalized electromagnetic null fields and geometrical optics. In *Perspectives in Geometry and Relativity,* ed. B. Hoffmann. Bloomington, IN: Indiana University Press.

Ehlers, J. (1971). General relativity and kinetic theory. In *General Relativity and Cosmology*, ed. R. K. Sachs. New York, London: Academic.

Ehlers, J. and Kundt, W. (1962). Exact solutions of the gravitational field equations. In *Gravitation*, ed. L. Witten. New York: Wiley.

Ehlers, J., Rosenblum, A., Goldberg, J. and Havas, P. (1976). Comments on gravitational radiation damping and energy loss in binary systems, *Astrophys. J.*, **208**, 77.

Einstein, A. (1917). Kosmologische Betrachtungen zur allgemeinen Relativitätstheorie, *Preuss. Akad. Wiss. Berlin Sitzber.*, 142.

Einstein, A. and Straus, E. G. (1945). The influence of the expansion of space on the gravitation fields surrounding the individual stars, *Rev. Mod. Phys.*, **17**, 120.

Ellis, G. F. R. (1971). Relativistic cosmology. In *General Relativity and Cosmology*, ed. R. K. Sachs. New York, London: Academic.

Ellis, G. F. R. and King, A. R. (1974). Was the big bang a whimper? *Commun. Math. Phys.*, **38**, 119.

Ellis, G. F. R. and Sciama, D. W. (1966). On a class of model universes satisfying the perfect cosmological principle. In *Perspectives in Geometry and Relativity*, ed. B. Hoffmann. Bloomington, IN: Indiana University Press.

Ellis, G. F. R. and Sciama, D. W. (1972). Global and nonglobal problems in cosmology. In *General Relativity*, ed. L. O'Raifeartaigh. Oxford: Clarendon Press.

Finkelstein, D. (1958). Past-future asymmetry of the gravitational field of a point particle, *Phys. Rev.*, **110**, 965.

Fischer, A. E. and Marsden, J. E. (1979). The initial value problem and the dynamical formulation of general relativity. In *General Relativity and Gravitation*, eds. S. W. Hawkins and W. Israel. Cambridge: Cambridge University Press.

Friedman, A. (1922). Über die Krümmung des Raumes, *Z. Phys.*, **10**, 377.

Gamow, G. (1948). The evolution of the universe, *Nature*, **162**, 680.

Geroch, R. (1971). Space-time structure from a global viewpoint. In *General Relativity and Cosmology*, ed. R. K. Sachs. New York, London: Academic.

Geroch, R. (1979). Asymptotic structure of space-time. In *Asymptotic Structure of Space-Time*, eds. F. P. Esposito and L. Witten. New York: Plenum.

Gibbons, G. W. (1979). Quantum field theory in curved spacetime. In *General Relativity*, eds. S. W. Hawking and W. Israel. Cambridge: Cambridge University Press.

Gödel, K. (1949). An example of a new type of cosmological solutions of Einstein's field equations of gravitation, *Rev. Mod. Phys.*, **21**, 447.

Greenberg, P. J. (1970). The general theory of space-like congruences with an application to vorticity in relativistic hydrodynamics, *J. Math. Anal. Appl.*, **30**, 128.

Hawking, S. W. (1975). Particle creation by black holes, *Commun. Math. Phys.*, **43**, 199.

Hawking, S. W. (1976). Breakdown of predictability in gravitational collapse, *Phys. Rev. D*, **14**, 2460.

Bibliography

Hawking, S. W. (1979). The path-integral approach to quantum gravity. In *General Relativity*, eds. S. W. Hawking and W. Israel. Cambridge: Cambridge University Press.

Hehl, F. W. (1973). Spin and torsion in general relativity I, *Gen. Rel. Grav.*, **4**, 333.

Hehl, F. W. (1974). Spin and torsion in general relativity II, *Gen. Rel. Grav.*, **5**, 491.

Hellings, R. W. (1984). Testing relativity with solar system experiments. In *General Relativity and Gravitation*, eds. B. Bertotti, F. de Felice and A. Pascolini. Dordrecht: Reidel.

Herlt, E. and Stephani, H. (1976). Wave optics of the spherical gravitational lens, *Int. J. Theoret. Phys.*, **15**, 45.

Hilbert, D. (1915). Die Grundlagen der Physik. *Nachr. Königl. Gesellsch. Wiss. Göttingen*, 395.

Israel, W. (1972). The relativistic Boltzmann equation. In *General Relativity*, ed. L. O'Raifeartaigh. Oxford: Clarendon Press.

Israel, W. (1985). General relativity: progress, problems and prospects, *Canad. J. Phys.*, **63**, 34.

Israel, W. and Stewart, J. M. (1980). Progress in relativistic thermodynamics and electrodynamics of continuous media. In *General Relativity and Gravitation II*, ed. A. Held. New York: Plenum.

Jordan, P., Ehlers, J. and Kundt, W. (1960). Strenge Lösungen der Feldgleichungen der Allgemeinen Relativitätstheorie, *Abh. Mainzer Akad. Wiss, math.-naturwiss Kl.*, No. 2.

Jordan, P., Ehlers, J., Kundt, W., Sachs, R. and Trümper, M. (1962). Beiträge zur Theorie der Strahlungsfelder, *Abh. Mainzer Akad. Wiss., math.-naturwiss Kl.*, No. 12.

Jordan, P., Ehlers, J. and Sachs, R. (1961). Beiträge zur Theorie der reinen Gravitationsstrahlung, *Abh. Mainzer Akad. Wiss., math.-naturwiss Kl.*, No. 1.

Jordan, P. and Kundt, W. (1961). Geometrodynamik im Nullfall, *Abh. Mainzer Akad. Wiss., math.-naturwiss Kl.*, No. 3.

Kaluza, T. (1921). Zum Unitätsproblem der Physik, *Sitzber. Preuss. Akad. Wiss.*, 966.

Kerr, R. P. (1963). Gravitational field of a spinning mass as an example of algebraically special metrics, *Phys. Rev. Lett.*, **11**, 237.

Killing, W. (1892). Über die Grundlagen der Geometrie, *J. Reine und Angew. Math.*, **109**, 121.

Kinnersley, W. (1975). Recent progress in exact solutions (to Einstein's equations). In *Proceedings of the 7th International Conference on General Relativity and Gravitation*, Tel Aviv, 23–28 June 1974. Chichester: Wiley.

Klein, O. (1926). Quantentheorie und fünf-dimensionale Relativitätstheorie, *Z. Phys.*, **37**, 895.

Komar, A. (1961), Covariant conservation laws in general relativity, *Phys. Rev.*, **113**, 934.

Kramer, D. (1969). Schwarzschildfeld im Friedmankosmos. *Wiss. Z. Friedrich-Schiller-Univ., Jena, math.-naturwiss, Reihe*, **1**, 155.

Kuchar, K. (1973). Canonical quantization of gravity, Banff Lectures. In *Relativity, Astrophysics and Cosmology*, ed. W. Israel. Dordrecht: Reidel.

Kundt, W. (1956). Trägheitsbahnen in einem von Gödel angegebenen kosmologischen Modell, *Z. Phys.*, **145**, 611.

Kundt, W. (1961). The plane-fronted gravitational waves, *Z. Phys.*, **163**, 78.

Lemaître, G. (1933). Condensations sphériques dans l'universe en expansion, *Compt. Rend. Acad. Sci. (Paris)*, **196**, 903.

Liang, E. P. T. and Sachs, R. K. (1980). Cosmology. In *General Relativity and Gravitation II*, ed. A. Held. New York: Plenum.

MacCallum, M. A. H. (1979). Anisotropic and inhomogeneous relativistic cosmologies. In *General Relativity*, eds. S. W. Hawking and W. Israel. Cambridge: Cambridge University Press.

Mehra, J. (1974). Origins of the modern theory of gravitation – the historical origins of the theory of relativity from 1907 to 1919, *Bull Soc. R. Sci. Liège*, **43**, 190.

Merat, P. (1974). Analysis of the optical data on the deflection of light in the vicinity of the solar limb, *Gn. Rel. Grav.*, **5**, 757.

Miller, J. C. and Sciama, D. W. (1980). Gravitational collapse to the black hole state. In *General Relativity and Gravitation II*, ed. A. Held. New York: Plenum.

Neugebauer, G. (1974). Einsteinsche Feldgleichungen und zweiter Hauptsatz der Thermodynamik, *Nova Acta Leopoldina Bd.*, **39**, No. 212.

Newman, E. T. and Penrose, R. (1962). An approach to gravitational radiation by a method of spin coefficients, *J. Math. Phys.*, **3**, 566.

Newman, E. P. and Todd, K. P. (1980). Asymptotically flat space-times. In *General Relativity and Gravitation II*, ed. A. Held. New York: Plenum.

Nobili, A. M. and Will, C. M. (1986). The real value of Mercury's perihelion advance, *Nature*, **320**, 39.

O'Brien, S., and Synge, J. L. (1952). Jump conditions at discontinuities in general relativity. *Communications of Dublin Institute of Advanced Studies A*, No. 9.

O'Connell, R. F. (1972). Present status of the theory of the relativity-gyroscope experiment, *Gen. Rel. Grav.*, **3**, 123.

Penrose, R. (1968). Structure of space-time. In *Batelle Rencontres. 1967 Lectures in Mathematics and Physics*. New York: Benjamin.

Penzias, A. A. and Wilson, R. W. (1965). A measurement of excess antenna temperature at 4080 Mcls, *Astrophys. J.*, **142**, 419.

Plebanski, J. and Mielnik, B. (1962). A study of geodesic motion in the field of Schwarzschild's solution, *Acta Phys. Polon.*, **21**, 239.

Pound, R. W. and Rebka, G. A. (1960). Apparent weight of photons, *Phys. Rev. Lett.*, **4**, 337.

Reasenberg, R. D. (1983). Present state of the experimental verification of the four-dimensional general theory of relativity. In *Unified Field Theories of More than Four Dimensions*, eds. V. de Sabbata and E. Schmutzer. Singapore: World Scientific.

Reissner, H. (1916). Über die Eigengravitation des elektrischen Feldes nach der Einsteinschen Theorie, *Ann. Phys.*, **50**, 106.

Robertson, H. P. (1936). Kinematics and the world structure, *Astrophys. J.*, **83**, 187.
Robinson, I. and Trautman, A. (1962). Some spherical gravitational waves in general relativity, *Proc. Roy. Soc.*, **265A**, 463.
Sachs, R. K. (1962). Waves in asymptotically flat space-time, *Proc. Roy. Soc.*, **270A**, 103.
Sachs, R. K. (1963). Gravitational radiation. In *Relativity, Groups and Topology*, eds. C. and B. de Witt. Les Houches. (1964, New York: Gordon & Breach.)
Sandage, A. and Tammann, G. A. (1971). Absolute magnitudes of cepheids III. Amplitude as a function of position in the instability strip: a period-luminosity-amplitude relation, *Astrophys. J.*, **167**, 293.
Sandage, A. and Tammann, G. A. (1974a). Steps towards the Hubble constant I, *Astrophys. J.*, **190**, 525; (1974b) II, *Astrophys. J.*, **191**, 603; (1974c) III, *Astrophys. J.*, **194**, 223; (1974d) IV, *Astrophys. J.*, **195**, 559; (1975a) V, *Astrophys. J.*, **196**, 313; (1975b) VI, *Astrophys. J.*, **197**, 265.
Schimming, R. (1973). Zur Geschichte der ebenen Gravitationswellen, *NTM-Schriftenreihe Gesch. Naturwiss. Technik u. Med., Leipzig*, **10**, 21.
Schrödinger, E. (1940). Maxwell's and Dirac's equations in the expanding universe. *Proc. Roy. Irish Acad.*, **46A**, 25.
Schücking, E. (1954). Das Schwarzschildsche Linienelement und die Expansion des Weltalls, *Z. Phys.*, **137**, 595.
Schwarzschild, K. (1916). Über das Gravitationsfeld eines Massenpunktes nach der Einsteinschen Theorie. *Sitz. Preuss. Akad. Wiss.*, 189.
Schwarzschild, K. (1916). Über das Gravitationsfeld einer Kugel aus inkompressibler Flüssigkeit nach der Einsteinschen Theorie, *Sitz. Preuss. Akad. Wiss.*, 424.
Sciama, D. W., Candelas, P. and Deutsch, D. (1981). Quantum field theory, horizons and thermodynamics, *Adv. Phys.*, **30**, 327.
Shapiro, I. I. (1980). Experimental tests of the general theory of relativity. In *General Relativity and Gravitation II*, ed. A. Held. New York: Plenum.
Stephani, H. (1966). Näherungsverfahren zur Lösung der Einsteinschen Feldgleichungen, *Wiss. Z. Univ. Jena*, **15**, 91.
Stephani, H. (1974). Physik in geschlossenen Kosmen, *Nova Acta Leopoldina*, **39**, 212.
Synge, J. L. (1964). The Petrov classification of gravitational fields. *Communications of Dublin Institute of Advanced Studies*, No. 15.
Taub, A. H. (1965). The motion of multipoles in general relativity. *Atti del convengo sulla relatività generale: Problemi dell'energia e onde gravitazionale*. Firenze: G. Barbera.
Thirring, H. and Lense, H. (1918). *Phys. Z.*, **19**, 156.
Thorne, K. S. *et al.* (1973). Foundations for a theory of gravitational theories, *Phys. Rev.*, **D7**, 3563.
Thorne, K. S. and Will, C. M. (1971a, b, c). Theoretical framework for testing relativistic gravity, *Astrophys. J.*, **163**, 595, 611; **169**, 125.
Trautman, A. (1962). Conservation laws in general relativity. In *Gravitation*, ed. L. Witten. New York: Wiley.

Walker, A. G. (1936). On Milne's theory of world-structure, *Proc. London Math. Soc.,* **42**, 90.
Weber, J. (1980). The search for gravitational radiation. In *General Relativity and Gravitation I,* ed. A. Held. New York: Plenum.
Wesson, P. S. (1974). Expanding clusters of galaxies as components of a relativistic hierarchical cosmology, *Astrophys. Space Sci.,* **30**, 95.
Westpfahl, K. (1967). Relativistische Bewegungsprobleme, *Ann. Phys.,* **20**, 113.
Weyl, H. (1917). Zur Gravitationstheorie, *Ann. Phys.,* **54**, 117.
Weyl, H. (1918). Gravitation und Elektrizität, *Sitz. Preuss. Akad. Wiss.,* 465.
Will, C. M. (1979). The confrontation between gravitation theory and experiment. In *General Relativity,* eds. S. W. Hawking and W. Israel. Cambridge: Cambridge University Press.
Zeldovich, Ya. B. (1972). The creation of particles and antiparticles in electric and gravitational fields. In *Magic without Magic: John Archibald Wheeler.* San Francisco: Freeman.

Index

acceleration parameter 255, 274
affine parameter 16
age of the universe 275
angular momentum 136, 144
    balance-equation 144
axially symmetric, static vacuum solution 213
axially symmetric, stationary vacuum solution 215

Bach bracket 33
background radiation, cosmic 275
baryon number 81
beginning of the universe 258, 272, 275
Bianchi identities 59
Bianchi type 204, 284
Bianchi-type I, solutions 284
Birkhoff's theorem 103
bivectors, self-dual 184
black holes 221, 230, 242, 247
Bogoliubov transformations 302
Boyer–Lindquist coordinates 212, 242
Brans–Dicke theory 294

Cartan theory 294
Cauchy problem 160
Christoffel symbols 5, 15, 16
    behaviour under transformation 19
classification of electromagnetic fields 185
    gravitational fields 189
closed spaces (universes) 219, 260
    and thermodynamical equilibrium 265
contraction 31
coordinate systems
    anholonomic 18
    co-moving 21
    Gaussian 20
    harmonic 114
    isotropic 114
    orthogonal 20
    rotating 71
    time-orthogonal 20

coordinate transformations 17, 25
Coriolis force 13, 71
cosmic background radiation 275
cosmological constant 87, 267
cosmological model 283
cosmological principle 251
covariant derivative 43
critical mass of a star 231
curl of a vector field 61
curvature, extrinsic 161
curvature scalar 57
curvature tensor 53, 57
    and global parallelism 52
    and geodesic deviation 11
    properties 56
    reduction formulae 162
    and second derivatives of the metric 56
    of a spherical surface 11

de Sitter universes 219, 268
deflection of light 108, 116
duality rotation 76
dualization 36
dust 83

Eddington–Finkelstein metric 225, 241
end of the universe 259
eigenvectors
    of the electromagnetic field tensor 184
    Weyl tensor 191
eikonal 78
Einstein tensor 87
Einstein universe 263, 267
embedding 210
energy–momentum complex 98
energy–momentum tensor of
    dust 83
    the gravitational field 98
    incoherent radiation 77
    matter 81, 97
    the Maxwell field 75
    a perfect fluid 83
entropy 81

ε pseudo-tensor 28, 32
equations of motion of test particles 91
ergosphere 244
event horizon of
   the Kerr metric 245, 247
   a Robertson–Walker metric 259, 273
expansion of
   a null-vector field 182
   a timelike vector field 175

far field 138
   of the linearized theory 132
Fermat principle 112
Fermi–Walker transport 47, 72, 94
four-velocity field 80
four-velocity of a point mass 70
Friedman universe 267, 270
fundamental metric form 3, 14

Gauss equation 163
Gauss's law 60, 65
geodesic deviation 8
geodesic equation 7, 15, 70
geometrical object 27
geometrical optics 78
geometrization of the gravitational force 14, 292
Gödel universe 288
Goldberg–Sachs theorem 193
gradient 61
gravitational collapse 221, 230, 235, 246
gravitational constant 91
gravitational lens 110
gravitational radius 104
gravitational waves 149
   plane (linearized theory) 151
   plane (exact solution) 154
groups of motion 200

Hawking radiation 306
heat current 82
Hubble constant 255, 274
   and the age of the universe 272

inertial system 13, 49, 69, 73
incoherent radiation 77
initial value problem, Cauchy 160
initial value problem, characteristic 166
integral laws 60, 65
intrinsic geometry 52

Kasner metric 287
Kerr metric 206, 212, 215, 242

Killing tensor 206
Killing vector 174, 194, 204
Kruskal diagram 226, 241
Kruskal metric 226

Lagrangian for the gravitational field 95
   Maxwell field 75
laws of conservation of baryon number 81
   of electrical charge 74
   of energy 147
   integral 66
   and Killing vectors 204
   of rest-mass density 81
Lemaître metric 225
Lense–Thirring effect 137
Levi-Civita symbol 28
Lie derivative 49, 196
light travel time 112, 118
limiting surface of stationarity 244
linearized gravitation theory 128
locally flat system 22
locally geodesic system 22, 69
locally Minkowskian system 21
Lorentz gauge 74

mass 136, 140
mass function 122
matter, concept of 86
matching conditions 123, 168
Maxwell equations 73, 78
metric tensor 3, 14
Minkowski space 12, 23
momentum-balance equation 144
monopole particle 91

neutron star 231
Newtonian approximation 89
null rotation 179
null fields, electromagnetic 77, 186
null geodesics 16, 78
null tetrad 178, 184
null vectors 31, 39, 178
   covariantly constant 155

observable 208
optical scalar 180

parallel displacement 45, 46
   along a curve 45
parallel propagator 30, 45
particle horizon 258, 272
particle mechanics 70
Penrose diagram 190

perfect fluid 83
perihelion precession 105, 115
Petrov classification 183, 190
plane waves 154, 213
polarization of
  an electromagnetic wave 79
  a gravitational wave 152
PPN formalism 295
pressure 81
principle of covariance 1, 69
principle of equivalence 69
projection tensor 81, 160, 180
projective field theory 293
proper time 16, 24
pseudo-Riemannian space 15, 22
pseudo-tensors 28

quadrupole moment 136
quadrupole radiation 147
quantization of the gravitational field 298
quantum mechanics 208
quantum theory 297

radiation universe 269
radius of the universe 253, 270
raising and lowering of indices 31
red shift
  of the cosmic background radiation 278
  in Robertson–Walker metrics 255, 256, 273
  in static gravitational fields 110, 117, 235
Reissner–Nordstrom solution 126
Ricci tensor 57
Riemannian space 12, 14
Robertson–Walker metric 251, 265
Robinson–Trautman solution 213
rotation of
  a coordinate system 71
  null vector field 182
  timelike vector field 175
  vector field 61, 172

scalar 25
scalar-tensor theory 294
Schwarzschild cavity 279
Schwarzschild radius 104
Schwarzschild solution 99, 193, 221, 239, 275
  in Eddington–Finkelstein coordinates 225
  interior 119
  in Lemaître coordinates 225
  maximal extension 226
shear of a null vector field 182
  timelike vector field 175
singularities 221, 289
singularity theorems 289
space of constant curvature 200, 216, 269
spinning particles 93
spinors 39
Stokes law 60, 65
structure constants 200
surface integral 62
symmetries, continuous 195

temperature 81, 207
tensor 27
  antisymmetric, of second rank 34
  completely antisymmetric 33
  symmetric, of second rank 34
tensor densities 27
tetrads 37
thermodynamics 80, 207
thermodynamics of black holes 306
Tolman solution 237

unified field theory 293

vector fields 171
  covariantly constant 174
  geodesic 173, 179
  hypersurface-orthogonal 172
  null 178
  timelike 175
vectors
  null 31, 39
  spacelike 31
  timelike 31
volume element 62
volume integral 62

Weyl class 213
Weyl tensor 57, 183
  classification 189
Weyl theory 293
white dwarf 231
worldlines, congruence of 172